Functional Dyes

Functional Dyes

Edited by

Sung-Hoon Kim
Department of Textile System Engineering
Kyungpook National University
Daegu, South Korea

ELSEVIER

Amsterdam • Boston • Heidelberg • London • New York • Oxford • Paris
San Diego • San Francisco • Singapore • Sydney • Tokyo

Elsevier
Radarweg 29, PO Box 211, 1000 AE Amsterdam, The Netherlands
The Boulevard, Langford Lane, Kidlington, Oxford 0X5 1GB, UK

First edition 2006

Library of Congress Cataloging-in-Publication Data
A catalog record for this book is available from the Library of Congress

British Library Cataloguing in Publication Data
A catalogue record for this book is available from the British Library

ISBN-13: 978-0-444-52176-7
ISBN-10: 0-444-52176-3

Printed and bound in The Netherlands

06 07 08 09 10 10 9 8 7 6 5 4 3 2 1

Preface

For a long time, organic dyes from our surroundings have been used to impart coloration effects to textiles and polymer substrates. These colored materials greatly influence our emotions and aesthetic satisfactions. Traditionally, organic dyes have been used for coloration, especially of textiles, by means of dyeing technology. However, nowadays, organic colorants have become increasingly important for high-technology applications in new fields such as electronic materials, devices, reprographics, and so on. These organic materials are called functional dyes.

There are a great number of books concerning dyes and pigments and they deal with a variety of subjects such as its synthesis, property, application, and analysis. However, few books have been written on functional dyes with respect to recent interesting research trends. For that reason, this book is considered to be necessary for the readers engaged in university, research institute, and industry so as to improve their fundamental knowledge and have wider appreciation for functional dyes. In addition, this book is to provide information on the latest developments and future directions in the functional dye chemistry.

In general, dyes showing brilliant and beautiful colors with higher fastness properties are of great demand because of their suitability for various end-uses. Various numbers of chromophore have been investigated to achieve satisfactory characteristics in terms of functional applications and specialized uses. In this context, the special effects of organic colorants have been reevaluated and employed in the new areas of information-recording materials, information-display media, or optoelectronic devices. These dyes can also be applied to organic photoconductors, solar-energy utilizations, sensitizers, biomedical probe, photocatalysts, and so on. These new trends of dye chemistry have been recently developed and these classes of dyes are of significant importance in high-technology industries.

This book addresses various knowledge and information about functional dyes. It reviews recent advances on synthesis and characteristics of the functional dyes and their applications in high-technology uses. Furthermore, it provides a broad and interesting introduction to the science and technology of the applications of functional dyes. It will be a valuable information source for readers in that this book was peer-reviewed and written by special dye chemists around the world. The first chapter of the book deals with the recent progress in phthalocyanine chemistry. The second chapter covers the structure and properties of cyanine dyes for solar cells and optical data storage. The third chapter presents the synthesis and characterization of photochromic naphthopyrans, and Chapter 4 deals with the synthesis and application of cyanine dyes as fluorescent non-covalent labels for nucleic acid research. Chapter 5 covers the surface plasmon resonance-based sensing system using functional dyes. In Chapter 6, the synthesis and application of squarylium dyes is described. The structures and physical properties of fluorine-containing dye are included in Chapter 7.

Thus, the main intention in preparing this book is to provide a fundamental understanding and an overview of the theories of dye chemistry for color chemists, organic chemists, and material scientists. It is hoped that it will also be useful for postgraduate students in chemistry and material science.

It is a pleasure to gratefully acknowledge the contributors of each chapter and their great enthusiasm in spite of their very busy schedules.

S.H. Kim

Contributors

Todor Deligeorgiev
University of Sofia,
Faculty of Chemistry,
Sofia, Bulgaria

Takamitsu Fukuda
Department of Chemistry, Graduate School of Science, Tohoku University,
Sendai, Japan

John D. Hepworth
Faculty of Science
University of Central Lancashire,
Preston, UK

B. Mark Heron
Department of Colour and Polymer Chemistry, University of Leeds,
Leeds, UK

Sung-Hoon Kim
Department of Textile System Engineering,
Kyungpook National University,
Daegu, Korea

Nagao Kobayashi
Department of Chemistry, Graduate School of Science, Tohoku University,
Sendai, Japan

Kwangnak Koh
College of Pharmacy,
Pusan National University,
Pusan, Korea

Masaki Matsui
Department of Materials Science and Technology, Faculty of Engineering,
Gifu University,
Gifu, Japan

Fanshun Meng
Laboratory for Advanced Materials and Institute of Fine Chemicals,
East China University of Science and Technology, Shanghai,
P.R. China

Hiroyuki Nakazumi
Department of Applied Chemistry,
Graduate School of Engineering,
Osaka Prefecture University,
Sakai, Japan

He Tian
Laboratory for Advanced Materials and Institute of Fine Chemicals, East China University of Science and Technology,
Shanghai, P.R. China

Aleksey Vasilev
University of Sofia,
Faculty of Chemistry,
Sofia, Bulgaria

Shigeyuki Yagi
Department of Applied Chemistry,
Graduate School of Engineering,
Osaka Prefecture University,
Sakai, Japan

TABLE OF CONTENTS

Functional Dyes
Sung-Hoon Kim (Editor)

Chapter 1

Recent progress in phthalocyanine chemistry: Synthesis and characterization

Nagao Kobayashi and Takamitsu Fukuda

Department of Chemistry, Graduate School of Science, Tohoku University, Sendai, Japan

1. INTRODUCTION

Phthalocyanine (Pc) was first synthesized [1] in 1907, so the first centenary of Pc research is fast approaching. Since the heteroaromatic Pc ligand has a strong blue color and is chemically stable, Pc complexes have been used extensively in pigments and dyestuffs for more than 70 years [2]. Pcs have also been used as catalysts for the removal of sulfur from crude oil, as charge-generation materials in xerography, in optical read/write discs, as photodynamic reagents for cancer therapy, in deodorants, germicides and anti-bacterial reagents, and as growth promoting and retarding agents of plants [3]. Other active fields of Pc-related research include applications in or as chemical sensors, electrochromism, conductors, semiconductors, photovoltaic elements for electricity generation, one-dimensional metals, non-linear optics, electrocatalysis, liquid crystals, Langmuir–Blodgett films, and electrophotography [4]. Although most research results have been published as patents, academic research has become very intensive over the last 20 years. The fact that over the last decade, more than 1,000 papers have been published annually on Pc-related research, including both academic papers and industrial patents, indicates that the Pc ligand remains of interest to many researchers. The *Society of Porphyrins and Phthalocyanines* was established in the year 2000 by leading researchers within the field to help foster future collaborative research. In this chapter, representative synthetic methods and spectroscopic characterizations are summarized.

2. PROGRESS IN Pc SYNTHESIS

Pcs (Fig. 1) can be prepared using phthalic acid, phthalic anhydride, phthalic imide, phthalamide, phthalonitrile, and isoindolediimine (isoindoline) in the presence and

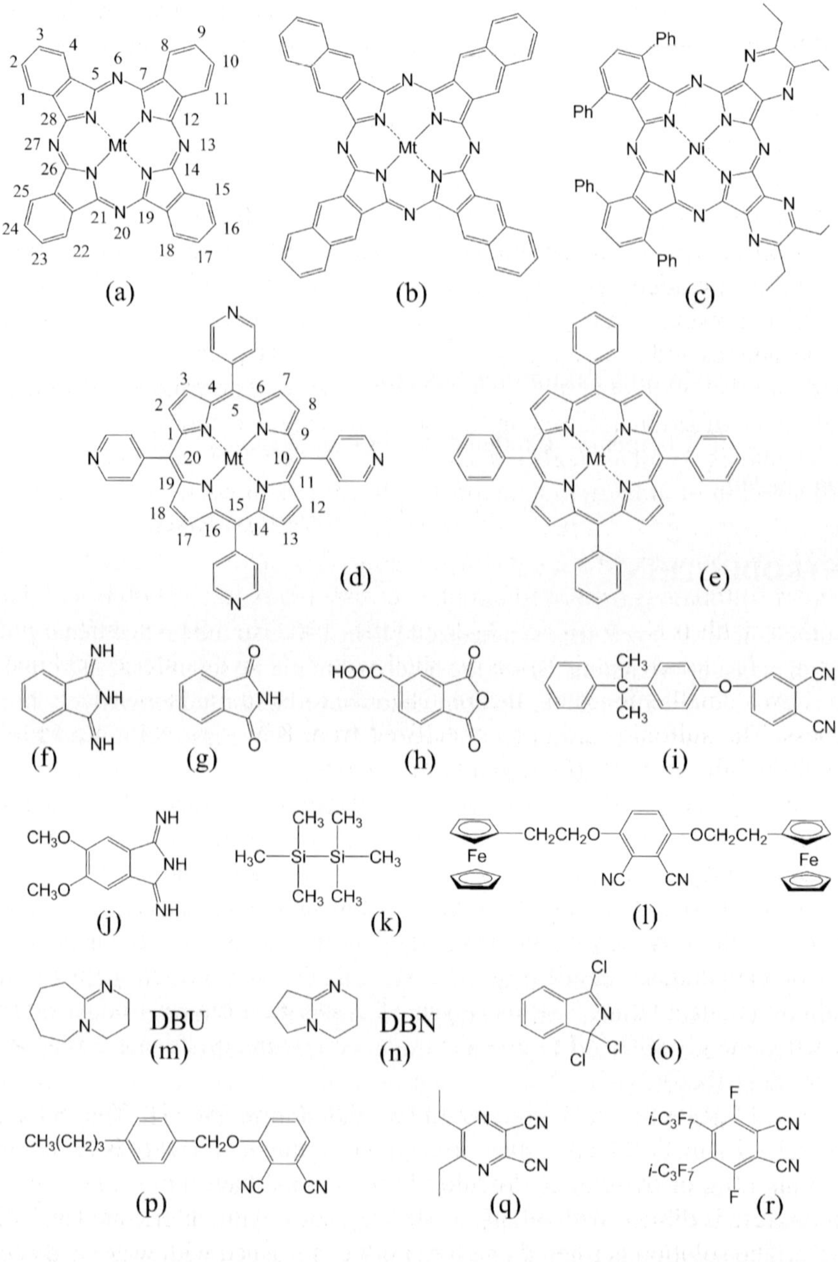

Fig. 1. Structures of some compounds appearing in this chapter: (a) Pc with numbering system, (b) Nc, (c) a C_{2v}-type NiPc derivative from Section 2.3.2(b), (d) 5,10,15,20-tetrapyridylporphyrin, (e) 5,10,15,20-tetraphenylporphyrin, (f) 1,3-diiminoisoinndoline, (g) phthalimide, (h) trimellitic anhydride, (i) 4–(cumylphenoxy)-4-phthalonitrile, (j) 5,6-dimethoxy-1, 3-diiminoisoinndoline, (k) HMDS, (l) 3,6-*bis*(2-ferrocenylethoxy)phthalonitrile, (m) DBU, (n) DBN, (o) trichloroisoindolenine, (p) 3-(*p*-*n*-butylbenzyloxy)phthalonitrile, (q) 2,3-dicyano-5,6-diethyl-1,4-pyrazine, and (r) perfluoro-(4,5-di-isopropyl)phthalonitrile.

absence of metal templates. Typical experimental procedures that have been developed for the synthesis of a wide range of different Pcs are presented below.

2.1. Water-soluble Pcs

2.1.1. Metal-free species

(a) Pc-2,9(or 10),16(or 17),23(or 24)-tetrasulfonic acid [5]. No Pc product can be obtained by simply heating triammonium 4-sulfophthalate (10 g) and urea (50–75 g) to 230 ºC. When, boric acid (0.5–1.5 g) and ammonium molybdate (1.0–2.5 g) are added as catalysts and the mixture is heated at 170–250 °C for 0.5~1.5 h, however, ammonium Pc-2,9(or 10),16(or 17),23(or 24)-tetrasulfonate can be synthesized in fairly good yields (8–18%). The reaction mixture is powdered and then extracted with water. The extract is then evaporated to dryness. The water extraction and evaporation steps are repeated approximately 10 times. The residue is then thoroughly washed with 70% ethanol and added into a saturated solution of ammonium carbonate, which is then warmed. The macrocycle is salted out to give ammonium sulfonate. The crude ammonium salt is thoroughly washed with 70% ethanol and then boiled in 95% ethanol. The tetraammonium sulfonate is converted into the corresponding tetrasulfonic acid through treatment with 6 N (or higher) hydrochloric acid. The blue crystalline product formed is collected by filtration, washed with 6 N hydrochloric acid and dissolved in a small amount of ethanol. The ethanol solution is then evaporated to dryness. The sulfonic acid is recrystallized from 8 N hydrochloric acid to give long thread-like crystals (final yield not reported).

(b) 2,3,9,10,16,17,23,24-octacarboxy*phthalocyanine* [6]. Diluted lithium propanoate ($CH_3CH_2CH_2OLi$) (0.45 mol) is added stepwise to a refluxing propanol solution containing tetracyanobenzene (1 mol). The reaction is continued for *ca.* 30 min after addition of the last portion of $CH_3CH_2CH_2OLi$. After the removal of the solvent with an evaporator, the residue is extracted using CCl_4 in a Soxhlet apparatus. The yield of 2,3,9,10,16,17,23,24-octacyanophthalocyanine dilithium is quantitative. The hydrolysis of cyano groups of this Pc is performed in triethylene glycol/KOH to give the target compound (yield and further details are not described).

(c) 2,3,9,10,16,17,23,24-octahydroxy*phthalocyanine* [7]. The corresponding 2,3,9,10,16,17,23,24-octamethoxyphthalocyanine (150 mg) is heated under reflux in 4.0 g of pyridine hydrochloride on a metal bath for 30 min. The reaction mixture is diluted with 40 mL of 10% aqueous hydrochloric acid and stirred for 1 h. The solution is filtered and the residue is washed with water and acetone under a nitrogen atmosphere. The product is then dried *in vacuo*. The yield is 99% of a black powder.

2.1.2. Metallated species

(a) The tetrasodium salt of cobalt(II) 2,9(or 10),16(or 17),23(or 24)-tetrasulfophthalocyanine 2-hydrate [8]. The monosodium salt of 4-sulfophthalic acid

(43.2 g, 0.162 mol), ammonium chloride (4.7 g, 0.09 mol), urea (58 g, 0.97 mol), ammonium molybdate (0.68 g, 0.6 mmol), and cobalt(II) sulfate 7-hydrate (13.6 g, 48 mmol) are thoroughly ground together. Nitrobenzene (40 mL) is added to a 500 mL three-necked flask fitted with a thermometer, a condenser, and a cork. The nitrobenzene is heated to 180 ºC. The solid mixture is then added slowly, with stirring, while keeping the temperature between 160 and 190 ºC. The heterogeneous mixture is heated for 6 h at 180 ºC. The crude product, a solid cake, is ground and washed with methanol until the odor of nitrobenzene can no longer be detected. The remaining solid is added to 1,100 mL of 1 N hydrochloric acid saturated with sodium chloride. This step is crucial for the removal of excess cobalt(II) from the product. The solution with the accompanying undissolved material is briefly heated to boiling, cooled to room temperature, and filtered. The resulting solid is dissolved in 700 mL of 0.1 N sodium hydroxide. The solution is heated to 80 ºC and the insoluble impurities are immediately separated by filtration. Sodium chloride (270 g) is added to the solution. At this point, some of the solid product precipitates. The slurry is heated and stirred at 80 ºC until the evolution of ammonia stops. The product is then obtained by filtration and the initial reprecipitation process is repeated twice more. The solid is separated and washed with 80% aqueous ethanol until the filtrate is chloride-free. This product is refluxed for 4 h. in 200 mL of absolute ethanol. The blue, pure product is filtered and dried overnight *in vacuo* over P_2O_5 with a yield of 80%.

(b) (2,3,9,10,16,17,23,24-octahydroxyphthalocyaninato)nickel(II), $Ni(OH)_8Pc$ [9]. $Ni(OMe)_8Pc$ (5.15 g, 6.4 mmol) is suspended in 100 mL of dichloromethane, and BBr_3 (24 mL, 254 mmol) is added under a nitrogen atmosphere. The mixture is stirred for 15 days, and 100 mL of methanol is added to the residue, and this procedure is repeated three additional times. $Ni(OH)_8Pc$ is separated from the solvent by filtration, to give a dark green precipitate in 47% yield (2.1 g). The complex can be recrystallized from pyridine, giving dark green solvated crystals of $Ni(OH)_8Pc{\cdot}2NC_5H_5$.

(c) Fe(III) and Co(II) Pc-2,9(or 10),16(or 17),23(or 24)-tetracarboxylic acid ($[(COOH)_4Fe^{III}Pc]^-$ or $(COOH)_4Co^{II}Pc$) [10]. 10 g (0.05 mol) of trimellitic anhydride (Fig. 1), 30 g (0.5 mol) of urea, 0.03 mol of metal salts ($FeCl_3$ or $CoCl_2$), 1 g (1 mmol) of ammonium molybdate and 150 mL of nitrobenzene, which is used as a solvent, are mixed in a mortar, and then placed in a 200 mL round-bottomed flask equipped with a reflux condenser. The reaction mixture is heated to 150–170 ºC for 3 h. A blue–green black solid is obtained, which is removed from the flask with 200 mL of methanol and collected by filtration. The filtrate is fully washed with methanol until no nitrobenzene can be detected, and dried for 24 h at 60ºC *in vacuo*. The yield of $(CONH_2)_4Fe^{III}Pc$ is 85% and that of $(CONH_2)_4Co^{II}Pc$ is 90%. The hydrolysis of these amides is carried out in NaOH-saturated boiling water for 12 h. After cooling, the solution is cautiously acidified by adding 6 N HCl to *ca.* pH 6. The blue solution is separated from the precipitated NaCl by decantation. This procedure is repeated several times. Finally, the precipitate of the target carboxylic acid is obtained once the pH of the solution is below pH 4. The yield of tetraacids is *ca.* 80%.

2.2. Water-insoluble Pcs

2.2.1. Metal-free species

(a) H_2Pc via Li_2Pc [11]. 60 g of phthalonitrile are added to a solution of lithium (4 g) in amyl alcohol (300 mL). A green color appears, and when the mixture is warmed, a vigorous exothermic reaction takes place, and the color changes to deep blue with the precipitation of some dilithium Pc (Li_2Pc). The mixture is boiled for 30 min, cooled, diluted to 1,000 mL with benzene, and left aside for 3 h. After filtration, the dull blue residue of Li_2Pc (36 g) is extracted by Soxhlet extraction with acetone and dried over sodium sulfate. The basic lithium compounds are left in the extractor. Evaporation under reduced pressure yields Li_2Pc as a crystalline deposit with a purple lustre (30 g; 50%). Water (2 mL) is added to a solution of Li_2Pc (3 g) in absolute alcohol (25 mL). A blue precipitate of H_2Pc is formed immediately is collected after 3 h and recrystallized from 1-chloronaphthalene. Li_2Pc can be hydrolyzed similarly by dilute acid. The yield is almost quantitative.

(b) Substituted H_2Pc using hydroquinone as a catalyst *Tetrakis*(cumylphenoxy)-Pc [12]. A mixture of 1.00 g (2.96 mmol) of 4-(cumylphenoxy)-4-phthalonitrile (Fig. 1) and 81 mg (0.74 mmol) of hydroquinone (purified by sublimation) is fused by gentle heating to melting point, cooled, sealed under vacuum, and then reacted at 180 ºC for 16 h and the yield is 0.49 g (49%).

(c) H_2Pc using DBU or DBN as a catalyst [13]. A mixture of phthalonitrile (2.56 g, 20 mmol) and 1,8-diazabicyclo[5.4.0]undec-7-ene (DBU) (or 1,5-diazabicyclo[4.3.0]non-5-ene (DBN)) (20 mmol) in ethanol or propanol (50 mL) is heated to reflux. As the reaction proceeds, H_2Pc is gradually deposited as a blue precipitate, which is collected by filtration, washed with ethanol, and purified by extraction with ethanol. The yield is time dependent. Based on the conditions outlined above, the yield is *ca.* 20% or 45% after 24 h, in the presence of DBU or DBN, respectively.

(d) $H_2(CN)_8Pc$ by solid phase, room temperature synthesis [14]. 90 mg (1.66 mmol) of sodium methoxide is added to a solution of 100 mg (0.526 mmol) of 1,2,4,5-tetracyanobenzene in 5 mL of dry THF. The resulting suspension is stirred for 5 days at room temperature, and the solution is then acidified with acetic acid to neutral pH. The residue of the target compound is filtered, washed with 2-propanol, and precipitated from DMF solution by the addition of water. The yield is 51 mg (50%).

(e) H_2Pc via isoindoline [15]. *Method A*: A mixture of phthalonitrile (m.p. 140.5–141.0 ºC, 25 g, 195 mmol) and 1-dimethylamino-2-propanol (b.p. 127–128 ºC, 100 mL) is placed in a four-neck, 500 mL round-bottomed flask, equipped with a mechanical stirrer, reflux condenser, thermometer, and gas inlet tube. The suspension is heated to 100 °C and the phthalonitrile is dissolved. A steady stream of ammonia gas is passed into the solution as the temperature is first raised to reflux (127–128 °C), which is then maintained for 7 h. A bluish-purple precipitate is collected by filtration from the hot solvent and washed thoroughly first with water or ethanol and then with acetone. The product is then air dried, with a yield of 22.5 g (90%). *Method B*: A mixture of 1,3-diiminoisoindoline (20.0 g, 138 mmol) and 2-dimethylaminoethanol (b.p. 130–135 °C, 100 mL) is refluxed with stirring for 7 h, during which time ammonia is evolved. The mixture is filtered hot and the purple crystals of Pc are washed thoroughly with ethanol and then with acetone. The product is dried *in vacuo* at 80 °C for 3 h, with a yield of 15.1 g (85.5%).

(f) H_2Pc from phthalimide using hexamethyldisilazane (HMDS) as a catalyst [16]. A glass tube is filled with phthalimide (100 mg, 0.68 mmol), *p*-TsOH·H_2O (13 mg, 0.07 mmol), HMDS (560 μl, 438 mg, 2.7 mmol, Fig. 1) and DMF (50 μl, 0.68 mmol) under an argon atmosphere and then sealed. Upon heating the mixture to 150 °C, a dark purple solid immediately appears. After heating for 10 h, the mixture is cooled and filtered. The solid is washed with methanol and then dissolved in concentrated H_2SO_4 (5.0 mL). The resulting solution is poured into H_2O (100 mL) and the blue precipitate is filtered and washed with H_2O. The solid is further purified by Soxhlet extraction with methanol to give 62 mg (58%) of H_2Pc as a blue solid.

(g) H_2Pc from phthalonitrile using oxime as a catalyst [17]. When 4 equiv. of oxime ($R^1R^2C = NOH$, where $R^1 = H$, $R^2 = Me$ or Et) are heated at 100 °C with 4 equiv. of phthalonitrile for 8 h, H_2Pc is obtained in *ca.* 60–65% yield (other details are not described).

(h) H_2Pc by electrosynthesis [18]. A 100 mL solution of either absolute ethanol or a mixture of ethanol and dimethyl acetamide, containing LiCl (3 g, 0.07 mol) is introduced into an electrolysis cell and deaerated at a fixed temperature (typically 75 °C). Phthalonitrile is added to the cathode compartment and a potential of $-1.6 - -1.9$ V is applied. With ethanol as the solvent, the initial uncolored solution becomes yellow, then blue–green after the passage of 20–40 C and finally turns into a viscous blue suspension. Electrolysis is stopped after a fixed amount of charge has been passed, typically $Q/Q_{theo} = 1$–1.5, where $Q_{theo} = (N \times F)/2$, and N is the initial mole number of phthalonitrile in the catholyte. Then the catholyte is poured into 100 mL of a 0.2 M H_2SO_4 solution. The resulting suspension is stirred for 30 min and then filtered. The blue solid is washed with water and acetone and then treated with warm acetone in a Soxhlet apparatus. The resulting blue solid is H_2Pc (α-form). The yield depends on the Q/Q_{theo} value and on the initial concentration of phthalonitrile in the catholyte. When the initial concentration of phthalonitrile is 50 mmol and electrolysis is stopped at $Q/Q_{theo} = 1$, the yield exceeds 70%.

2.2.2. Metallated species

(a) *Tert*-butylated CoPc by the anhydride method [19]. 4-*tert*-butylphthalic anhydride (13 g, 64 mmol), 13.5 g (225 mmol) of urea (water free), and 2.4 g (19 mmol) of cobalt(II) chloride are suspended in 50 mL of trichlorobenzene, treated with 0.5 g of ammonium molybdate, and heated at 190 °C for 4 h. After the mixture is cooled, 50 mL of petroleum ether (60–90 °C) is added and the mixture is vacuum filtered. The residue on the filter is discarded, and the filtrate is concentrated. The precipitate is purified as described above with 1 N HCl and 1 N NaOH solutions to give a yield of 4 g (31%).

(b) CuPc from phthalimide using HMDS as a catalyst [20]. A glass tube is filled with phthalimide (100 mg, 0.68 mmol), $CuCl_2$ (23 mg, 0.25 mmol), *p*-toluenesulfonic acid monohydrate (13 mg, 0.07 mmol), and HMDS (560 μl, 438 mg, 2.7 mmol) and the mixture is stirred at 100 °C for 1 h under an argon atmosphere. The mixture is cooled and then DMF (50 μl, 0.68 mmol) is added. The tube is then sealed and the mixture is heated at 150 °C. A dark purple solid immediately appears and the tube is heated for a further 10 h. The mixture is cooled and filtered. The solid is washed with methanol and then dissolved in concentrated H_2SO_4 (5 mL). The solution is poured into 100 mL of water. The resulting blue precipitate is filtered and washed successively with dilute H_2SO_4, water, and methanol. The solid is further purified by Soxhlet extraction with methanol to give 62 mg of CuPc (63%) as a blue solid.

(c) ZnPc from substituted phthalonitrile using DBU in alcohol. [1,4,8,11,15,18,22,25-*Octakis*(2-ferrocenylethoxy)phthalocyaninato]zinc(II) [21]. A mixture of 3,6-*bis*(2-ferrocenylethoxy)phthalonitrile (100 mg, 0.17 mmol, Fig. 1) and $Zn(OAc)_2{\cdot}2H_2O$ ((22 mg, 0.10 mmol), Ac = anthracocyanine) in *n*-pentanol (3 mL) is heated to 90 °C. DBU (0.01 mL, 0.07 mmol) is then added. The mixture is stirred at 150 °C overnight and then poured into a mixture of methanol and water (1:1, 50 mL). The precipitate formed is filtered off and chromatographed on a basic alumina column using $CHCl_3$/THF (5:1) as the eluent to give ZnPc as a green powder (54 mg, 53%).

(d) NiPc from phthalonitrile using hydroquinone in quinoline [22]. A 2.00 g (15.6 mmol) sample of phthalonitrile is reacted with 0.76 g (4.3 mmol) of 99.999% pure anhydrous $Ni(OCOCH_3)_2$ and 0.48 g (4.4 mmol) of hydroquinone in 6 mL of quinoline. The solution is refluxed under nitrogen for 5 h and then cooled. The resulting precipitate is collected and washed with hot water and acetone. Purification by sublimation yields 0.47 g (21%) of NiPc.

(e) ZnPc from substituted phthalonitrile and zinc dust [*Tetrakis*(cumylphenoxy)Pc]zinc [12]. A mixture of 676 mg (2.00 mmol) of 4-(cumylphenoxy-4-phthalonitrile (Fig. 1) and 262 mg (4.00 mmol) of Zn powder (HCl etched, washed, and dried) is reacted at 280 °C for 70 h resulting in a yield of 397 mg (56%).

(f) MgPc from substituted phthalonitrile and magnesium powder [*Tetrakis*(cumylphenoxy)Pc]magnesium [12]. A mixture of 0.50 g (1.48 mmol) of

4- (cumylphenoxy)-4-phthalonitrile (Fig. 1) and 0.14 g (5.92 mmol) of Mg powder (lightly etched with dilute HCl) is reacted at 280 °C for 70 h to yield 72 mg (14%).

(g) PbPc from substituted phthalonitrile and PbO [*Tetrakis*(cumylphenoxy)Pc]lead [12]. A mixture of 500 mg (1.48 mmol) of 4-(cumylphenoxy)-4-phthalonitrile (Fig. 1) and 250 mg (1.12 mmol) of PbO is reacted at 210 °C for 14 h resulting in a yield of 280 mg (49%).

(h) O=TiPc from phthalonitrile and $Ti(OBu)_4$ [23]. *Method A*: A mixture of phthalonitrile (5.0 g, 39 mmol), $Ti(OBu)_4$ (3.65 g, 10.7 mmol), urea (1.17 g, 19.5 mmol), and 1-octanol (6.0 g, 46.1 mmol) is heated at 150 °C under nitrogen for 6 h. After addition of methanol (30 mL) to the reaction mixture followed by refluxing for 30 min, the fine blue crystals are collected by filtration, washed with toluene, methanol, and water, and then dried at 100 °C and at a pressure of 1 Torr for 3 h to give 4.92 g of O=TiPc (87.6%). *Method B*: A mixture of phthalonitrile (64.07 g, 0.5 mol), $Ti(OBu)_4$ (46.75 g, 138 mmol), urea (15.15 g, 0.25 mol), and 1-octanol (76.8 g, 0.59 mol) is heated at 150 °C under stirring for 6 h. After addition of methanol (100 mL) followed by refluxing for 30 min, the suspension is filtered. The collected solid is washed with toluene (3×250 mL), methanol (3×100 mL), and water (500 mL) and dried at 100 °C and a pressure of 1 Torr for 6 h to give 62.1 g of O=TiPc (86.2%). Although $TiCl_4$ had long been used for this procedure, the use of $Ti(OBu)_4$ makes the synthesis much easier, since $Ti(OBu)_4$ is much more stable against moisture.

(i) AlPc from phthalonitrile and $Al(OBu)_3$ [23]. A mixture of phthalonitrile (1.0 g, 7.8 mmol), $Al(OBu)_3$ (0.58 g, 2.3 mmol), urea (0.47 g, 7.8 mmol), and 1-butanol (1.5 g, 20 mmol) is heated at 140 °C for 6 h. 25 mL of dilute H_2SO_4 (3 M) is added to the reaction mixture, and the suspension is then stirred at 100 °C for 2 h. The blue solid is collected by filtration, successively washed with water, 28% ammonia solution, water, and methanol, and then dried to give 0.73 g (65%) of $(OH)AlPc{\cdot}H_2O$.

(j) FePc from phthalonitrile and $Fe(CO)_5$. 1,2,3,4,8,9,10,11,15,16,17,18,22, 23,24,25-hexadecamethyl-phthalocyaninato iron(II) [24]. A solution of $Fe(CO)_5$ (0.2 g, 1 mmol) dissolved in 1-chloronaphthalene (2 mL) is added dropwise over 5 min to a solution of tetramethylphthalonitrile (0.73 g, 4 mmol) in 1-chloronaphthalene (6 mL) under reflux in a nitrogen atmosphere. The black suspension is held at 270 °C for a further 3 h and then cooled, after which the precipitate is suction filtered, and then washed with chloroform, toluene, and acetone. The product is first boiled in dilute HCl followed by water, placed in the thimble of a Soxhlet extractor, washed overnight with acetone, and then the following night with chloroform. After drying *in vacuo*, a blue–black powder of the target compound is obtained (0.13 g, 17%). The use of metallocarbonyl compounds is the preferred method when the reactivity of phthalonitrile derivatives is not high. CoNc [25] (Nc = naphthalocyanine, Fig. 1) and FeNc [26] have been synthesized by this metal-carbonyl method.

(k) Cl_2SiPc from isoindoline [27]. In a flask equipped with a water condenser, a mechanically stirred mixture of 36.5 g (0.25 mol) of 1,3-diiminoisoindoline, 41.5 mL (0.36 mol) of silicon tetrachloride, and 415 mL of quinoline is slowly brought to reflux (219 °C). It is maintained at this temperature for 30 min and then cooled to 184 °C and filtered. The purple crystalline product is washed with quinoline, benzene, methanol, and acetone and then dried at 110 °C to yield 27.4 g (71% based on 1,3-diiminoisoindoline). This Cl_2SiPc can be recrystallized from 1-chloronaphthalene (*ca.* 60–70%) and can be hydrolyzed in H_2O/MeOH containing NaOH, pyridine/aqueous ammonia, or dilute H_2SO_4 to produce $(OH)_2SiPc$.

2.3. Phthalocyanines with specific symmetry

2.3.1. Metal-free species

(a) A D_{2h} symmetry H_2Pc from trichloroisoindolenine and isoindoline 2,16(or 17)-*bis*(4-*tert*-butylphenoxy)-9(or 10),23(or 24)-dinitroPc [28]. 100 mg (0.34 mmol) of 6-(4-*tert*-butylphenoxy)diiminoisoindoline and 69 mg (0.068 mmol, 1.5 equiv.) of triethylamine are placed in a dried 100-mL round-bottomed flask equipped with a magnetic stirrer and sealed with a septum. Dry nitrogen is swept through the flask by the two-needle method; and 30 mL of freshly dried THF is added with a syringe, and the mixture is stirred to dissolve the diiminoisoindoline. The mixture is then cooled to approximately 0 °C in a salt and ice bath. A solution of 90 mg (0.34 mmol) of 6/7-nitro-1,3,3-trichloroisoindolenine (Fig. 1) in 10 mL of dry THF is gradually added by means of a syringe over a 15-min period. The reaction mixture is then stirred for 1 h at approximately 0 °C and allowed to slowly warm to room temperature with stirring over a 6-h period. During this time, the solution turns from yellow to a yellow–green color. The reaction mixture is then filtered to remove the triethylamine hydrochloride and is then returned to the reaction flask. 37 mg (0.34 mmol) of hydroquinone and 54 mg (1 mmol) of sodium methoxide are added to the reaction vessel, which is equipped with a reflux condenser and is refluxed under nitrogen for 6 h. The solution is cooled to room temperature and filtered, leaving a dark blue–black residue. The residue is washed by boiling in water and filtering until the filtrate is clear; this procedure is then repeated with ethanol. The yield of the target compound is 48%.

(b) D_{2h} and C_{2v} symmetry H_2Pc from *bis*-phthalonitrile (Fig. 2, [29]). *Bis*-phthalonitriles ((*S*)-**1** and (*R*)-**1**) prepared from (*S*)-(-)-dihydroxy-1,1'-binaphthyl and (*R*)-(+)-dihydroxy-1,1'-binaphthyl and 3-nitrophthalonitrile are converted into the isoindoline derivatives by bubbling ammonia gas in dry methanol followed by reaction with 2 equiv. of 5,6-dimethoxyisoindoline [19] in refluxing *N, N*-dimethylaminoethanol for 3 h. After evaporation of the solvent, the residue is separated on a basic alumina column (Act III) using CH_2Cl_2/methanol (9:1 v/v) and then CH_2Cl_2/pyridine (1:1 v/v). The blue–green fraction is collected and separation is first attempted by gel-permeation chromatography using Bio-beads SX-2 (Bio-rad)

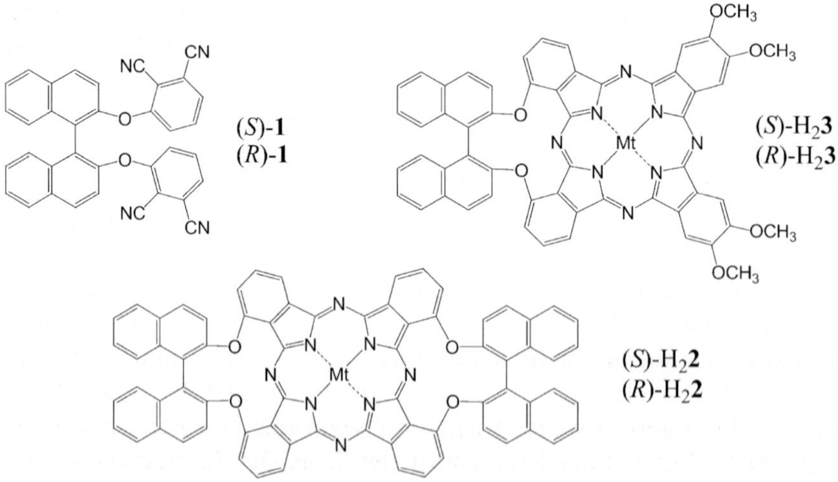

Fig. 2. Structures of (*S*)-**1**, (*R*)-**1**, (*S*)-H_2**2**, (*R*)-H_2**2**, (*S*)-H_2**3**, and (*R*)-H_2**3**.

and CH_2Cl_2/methanol (9:1 v/v). Three blue–green bands appear, but the separation between the first two bands is usually too small to enable a full separation. The third band (blue) has been identified as (*S*)-H_2**2** and (*R*)-H_2**2** with two binaphthyl units via fast atom bombardment (FAB) mass spectroscopy (5–7%). The first and second bands can be mixed together and then separated on a column of Bio-beads SX-8 (Bio-rad) using THF as the eluent. The first band is collected and recrystallized from CH_2Cl_2/EtOAc and then THF/EtOAc to give optically active non-centrosymmetric (*S*)-H_2**3** and (*R*)-H_2**3** as a blue–green powder in 21–26% yield.

(c) C_{4h} symmetry H_2Pc using phthalonitrile substituted with a bulky group at the 3 position. 1,8,15,22-*Tetrakis*(*p*-*n*-butylbenzyloxy)Pc [30]. Lithium (0.30 g) is suspended in 30 mL of 1-octanol. The suspension is heated to 170 ºC and stirred for 4 h. 0.35 g (1.2 mmol) of 3-(*p*-*n*-butylbenzyloxy)phthalonitrile (Fig. 1) in 4 mL of dried THF is added to the homogeneous solution, which is then cooled to 40 ºC. The temperature is raised to 60 ºC, and the mixture stirred for 12 h. The temperature is raised again to 120 ºC, and the solution is stirred for another 2 h. The mixture is cooled to room temperature, and the reaction is quenched with methanol and water (1:1 v/v) to form a blue precipitate. The precipitate is collected by centrifugation, washed successively with water, methanol, and hexane, and dried to give 126.7 mg (40%) of a dark green solid.

2.3.2. MetalloPcs

(a) A D_{2h} symmetry ZnPc [31]. Zinc acetate (1 equiv.), 4,5-dimethoxyphthalonitrile (1 equiv.), and 3,6-diphenylphthalonitrile (3 equiv.) are mixed thoroughly and heated at 250–270 ºC for *ca.* 20 min. After cooling, the reaction mixture is washed with water and ethanol and chromatographed on basic alumina with chloroform as the eluent. The first band is collected and further separated by gel-permeation chromatography using a Bio-bead SX-8 (Bio-rad) column with THF as the eluent. The initial band observed is a slight trace of ZnPc (>0.01%) originating from three 3,6-diphenylphthalonitrile and one 4,5-dimethoxyphthalonitrile. The second green band (yield of 9% after two recrystallizations from toluene) is the D_{2h} symmetry ZnPc target compound.

(b) Adjacently substituted NiPc [32]. Lithium (10 mg, 1.4 mmol) is heated at 100 °C in 1-hexanol (3 mL) until all of the metal is dissolved. After cooling to room temperature, 3,6-diphenylphthalonitrile (1 g, 3.57 mmol) is added and heated to 170 °C for *ca.* 3 min, to give a yellowish-green solution. With the temperature maintained at 170 °C, 2,3-dicyano-5,6-diethyl-1,4-pyrazine (1 g, 5.37 mmol, Fig.1) dissolved in hot 1-hexanol (*ca.* 3 mL) is added drop wise via a syringe. The solution immediately turns dark blue, and the reaction is continued for a further 1 h. After boiling off most of the solvent, DMF (20 mL) and $NiCl_2{\cdot}2H_2O$ (3.7 g, 22 mmol, 10 equiv.) is added, and the solution is kept at 170 °C for 30 min. The mixture is then poured into water, and the resulting solid filtered off, washed with water and methanol, and dried *in vacuo*. The residue is briefly passed through a short column (silica, $CHCl_3$/methanol (20:1 v/v)) and the blue mixture obtained is further purified by column chromatography (silica, toluene/methanol (1:0–30:1 v/v)) to give 151 mg of the NiPc derivative target compound.

(c) An adjacently substituted C_{2v} symmetry ZnPc via a half-Pc intermediate 2,3,23,24-tetramethoxy-Pc [33]. 0.07 g of lithium (10 mmol) and 2.56 g (20 mmol) of phthalonitrile is added to 20 mL of freshly distilled methanol. The solution is refluxed for 2 h under argon. 1.84 g (10 mmol) of 4,5-dimethoxyphthalonitrile in 10 mL of octanol is added to 2 mL (2 mmol of phthalonitrile) of the dark green solution. The mixture is heated to 100 °C and stirred overnight. Zinc acetate is added to the mixture, which is stirred for a further 8 h. The mixture is poured into 50 mL of methanol/water (1:1 v/v) to give a blue precipitate, which is collected by means of a centrifuge. The residue is passed through a silica gel column to remove some impurities by eluting with THF/hexane (1:1 v/v). Further purification is performed by GPC on Bio-beads gel and the ZnPc and 2,3-dimethoxyZnPc products are collected in 28% and 22% yields, respectively.

2.4. Deformed Pcs

(a) [1,4,8,11,15,18,22,25-octaphenylphthalocyanininato]nickel [34]. Lithium (100 mg, 15 mmol) is heated at 100 °C in 1-hexanol (5 mL) until all of the metal is dissolved. After cooling to room temperature, 3,6-diphenylphthalonitrile (1 g, 3.6 mmol) is added and reacted for 1 h at 170 °C. Most of the solvent is boiled off, DMF (5 mL) is added, and the mixture is poured into 200 mL of water. The precipitate is filtered off, washed with water and methanol, and dried under reduced pressure. The crude products are further purified using column chromatography (silica, toluene/cyclohexane (1:1 v/v)) to give 40 mg (4.0%) of 1,4,8,11,15,18,22,25-octaphenylPc (R_f = 0.7) as a yellowish-green solid. This compound has a saddle-shaped D_{2d} symmetry deformation due to steric repulsion between the bulky phenyl groups.

(b) 1,4,8,11,15,18,22,25-octafluoro-2,3,9,10,16,17,23,24-octaperfluoro(isopropyl)Pc [35]. Perfluoro-(4,5-diisopropyl)phthalonitrile (1020 mg, 2.04 mmol, Fig. 1) and $Fe(OAc)_2$ (139 mg, 0.799 mmol) are placed in a pressure vessel, evacuated, and then heated to 210 °C for 6 h. The greenish-blue mixture is allowed to cool, washed with toluene, and sublimed at 100 °C to remove unreacted phthalonitrile. The crude material is purified by silica-gel chromatography with 20–30% acetone–hexane as the eluent. The target fluorinated Pc is obtained in 22.4% yield, together with F_{64}PcFe (50.2%). The metal-free species exhibits a dome-like molecular distortion.

2.5. Cofacial homodimers

2.5.1. μ-(O, N, or C) dimers

(a) $(SiPc)_2O$ [36]. A mixture of $(OH)_2SiPc$ (785 mg) and $(OH)SiPc(OSi(CH_3)(OSi(CH_3)_3)_2$ (543 mg) is finely ground, heated at *ca.* 300 °C under vacuum for 20 h, and then subjected to a 3-h carbon tetrachloride extraction (250 mL). The extract is evaporated to dryness and the product chromatographed on Woelm alumina of activity II with petroleum ether and benzene as eluents. This gives fractions containing one-ring (22 mg), two-ring (50 mg), three-ring (25 mg), four-ring(28 mg), and some five-ring oligomers (10 mg). Pure crystals of the one-ring oligomer can be obtained via recrystallization from an ethanol/*n*-hexane solution (2:3). Red-reflecting, blue-transmitting, NMR-pure crystals of the two-ring oligomer can be obtained via recrystallization from *n*-heptane.

(b) $(FePc)_2O$ [37]. Finely ground FePc (200 mg) is added to a solution of 2-propylamine (0.75 mL) in 1-chloronaphthalene (20 mL), the mixture is stirred for 30 min, and then filtered. The solution obtained is stirred in air for 16–20 h. The use of a wide-necked beaker facilitates the complete evaporation of 2-propylamine. The solid formed is separated from the mother liquor by filtration, washed with methanol, and dried *in vacuo* to yield *ca.* 100 mg.

(c) $(FePc)_2N$ [38]. NaN_3 (1 g) is added to a 1-chloronaphthalene solution (20 mL) containing FePc (1 g) at 265 °C with constant stirring for 1 h. During this period, N_2 gas is evolved and the color of the solution changes from the typical blue–green of FePc to the dark blue–purple of the suspended product. Since the reaction is heterogeneous, it is difficult to determine visually when it is complete. Progress can be monitored by taking small aliquots of the reaction mixture, filtering the sample, and washing the filtrate with water and methanol. Once the filtrate is dried, the IR spectrum can be examined for the disappearance of the N_3^- band and the appearance of the anti-symmetric Fe–N–Fe band (915 cm^{-1}). Once the reaction is completed, the product is filtered, washed with $CHCl_3$ (to remove 1-chloronaphthalene) and alcohol, and finally with water (to remove any excess N_3^- and other water-soluble impurities), and then dried under vacuum to yield 90%.

(d) $(RuPc)_2N$ [39]. RuPc (300 mg), or (RuPc)(pyridine)$_2$ (350 mg), and NaN_3 (300 mg) are suspended in 1-chloronaphthalene and heated with stirring at 280–290 °C for 20 h. After cooling to room temperature, the suspension is filtered, the bluish solid $(RuPc)_2N$ is washed with water and acetone, and then heated for 30 min at 250 °C under vacuum (10^{-2}–10^{-3} mmHg) to give the product, which is indefinitely stable to air, and practically insoluble in non-donor or weakly donor solvents. The yield is not reported. The oxidation state of Ru in this complex is both (III) and (IV) (i.e. this is a mixed-valence dimer).

(e) $(FePc)_2C$ [40]. A mixture of FePc (500 mg), CI_4 (500 mg), and sodium dithionite (500 mg) in 1-chloronaphthalene (10 mL) is heated at 140–150 °C for 30 min in air with stirring. After cooling and filtration, the solid residue is washed with water (to eliminate excess dithionite) and then with acetone, and dried at reduced pressure (10^{-2} mmHg) at room temperature for 2 h to give 200 mg of the product.

(f) $(RuPc)_2C$ [41]. H[Cl_2RuPc] (1 g) is dissolved in 0.2 M KOH in 2-propanol (50 mL) and boiled, with vigorous evaporation of the alcohol, until the color of the solution changes (<10 min). Chloroform (3 mL) is added cautiously to the boiling solution, and after the vigorous reaction has ceased, the precipitation of the μ-carbido complex is completed through the addition of water (50 mL). The crude product is filtered off, washed thoroughly with copious amounts of water, and dissolved in a mixture of acetone/methanol (3:1, 60 mL) and KOH (200 mg), and then heated to 40 °C for 1 h. The residue is filtered off and washed with water. After drying, it is dissolved in warm (60 °C) pyridine. The pyridine adduct [{(pyridine)RuPc}$_2$(μ-C)] is precipitated with *n*-pentane and, after isolation, heated to 200–250 °C under low pressure to remove the pyridine (yield not reported).

2.5.2. Sandwich dimers

(a) $(Ln^{III}Pc_2(NBu)_4$ (reduced non-radical form) [42]. A mixture of $Ln(CHCOO)_3 \cdot nH_2O$ (1 mmol), phthalonitrile (0.8 mmol), and Na_2CO_3 (0.5 mmol) is placed in a Pyrex tube (4 cm in diameter and 20 cm in length) equipped with a glass tube and a thermocouple. The test tube is placed in a mantle heater equipped with a programmable temperature controller. The mixture is heated from room temperature to 150 °C at a constant rate over 20 min, and then it is kept at 150 °C for 30 min. The mixture melts and develops a green color. The temperature is then raised again to 280 °C over 15 min and is kept at 280 °C for a further 20 min. The mixture changes color to dark blue and then solidifies. The glass tube is connected to a vacuum pump, and the temperature is raised to 340 °C over 20 min *in vacuo* and is then kept at 340 °C for 30 min in order to remove any unreacted phthalonitrile. The dark blue crude reaction product is cooled to room temperature, ground to a powder, dissolved in a small amount of DMF under sonication, and the resulting solution (*ca.* 15 mL) is filtered to remove any insoluble residue. The solution is placed on an alumina column and eluted with methanol containing 2% CH_3ONa. The deep blue band that elutes after the green band for $[Ln^{III}Pc_2]^0$ is collected, an excess amount of tetrabutylammonium bromide is added, the resulting solution is concentrated to a volume of 30 mL on a rotary evaporator, and then cooled in a refrigerator prior to filtration. After recrystallization of the dark crude filtrate from acetone, the target product is obtained at a yield of 36–41%.

(b) $Ln^{III}Pc_2$ (radical form) by the nitrile method [43]. Excess phthalonitrile (recrystallized from methanol) is mixed in an approximately 14:1 molar ratio with the appropriate lanthanide acetate and vigorously ground together with a mortar and pestle. The mixture is then placed in a large reaction tube equipped with a water-cooled condenser, heated in a tin bath at 280–300 °C for 4–5 h, and cooled slowly. Unreacted phthalonitrile is collected on the sides of the reaction tube and the condenser. The dark green product that forms is separated from the unreacted phthalonitrile and reground. The product is then returned to a clean, dry reaction tube and heated in a tin bath at 300 °C *in vacuo* for 12–24 h. The product is washed with acetic anhydride and cold acetone, dissolved in a minimum amount of chloroform, and gravity filtered to remove any H_2Pc that may have formed. The solution is separated on a 15 × 2 in basic alumina column using a 5% methanol/toluene solution. In most cases, a green fraction is eluted first, followed by a blue fraction. The portion containing the green product is evaporated to dryness on a rotary evaporator and then dried in a vacuum oven to give the target radical form of $Ln^{III}Pc_2$. The ratio of blue and green products obtained depends on the lanthanide used, with the blue product being increasingly favored toward the left of the periodic table.

(c) $Ln^{III}Pc_2$ (radical form) by the use of Li_2Pc [44]. A mixture of Li_2Pc (53 mg, 0.1 mmol) and $Ln(acac)_3{\cdot}nH_2O$ (150 mg, 0.3 mmol) is refluxed in trichlorobenzene (10 mL) for 10 h under nitrogen. The resulting dark blue solution is cooled and evaporated under reduced pressure to give a residue, which is chromatographed on a silica gel column with $CHCl_3$ as the eluent. A pale blue band containing a trace amount of H_2Pc is observed, followed by a greenish-blue band, which is collected and concentrated. The crude product is purified by recrystallization from a mixture of $CHCl_3$ and MeOH. The yield for the Eu complex is 19 mg (32%) and for the Gd complex is 15 mg (25%).

(d) Soluble $Ln^{III}Pc_2$ (radical form) in the presence of DBU [44]. Under a slow stream of nitrogen, a mixture of 4,5-*bis*(pentyloxy)phthalonitrile (300 mg, 1.00 mmol) "DBU" (78 mg, 0.51 mmol) and $Ln(acac)_3{\cdot}nH_2O$ (30 mg, 0.06 mmol) is heated in amylalcohol (4 mL) under reflux for 8 h to give a dark greenish-blue solution. The volatiles are removed *in vacuo* and then the residue is chromatographed with $CHCl_3$/hexane (2:3 v/v) as the eluent. The intial green band contains a small amount of metal-free Pc. A dark greenish-blue band containing the sandwich compound is then collected and evaporated under reduced pressure. The crude product obtained is purified further by the same chromatographic procedures. The yield for Eu is 121 mg (79%).

(e) $SnPc_2$ [45]. A freshly ground mixture of Cl_2SnPc (6.86 g, 9.8 mmol) and Na_2Pc (6.45 g, 11.6 mmol) is placed in a 200-mL round-bottomed flask and dried in an oven for 3 h at 110 °C. Technical grade 1-chloronaphthalene (140 mL) is added and the resulting solution is refluxed for 90 min in a nitrogen atmosphere. The solution is allowed to cool for 1 h and is then filtered. The precipitate is a purple gum-like solid which when peeled away yields a layer of red crystals mixed with green powder. The red crystals and green powder are washed with ether and the slurry containing a light green precipitate is decanted, leaving the crystals of the target compound behind. The yield is about 0.5 g or 4.5%.

2.5.3. Metal–metal directly linked dimers

(a) MoPc–MoPc [46]. *Method A*: A mixture of O = MoPc (*ca.* 1 g) and triphenylphosphine (*ca.* 10 g) is heated at 380 °C for 3 h. After cooling, the residue is washed with CH_2Cl_2, leaving a blue powder, which is purified further by sublimation at 550 °C under high vacuum (yield not reported). *Method B*: A mixture of $Mo(CO)_6$ (*ca.* 1 g) and Li_2Pc (*ca.* 5 g) is refluxed in decaline (50 mL) for 10 h. The residue is washed with copious amounts of acetone and dried (yield not reported).

(b) IrPc–IrPc [47]. ClIrPc coordinated by a phthalonitrile molecule as an axial ligand (2 g, 2.2 mmol) and $NaBH_4$ (2.5 g) are refluxed in THF for 12 h. The hot solution is poured into water (250 mL) and the target compound is collected by filtration, washed thoroughly with hot water, and dried (yield not reported).

(c) RhPc–RhPc [48]. *Method A*: $H(X)_2$ RhPc (200 mg, X = Cl, Br) is kept at 300 °C under reduced pressure for 24 h. The dark red solid obtained is the target compound and requires no further purification (yield not reported). *Method B*: $H(X)_2$ RhPc (200 mg, X = Cl, Br) is refluxed in 1-chloronaphthalene (25 mL) for 12 h. The dark green solution is filtered, the residue washed several times with ether, and then with acetone to remove traces of soluble mononuclear RhPc (yield not reported).

2.6. Cofacial homotrimers and homooligomers

(a) μ-oxo-SiPc trimer and tetramer [36]. In the synthesis described above for μ-oxo SiPc dimer (Section 2.5.1(a), the trimer (25 mg), tetramer (28 mg), and pentamer (10 mg) are obtained during column chromatography using alumina with petroleum ether and benzene as eluents. The trimer can be recrystallized from toluene/*n*–heptane (1:4) solution.

(b) Soluble sandwich-type (Pc)Y(Pc)Y(Pc) and (Pc)Dy(Pc)Dy(Pc) [49]. A powder of $Yb(CH_3COO)_3{\cdot}4H_2O$ or $Dy(CH_3COO)_3{\cdot}4H_2O$ (0.26 mmol) is added to a 1-octanol suspension (2.5 mL) of metal-free octabutoxyPc (110 mg, 0.1 mmol) under reflux. After a further 4 h of refluxing, the reaction mixture is added to methanol (100 mL). The resulting precipitate is purified by column chromatography using silica gel with $CHCl_3$ as the eluent, and then by size-exclusion chromatography (Bio-beads SX1, Bio-rad) using benzene. Finally, the dark blue solid is recrystallized from $CHCl_3$ and methanol (yield not reported).

(c) (Pc)In(Pc)In(Pc) [50]. This compound is prepared from $InSn_4$ filings ($InSn_4$ alloy is prepared by melting indium and tin, in stoichiometric proportions, both 99.999% pure, *in vacuo* in a glass ampoule) and phthalonitrile, in 1:2 weight proportion, mixed together, and pressed into pellets. The pellets are inserted into an evacuated glass ampoule and sealed. The ampoule is heated at 480 K for 2 days resulting in the formation of two kinds of well-formed crystals. One set of crystals has been identified as SnPc, while the second set has been identified as (Pc)In(Pc)In(Pc) (yield not reported).

2.7. Cofacial heterodimers and heterotrimers

(a) U (OEP)(Pc) [51]. $Cl_2U(OEP)$[52] (OEP = 2,3,7,8,12,13,17,18-octaethylporphyrin) and Na_2Pc[53] are used as starting materials for the synthesis of U(OEP)(Pc). These compounds are prepared according to literature procedures after which they are heated *in vacuo* at 110 °C for 3 h prior to use. A 1-chloronaphthalene solution (30 mL) containing 0.38 g of $Cl_2U(OEP)$ (0.59 mmol) is added to 1.0 g of Na_2Pc (2.68 mmol). The mixture is refluxed for 40 h under argon and the progress of the reaction is monitored by UV–visible spectroscopy. The mixture is then cooled and precipitated by addition of hexane. The crude product is filtered and dissolved in $CHCl_3$, and the solution is concentrated and purified by column chromatography (SiO_2, 30 × 5 cm) with CH_2Cl_2/heptane (1:1) as the eluent. A fraction containing trace H_2OEP is eluted first. A second fraction containing U(OEP)(Pc) is then collected, and finally a trace of UPc_2 is observed. No $U(OEP)_2$ has been detected. After evaporation, the crude solid is recrystallized through the addition of heptane to yield 0.34 g (45%) of U(OEP)(Pc).

(b) Li[Eu(Pc)(TPyP)] [54]. A mixture of 210 mg (0.43 mmol) of $Eu(acac)_3{\cdot}nH_2O$ and 90 mg (0.15 mmol) of H_2TPyP (TpyP = 5,10,15,20-tetrapyridylporphyrin, Fig. 1) is refluxed in 10 mL of 1,2,4-trichlorobenzene under nitrogen for 6 h. The resulting dark cherry-red solution is cooled to room temperature and then 120 mg (0.23 mmol) of Li_2Pc is added. The whole mixture is refluxed for a further 12 h, then the volatiles are removed *in vacuo*, and the residue is chromatographed on a silica gel column with $CHCl_3$ as the eluent. A small amount of unreacted H_2TpyP and the triple-decker complex $Eu_2(Pc)_2(TPyP)$ are collected as the first fraction. The column is then eluted with $CHCl_3$/methanol (6:1) to give a mixture containing Li[Eu(Pc)(TPyP)] and Eu(acac)(TPyP), which are separated by repeated chromatography with $CHCl_3$/methanol (20:1) as the eluent. The solution is evaporated to give the target product as a dark violet solid, which is dried *in vacuo*. The yield is 91 mg (49%).

(c) Eu(Pc)(TPyP) [54]. A solution of 30 mg (0.023 mmol) of Li[Eu(Pc)(TPyP)] in 10 mL of $CHCl_3$ and 5 mL of methanol is stirred at room temperature in air for 3 days. The color changes gradually from dark blue to dark brown. The volatiles are then removed *in vacuo* and the residue is chromatographed with $CHCl_3$/methanol (30:1) as the eluent. A yellowish-brown band is observed which is collected and evaporated to give the Eu(Pc)(TPyP) product as a dark purple solid (19 mg, 65%), which can be recrystallized from $CHCl_3$/methanol. The column is then eluted with $CHCl_3$/methanol (6:1) to recover the unreacted Li[Eu(Pc)(TPyP)] (10 mg, 35%).

(d) Lu(Nc)(Pc) [55]. Lutetium acetate (41 mg, 0.1 mmol) is added, under nitrogen, to a solution of Li_2Pc (52 mg, 0.1 mmol) and Li_2Nc (72 mg, 0.1 mmol) in 20 mL of 1-chloronaphthalene. The mixture is stirred at 245 °C for 4 h. After cooling, the reaction mixture is poured onto an alumina column. The small amount of $Lu(Nc)_2$ produced during the synthesis is eluted with $CHCl_3$ and emerges from the column with the 1-chloronaphthalene; a subsequent blue band contains the Lu(Nc)(Pc) target product. A green band which contains $Lu(Pc)_2$ is finally eluted with CH_2Cl_2. Lu(Nc)(Pc) is isolated as a dark blue powder after evaporation of the solvent. The yield is 23.8 mg (17%). R_f (silica gel, CH_2Cl_2) = 0.6 (R_f = 0.4 for $Lu(Pc)_2$ under the same conditions).

(e) Ln_2Pc_2[T(4-OCH_3)PP] [56]. A 1,2,4-trichlorobenzene solution of H_2T(4-OCH_3)PP (T(4-OCH_3)PP = T*etrakis*(*p*-methoxyphenyl)porphyrin) and $Ln(acac)_3{\cdot}nH_2O$ (Ln = Nd, Eu, Gd) is refluxed for 4 h under a slow stream of argon to form the mono(porphyrinato)lanthanide. Li_2Pc is added to this solution, after cooling under argon, the mixture is refluxed again for 2 h. Green microcrystalline powders are obtained after cooling, addition of hexane, redissolving of the precipitate in CH_2Cl_2, chromatography on a basic alumina column with a 2% methanol/CH_2Cl_2 mixture as the eluent, concentration, and addition of hexane. All three compounds give analytical and spectroscopic data consistent with a Ln_2Pc_2[T(4-OCH_3)PP] formulation (Ln = Nd(III), Eu(III), and Gd(III)).

(f) $Ln_2(Pc)(TPP)_2$, $Ln_2(Pc)_2(TPP)$, and Ln(Pc)(TPP) [57]. A mixture of H_2TPP (62 mg, 0.10 mmol) (TPP = 5,10,15,20-tetraphenylporphyrin, Fig. 1) and $Ln(acac)_3{\cdot}nH_2O$ (Ln = Sm, Eu, and Gd) (150 mg, ~0.3 mmol) is refluxed in 1,2,4-trichlorobenzene (20 mL) for 4 h under a slow stream of nitrogen. Li_2Pc (80 mg, 0.15 mmol) is added once the resulting dark cherry-red solution is cooled to room temperature. The mixture is refluxed for a further 3 h and then the solvent is removed under vacuum. The residue is chromatographed on a silica gel column (Merck, 70–230 mesh) with CH_2Cl_2 as the eluent to give an olive–brown band which primarily contains the $Ln_2(Pc)(TTP)_2$ triple-decker complexes with a small amount of unreacted H_2TPP. The column is further eluted with CH_2Cl_2/methanol (10:1) to give a dark greenish blue band which contained the triple-decker complex $Ln_2(Pc)_2(TPP)$ and a trace amount of the Ln(Pc)(TPP) double-decker complex. The first fraction is concentrated and chromatographed again with CH_2Cl_2/hexanes (1:1) as the eluent to remove the H_2TPP. The solution containing the Ln_2(Pc)$(TPP)_2$ triple-deckers is evaporated to give a violet–blue solid, which is recrystallized from a CH_2Cl_2/methanol mixture. Similarly, the second fraction is concentrated and loaded onto a silica gel column with CH_2Cl_2/methanol (20:1) as the eluent. The initial yellowish-brown solid contains Ln(Pc)(TPP) (2–5 mg). The column is then further eluted with CH_2Cl_2/methanol (10:1) to give a blue band containing $Ln_2(Pc)_2$(TPP). The crude product is purified by chromatography using similar conditions, followed by a recrystallization from a mixture of CH_2Cl_2/methanol.

2.8. Co-planar dimers

(a) Co-planar Pc dimer sharing a common benzene ring [58]. Crude 5-neopentoxy-1,3-diiminoisoindoline obtained from 5.70 g (26 mmol) of 5-neopentoxyphthalonitrile and *bis*(1,3-diiminoisoindoline) obtained from 170 mg (0.96 mmol) of 1,2,4,5-tetracyanobenzene are refluxed for 44 h in *N*, *N*-dimethylaminoethanol (25 mL) (Fig. 3(a)). The reaction mixture is allowed to cool and is then poured into 200 mL of water. The resulting mixture is filtered and washed with water until the filtrate becomes clear. The residue is washed again with methanol and then dried overnight in an oven at 60 °C to give 4.0 g of crude product. Flash chromatography of this product on silica gel using a 5 cm diameter column and 1:1 (v/v) toluene/hexane as the eluent results in 3.0 g of mononuclear tetraneopentoxyPc in 54% yield. Further elution using toluene followed by 2-methoxyethanol/toluene(2:98 v/v) yields a fraction consisting largely of the binuclear Pc target compound. This binuclear fraction is further purified by size-exclusion chromatography (Bio-beads SX1, Bio-rad) with THF as the eluent. Final purification of the binuclear Pc is achieved by vacuum liquid chromatography using toluene as the eluting solvent followed by gradually increasing the 2-methoxyethanol/toluene ratio (1:100–5:100), to give 196 mg (12% yield) of the blue powder of the binuclear H_2Pc target compound.

(b) Planar Pc–pyrazinoporphyrazine heterodinucleate [59]. 2[2],7[2],12[2]-Tri(*tert*-butyl)benzo-[*b,g,l*]-17[2],17[3]-dicyanobenzo-[*q*]-5,10,15,20-tetraazaporphine (200 mg, 0.268 mmol) obtained from 5-*tert*-butyldiiminoisoindoline and 5,6-dicyanodiiminoisoindoline by mixed condensation, is treated with ammonia gas in the presence of sodium methoxide in dry dioxane/methanol (3:1 v/v, 22 mL). The resulting diiminoisoindoline derivative and 2-*tert*-butyl-5,7-diimino-6*H*-pyrrolo[3,4-*b*]-pyrazine (240mg, 1.338 mmol) are mixed well and heated to *ca.* 230–250 ºC under a nitrogen atmosphere (Fig. 3(b)). The crude product is imposed on a basic alumina column using CH_2Cl_2 and a blue–green fraction is further purified by gel-permeation columns of Bio-beads SX-2, with THF as the eluent. The initial dinuclear fraction is collected and recrystallized from CH_2Cl_2/hexane to give the planar heterodimer target compound as a bluish powder (40 mg, 12%).

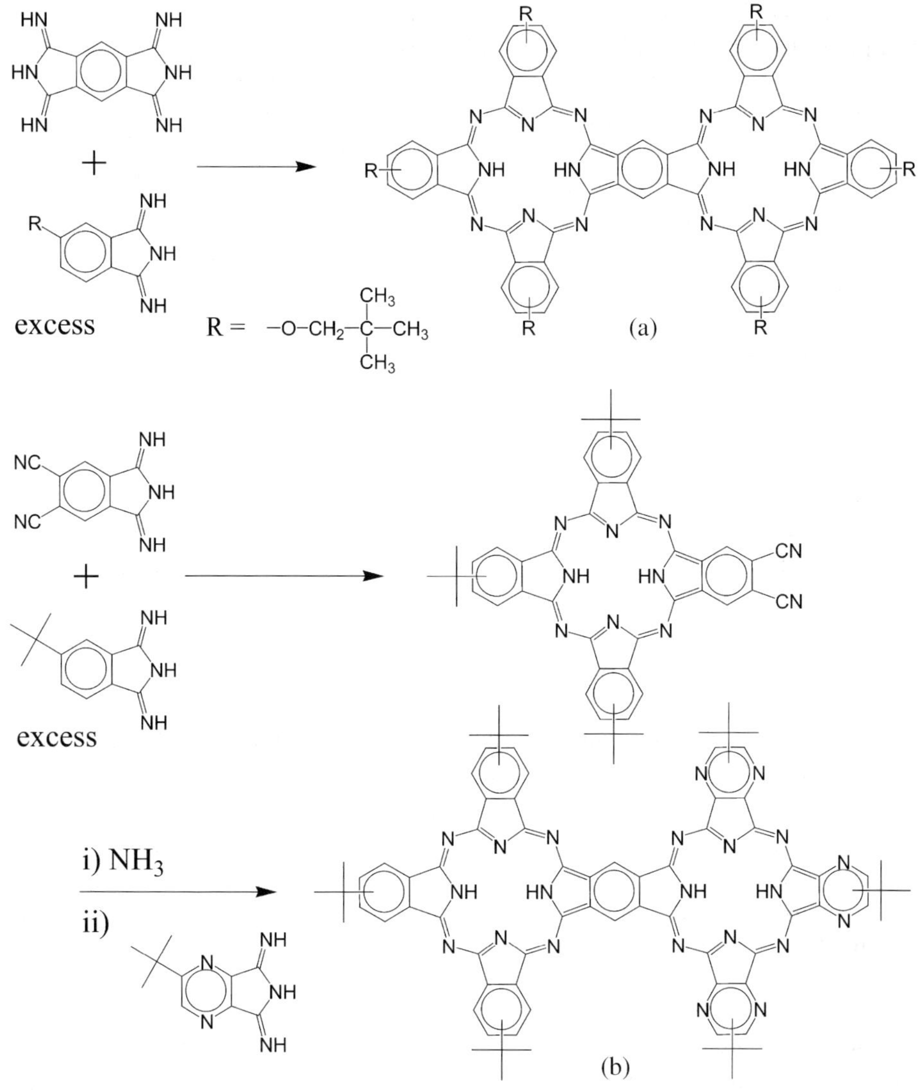

Fig. 3. Synthetic route to a planar homodinuclear Pc (a) and to a heterodinuclear Pc (b).

3. PROGRESS IN THE CHARACTERIZATION

Recent advances in synthetic methods and in chromatographic separation techniques have made a wide range of new Pc derivatives available for spectroscopic study. The optical and/or electrochemical properties of Pcs depend largely on the nature of their chemical structures. For example, excited states of Pcs are influenced by the in-plane π-structures, and peripheral substituents control the oxidation and reduction potentials. Axial substituents and non-planar π-distortions are also potentially important factors for controlling these properties. As a consequence, more rapid and sophisticated characterization methods are required to determine the chemical structures of newly synthesized Pcs. Modern instrumental analytical methods are satisfactory for these requirements. In this section, first, the application of mass spectrometry and infrared spectroscopy to Pcs is reviewed. Secondly, the importance of electronic absorption and magnetic circular dichroism (MCD) spectroscopy in the Pc chemistry is described. Finally, the progress in electrochemical characterization methods is briefly introduced.

3.1. Mass spectrometry

A mass spectrum is obtained by converting analytes into gaseous ions, which are then separated on the basis of their mass-to-charge ratios. Mass spectrometry (MS) is capable of providing a direct measurement of the molecular weight of a complex. Although inherently a destructive measurement, the required sample amounts are very small and all metallo Pcs can be potentially analyzed, unlike the case with NMR spectroscopy in which paramagnetic species are out of the subjects. From the 1960s to the 1980s, electron impact (EI) was the dominant MS method. However, since samples are required to have moderate vapor pressures under the measurement conditions, EI–MS could only be employed for a limited range of Pcs. The development of the FAB and secondary ion mass spectrometry (SIMS) techniques improved the situation to some extent. Many peripherally or axially substituted Pcs have been characterized through the use of these methods. More recently, soft ionization techniques such as electrospray ionization (ESI) or matrix-assisted laser desorption ionization (MALDI) finally enabled the analysis of even peripherally substituted, supramolecular, and self-organized Pcs. Discrimination of some structural Pc isomers can be achieved by employing the MS^n technique. Within this chapter, we focus primarily on the more modern ionization methods. Only selected references are cited in the case of EI [60–68] and field desorption (FD) studies [69].

3.1.1. FAB and SIMS methods

The FAB and SIMS techniques are usually employed in the case of high molecular weight species. The sample is mixed with a supporting matrix and is ionized by bombardment of energetic xenon, argon, or cesium atoms. Although FAB and SIMS tend to induce more sample fragmentation than soft ionization methods, satisfactory results can often be obtained for rigid Pc compounds. Glycerin and *m*-nitrobenzylalcohol (*m*-NBA) are the most frequently used matrices. Freas and Campana [70] examined the utility of mixtures of concentrated sulfuric acid and 85% phosphoric acid (1:2 v/v, H_2SO_4/H_3PO_4 matrix), and butyl benzyl phthalate as matrices. Partial decomposition of the Pc occurs when concentrated H_2SO_4 is used as the matrix, and as a consequence, $m/z = 129$ and 130 signals (corresponding to phthalonitrile +1 and +2, respectively) are detected. H_2SO_4/H_3PO_4 matrices generate no such fragments. The results of FAB analyses for unsubstituted metallo Pcs are summarized in Table 1. All but the PbPc mass spectra contain oxygen atom adducts, in addition to the M^+ and $[M + H]^+$ peaks, which are consistent with the existence of the Pc–oxygen interaction proposed by the ESR study [71]. In the case of cobalt and lead Pcs, demetallated $[Pc + 3H]^+$ peaks can also be detected and demetallated fragments appear in the case of PbPc. The lead ion is probably too large to remain in the Pc core during the ionization processes [72]. Table 2 summarizes FAB mass data for *tetrakis*(cumylphenoxy)Pc (Fig. 1, $Pc(CP)_4$). Only M^+ and $[M + 1]^+$ ions were observed. Although the result from vapor pressure osmometry (VPO) indicates that metallo $Pc(CP)_4$s form dimer and oligomer species in solution, the FAB data show no evidence of aggregation. More detailed study on $Pc(CP)_4$ was performed using laser ablation method as outlined in the subsequent section. No molecular ions were detected for platinum and palladium $Pc(CP)_4$s.

3.1.2. Laser ablation method

Laser ablation time-of-flight (TOF) MS is a less frequently used method for Pc compounds. Some noteworthy results were reported on the chemical structure of Pcs [73]. The ionization mechanism is based on laser irradiation of frozen

Table 1
FAB–MS (matrix: H_2SO_4/H_3PO_4) results of unsubstituted metallo Pcs

Metal	M^+	Observed ion peaks
H_2	514	$[H_2Pc]^+$, $[H_2Pc + H]^+$, $[H_2Pc + O]^+$, $[(H_2Pc + O) + H]^+$, $[H_2Pc + H_2O]^+$, $[H_2Pc + O_2]^+$, $[(H_2Pc + O_2) + H]^+$
Fe	568	$[FePc]^+$, $[FePc + H]^+$, $[FePc + O]^+$, $[(FePc + O) + H]^+$, $[FePc + H_2O]^+$, $[FePc + O_2]^+$, $[(FePc + O_2) + H]^+$
Co	571	$[CoPc]^+$, $[CoPc + H]^+$, $[CoPc + O]^+$, $[(CoPc + O) + H]^+$, $[CoPc + H_2O]^+$, $[CoPc + O_2]^+$, $[(CoPc + O_2) + H]^+$, $[Pc + 3H]^+$
Ni	570	$[NiPc]^+$, $[NiPc + H]^+$, $[NiPc + O]^+$, $[(NiPc + O) + H]^+$, $[NiPc + H_2O]^+$, $[NiPc + O_2]^+$, $[(NiPc + O_2) + H]^+$
Cu	575	$[CuPc]^+$, $[CuPc + H]^+$, $[CuPc + O]^+$, $[(CuPc + O) + H]^+$, $[CuPc + H_2O]^+$, $[CuPc + O_2]^+$, $[(CuPc + O_2) + H]^+$
Zn	576	$[ZnPc]^+$, $[ZnPc + H]^+$, $[ZnPc + O]^+$, $[(ZnPc + O) + H]^+$, $[ZnPc + H_2O]^+$, $[ZnPc + O_2]^+$, $[(ZnPc + O_2) + H]^+$
Pb	720	$[Pc + 3H]^+$

Table 2
FAB-MS results of metallo $Pc(CP)_4s$

Metal	M^+	Observed ion peaks
H_2	1354	$[H_2Pc(CP)_4]^+$, $[H_2Pc(CP)_4 + H]^+$
Be	1361	$[BePc(CP)_4]^+$, $[BePc(CP)_4 + H]^+$
Mg	1376	$[MgPc(CP)_4]^+$, $[MgPc(CP)_4 + H]^+$
Mn	1407	$[MnPc(CP)_4]^+$, $[MnPc(CP)_4 + H]^+$
Fe	1408	$[FePc(CP)_4]^+$, $[FePc(CP)_4 + H]^+$
Co	1411	$[CoPc(CP)_4]^+$, $[CoPc(CP)_4 + H]^+$
Ni	1410	$[NiPc(CP)_4]^+$, $[NiPc(CP)_4 + H]^+$
Cu	1415	$[CuPc(CP)_4]^+$, $[CuPc(CP)_4 + H]^+$
Zn	1416	$[ZnPc(CP)_4]^+$, $[ZnPc(CP)_4 + H]^+$
Pt	1458	
Pd	1547	
Pb	1560	$[PbPc(CP)_4]^+$, $[PbPc(CP)_4 + H]^+$, $[Pc(CP)_4 + 3H]^+$
Bi	1561	$[BiPc(CP)_4]^+$, $[BiPc(CP)_4 + H]^+$, $[Pc(CP)_4 + 3H]^+$

sample solutions, which causes ablation followed by ionizations of the sample [74]. McMillan et al. [73] performed the laser ablation mass analysis for metallo $Pc(CP)_4s$ at 77 K in benzene solution. As shown in Fig. 4, the M^+ ion is observed predominantly for lead (b), copper (c), and metal-free (d) Pcs, and less intense cluster ions such as M_2^+ and M_3^+ are associated with the molecular ions. Compared with the lead complex, the copper complex has a slightly more intense cluster signal, indicating a higher tendency to aggregation. The mercury complex shows far more intense cluster signals. The VPO study also supported these results. The proposed oligomeric structure of $HgPc(CP)_4$ is shown in Fig. 5.

3.1.3. MALDI method

MALDI is one of the emerging soft ionization techniques, which has enabled the mass determination of large bioorganic and synthetic organic molecules with molecular weights as high as 200,000 Da without marked decomposition of the analyte. The MALDI technique is based on placing the sample on an ultraviolet absorbing matrix. The matrix and sample are mixed and placed onto a sample plate. Under vacuum conditions, the sample disperses homogeneously

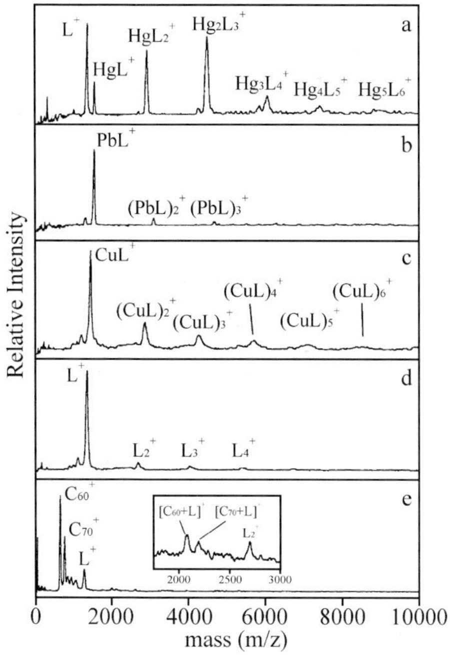

Fig. 4. Laser ablation TOF-MS spectra of (a) $HgPc(CP)_4$, (b) $PbPc(CP)_4$, (c) $CuPc(CP)_4$, (d) $H_2Pc(CP)_4$, and (e) C_{60}/C_{70} + $H_2Pc(CP)_4$ for internal standard. L represents $Pc(CP)_4$.

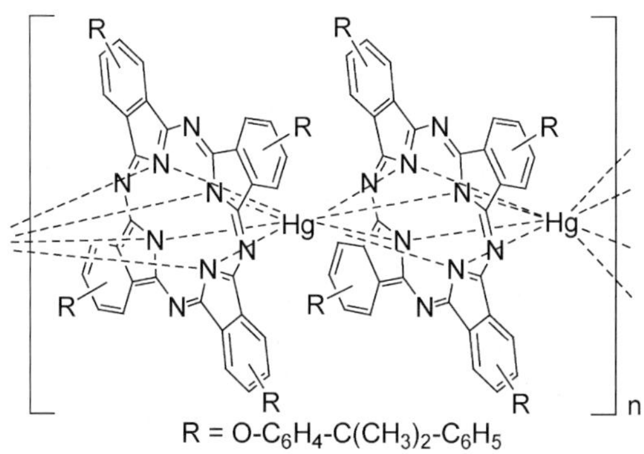

Fig. 5. Proposed oligomeric structure of $HgPc(CP)_4$.

within the matrix molecules. A pulse laser shot excites the matrix molecules, which is followed by rapid vaporization and charging of the sample molecules. Matrix selection is an important factor in obtaining clear molecular ion signals. Table 3 summarizes the practical examples of matrices suitable for Pcs.

Table 3
Examples of MALDI matrices applicable to Pcs

Matrix	Structure
Cyano-4-hydroxy cinnamic acid (CHCA)	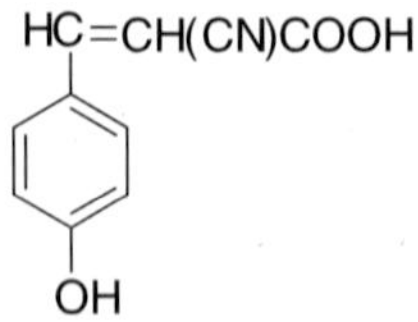
trans-4-Hydroxy-3-methoxycinnamic acid (ferulic acid)	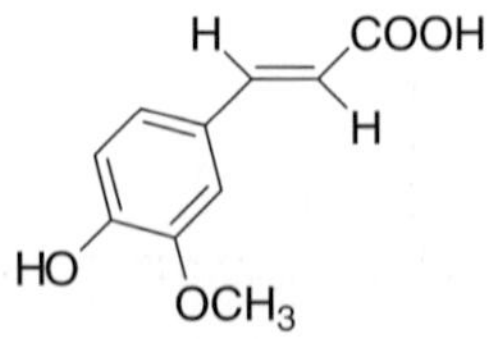
1,8,9-Anthracenetriol (dithranol)	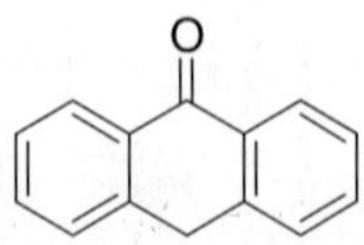
1,4-*bis*(5-Phenyloxazol-2-yl)benzene (POPOP)	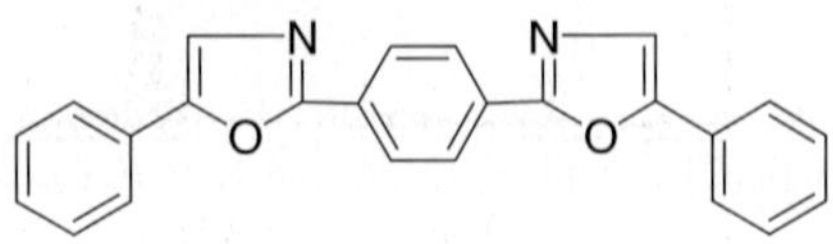
1,4-Benzoquinone (BQ)	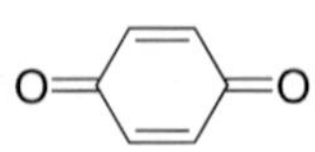
2,3,5,6-Tetrachloro-1,4-benzoquinone (TCQ)	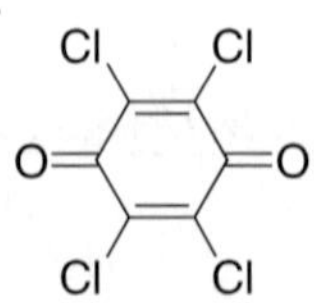

Fig. 6 shows the MALDI–MS spectra of a series of alkyne-bridged Pc oligomers [75]. In the case of monomer species, m/z = *ca.* 1,800 peaks (corresponding to dimer species) were observed in addition to M^+. This is due to the recombination of the ionized monomer species under the measurement conditions. Similarly, the dimer spectrum detected tetramer peaks. No tetramer ions were detected by FAB–MS. MALDI is capable of providing the molecular weight of relatively sensitive Pc oligomers.

Lindsey et al. [76] applied negative ion detection mode to Pcs. However, unusual fragmentation patterns complicate the results. Generally, the positive ion

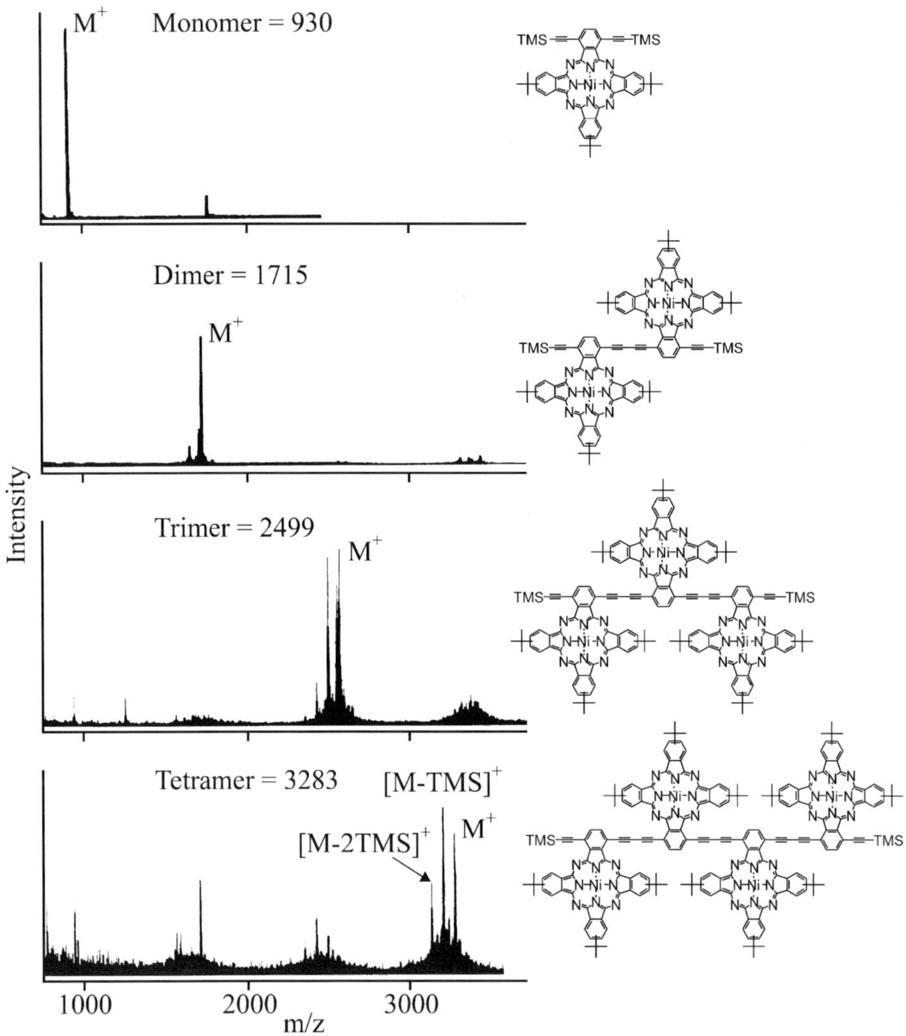

Fig. 6. MALDI–MS spectra of a series of acetylene-bridged Pc monomer, dimer, trimer, and tetramer (from top to bottom). Dithranol was used as the matrix. TMS represents the trimethylsilyl group.

mode is the preferred technique. Another example of MALDI studies is provided in References [77,78].

3.1.4. ESI method

ESI tends to be the soft ionization technique of choice within Pc chemistry. A solution containing the sample is forced through a capillary tip in the presence of an electric field. As the liquid becomes charged, the sample molecules begin to repel each other to form a fine mist of charged droplets. The solvent is evaporated, and the concentrated charged analyte explodes into multiple charged molecular ions. The ESI method tends to suppress fragmentation, and requires no matrix. Protic solvents tend to be preferred. Chloroform-containing methanol is used frequently as the solvent for Pc studies.

Fig. 7 contains the ESI–Fourier transform ion cyclotron resonance (FTICR) spectrum of a triple-decker $Eu_2(TPP)(Pc)_2$ [79]. The observed isotropic peaks (solid line) are in close agreement with the calculated pattern (broken line). However, the detection of the M^+ ion is insufficient to identify whether the chemical structure is (TPP)Eu(Pc)Eu(Pc) or (Pc)Eu(TPP)Eu(Pc). The M^+ ion could only be detected using the sustained off-resonance irradiation collision-induced dissociation (SORI–CID) technique. The M^+ fragmented into $[(TPP)Eu(Pc)^+]$ and $[(Pc)Eu(Pc)^+]$ and the structure was finally determined to be (TPP)Eu(Pc)Eu(Pc).

3.2. IR spectroscopy

IR spectroscopy is one of the most important analytical methods for molecular structure characterization. Unsubstituted metallo Pcs with D_{4h} symmetry are composed of 57 atoms and therefore have 165 vibrational freedoms. However, many of these are forbidden bands due to selection rules. In-plane vibrations have been classified as $X\text{(in-plane)} = 14A_{1g} + 13A_{2g} + 14B_{1g} + 14B_{2g} + 28E_u$, of which the E_u mode alone is IR-active. The out-of-plane vibrations can be represented as $X\text{(out-of-plane)} = 6A_{1u} + 8A_{2u} + 7B_{1u} + 7B_{2u} + 13E_g$, of which the A_{2u} mode is IR-active. For metallo Pcs containing the first-row transition elements, several vibrations are known to shift from high to low frequencies in the order Ni > Co > Fe > Cu > Zn [80]. However, the assignments of most of the bands can be achieved by comparing

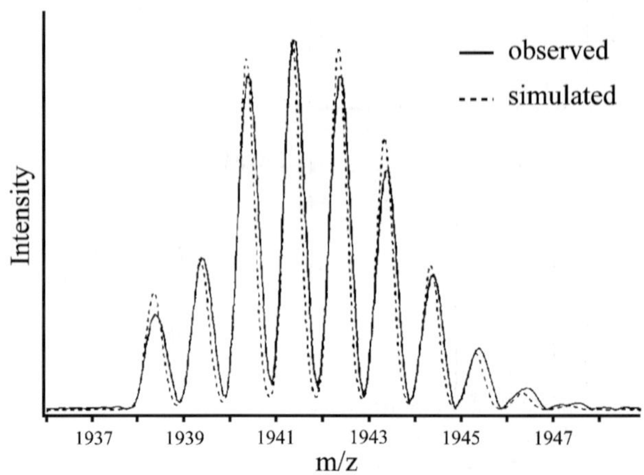

Fig. 7. Observed (solid line) and calculated (broken line) ESI–FTICR–MS spectra of $Eu_2(TPP)(Pc)_2$.

experimental spectra with density functional theory (DFT) level calculated frequencies. With the development of supercomputer resources, it has recently become possible to fully assign the IR bands of Pc derivatives [81–86]. Semi-empirical methods are known to be ineffective for frequency calculations. The combination of the B3LYP hybrid functional and 6–31G(d) basis set with an optimum scaling factor of 0.9613 gives practical results [87]. DFT calculations are also powerful for determining the chemical structure of low symmetry Pcs [88, 89].

Fig. 8 shows experimental IR spectra (solid lines) and calculated vibrational frequencies (solid bars) of nickel tribenzotetraazachlorin (TAC), dibenzotetraazabacteriochlorin (TABC), and -isobacterochlorin (TAiBC) (from top to bottom) [88].

Geometry optimization calculations predict nearly planar aromatic structures. The calculated modes were assigned to experimental bands based on both

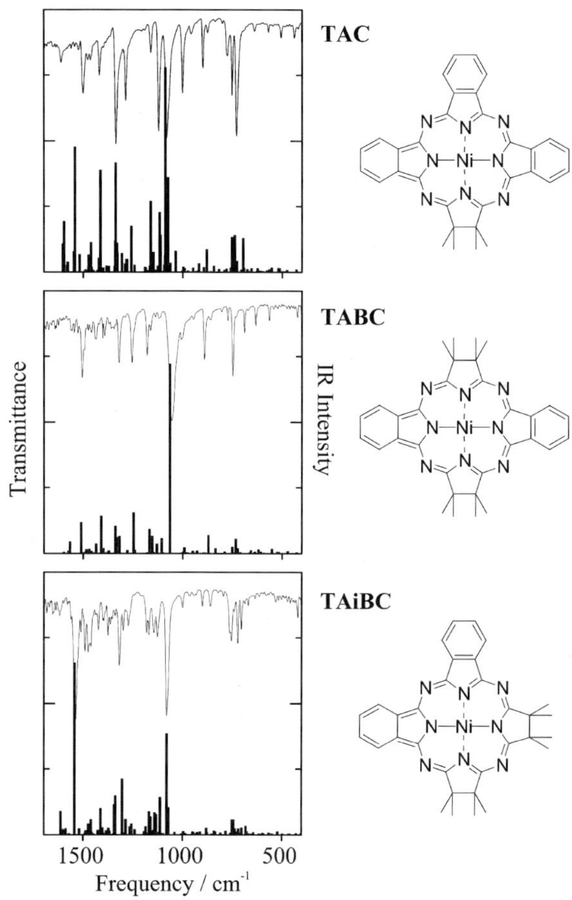

Fig. 8. Experimental IR spectra (solid lines) and calculated vibrational frequencies (solid bars) of the nickel complexes of TAC, TABC, and TAiBC (from top to bottom).

the frequency and intensity data. Some of the vibrational mode assignments are illustrated in Figs. 9–11.

The correspondence between the experiment and calculation is well brought out. IR spectral features reflect the molecular symmetry. In particular, differences between isomeric D_{2h} TABC and C_{2v} TAiBC were well reproduced by the calculations. Out-of-plane motion of the fused aromatics appeared in the energy region below 1,000 cm^{-1}. However, the intensity of these peaks tends to be low, and therefore, it is hard to utilize these for structural characterization. Most of the assigned modes result from in-plane motion of the peripheral hydrogen atoms. This type of mode spreads over the whole spectral region shown (400–1,700 cm^{-1}). The bands at 1,161 and 1,249 cm^{-1} of TAC and TABC, respectively, are of this type. The vibrational modes localized within the inner 18-π system (the tetraazaporphyrin (TAP) skeleton) were rarely recognized and generally had low intensity. However, medium-intensity bands at 1,543 and 1,542 cm^{-1} of TAiBC and TAC, respectively, were significantly affected by the inner skeletal vibrational modes. In the case of D_{2h} TABC, this kind of vibrational mode was hard to assign due to their low intensity. The above results illustrate that the DFT calculations can satisfactorily reproduce the experimental IR spectra and can enable the identification of the IR spectra of different isomeric species.

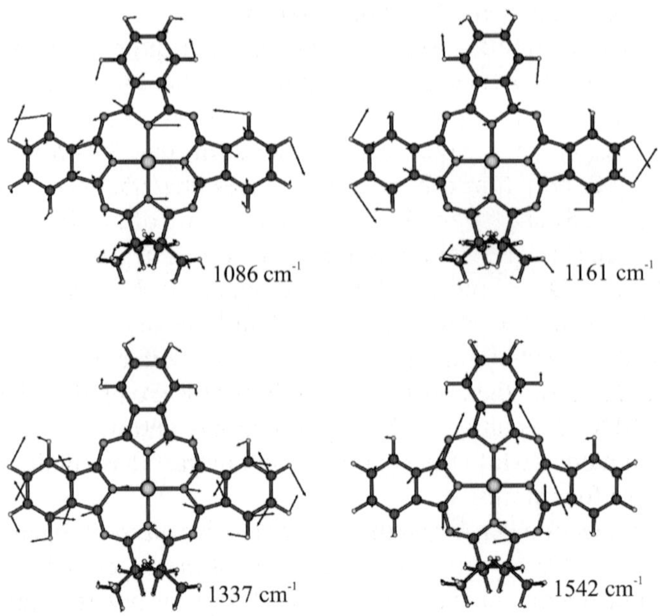

Fig. 9. Atomic movements at selected predicted frequencies for TAC.

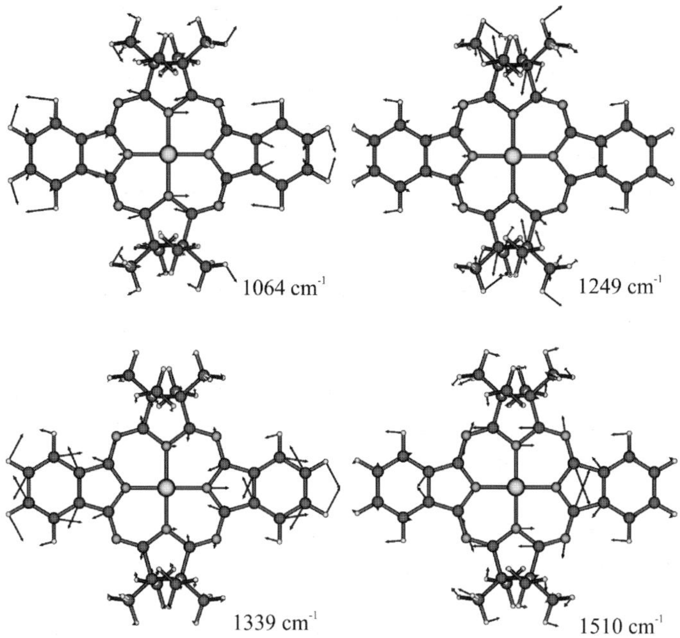

Fig. 10. Atomic movements at selected predicted frequencies for TABC.

3.3. Electronic absorption and MCD spectroscopy

Electronic absorption spectra provide spectral information related to the most fundamental properties of Pcs. Typical metallo Pcs with D_{4h} symmetry have an unsplit lowest energy band (Q band) in the visible region (*ca.* 650–700 nm), and a less-intense Soret band located in the 300–500 nm region. Gouterman's [90, 91] four-orbital model states that the two lowest singlet excited states, i.e. S_1 and S_2, of D_{4h} Pcs consist mainly of the linear combination of the $^1(a_{1u}e_g)$ and $^1(a_{2u}e_g)$ configurations, where a_{1u}, a_{2u}, and e_g correspond to the HOMO, HOMO−1, and LUMO of Pc, respectively. Since the LUMO is degenerate, the S_1 and S_2 states are also orbitally degenerate. Similarly, Soret bands can be described by the same configurations with a different linear combination. Fig. 12 shows the four frontier orbitals of ZnPc predicted by the semi-empirical ZINDO/S technique. The a_{1u} orbital has anti-nodes at the diagonal *meso*-nitrogen atoms, while the a_{2u} orbital, in contrast, has significant electron density at these four atoms.

The large electron density on the electronegative *meso*-nitrogen atoms stabilizes the energy of the a_{2u} orbital relative to that of the a_{1u} orbital. Therefore, the configuration interaction between $^1(a_{1u}e_g)$ and $^1(a_{2u}e_g)$ is not as significant as it is for the porphyrins. In the case of the Q band of Pcs, the $^1(a_{1u}e_g)$ configuration accounts for *ca.* 87%. The envelope of the Q band is sensitive to the molecular

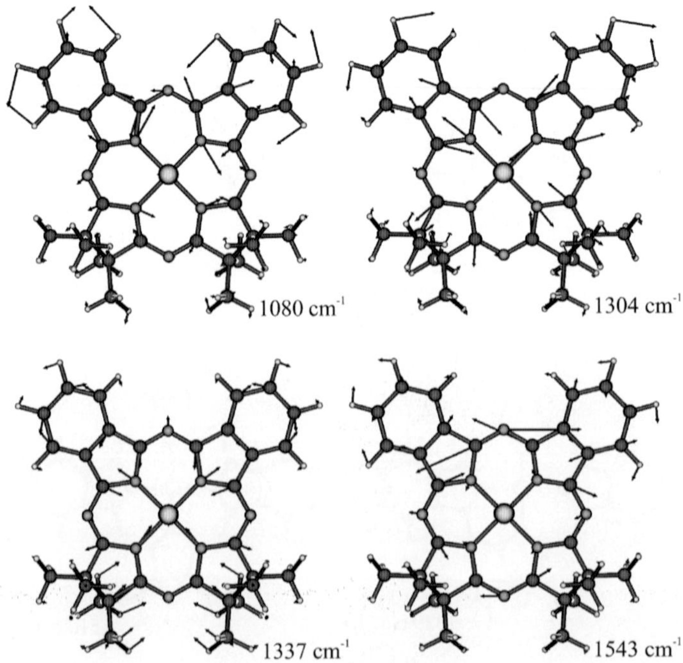

Fig. 11. Atomic movements at selected predicted frequencies for TAiBC.

symmetry. Hence, the Q band of low-symmetry Pcs often splits into separate x- and y-polarized component bands.

Magnetic circular dichroism (MCD) spectroscopy is complementary to absorption spectroscopy in providing information about both ground and excited state degeneracies that is essential in understanding the electronic structure of highly symmetrical molecules. The MCD signal arises from the same transitions as those seen in the UV–visible absorption spectrum, but the selection rules are different because the intensity mechanism depends on the magnetic dipole moment in addition to the electronic dipole moment. MCD spectra are comprised of three specific spectral features, called the Faraday A, B, and C terms, and is expressed in terms of magneto-molar ellipticity $[\theta]_M$ (deg mol^{-1} dm^3 m^{-1} G^{-1}) versus incident light energy, i.e. $[\theta]_M = -21.35\ (f_1 A + f_2(B + C/kT))$, where A (D$^2\beta_e$), B (D$^2\beta_e$/cm^{-1}), and C are the Faraday terms, and f_1 and f_2 are the dispersion and Gaussian-type shape functions, respectively. The derivative-shaped Faraday A term is temperature independent and identifies degenerate excited states, which is important in the case of the Pcs, since the LUMO of typical D_{4h} symmetry complexes are orbitally degenerate. The Gaussian-shaped C terms arise from orbital and/or spin degeneracy in the ground state, and their intensity is proportional to the inverse of the absolute temperature. Their spectroscopic

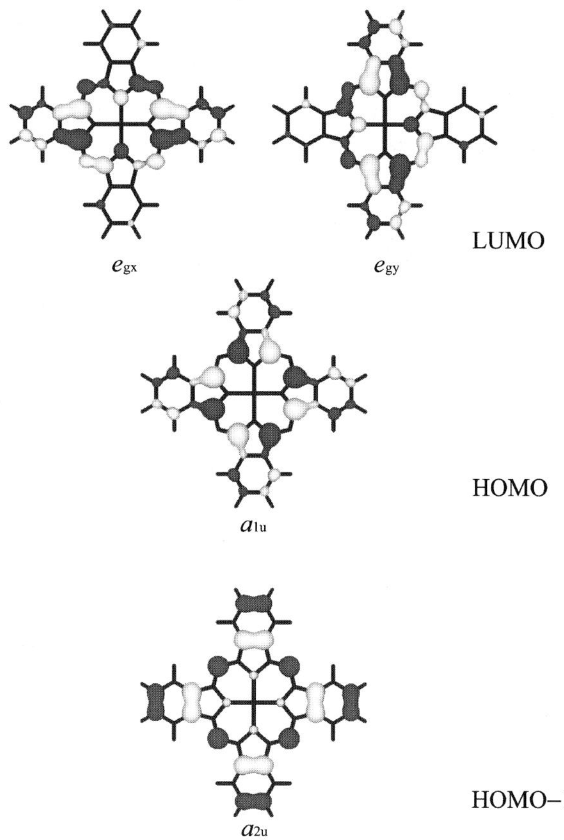

Fig. 12. Four frontier orbitals of ZnPc predicted by ZINDO/S.

center is, in the real system, close to that of the corresponding absorption peak. Gaussian-shaped, temperature independent *B* terms arise from mixing between closely related states linked by a magnetic dipole transition moment. The extent of mixing is proportional to the reciprocal of energy separation. Although it is usually difficult to extract physical meanings from the *B* terms, they play an important role for band deconvolution analysis, which isolates the required bands from the spectra by fitting each band with a Gaussian line shape.

In this section, several examples of absorption and MCD spectra are provided in order to illustrate the relationship between molecular structure, oxidation state, and the orbital degeneracy of the excited states.

3.3.1. Neutral species

3.3.1.1. D_{4h} metallo Pcs The UV–visible spectra of D_{4h} symmetry Pcs are dominated by the *x/y*-polarized Q band. The Q band can be observed at 674 nm in the case of ZnPc in pyridine (Fig. 13 (bottom)). The corresponding MCD signal (Fig. 13 (top)) is a derivative-shaped Faraday *A* term, indicating that the Q band arises from a transition to an orbitally degenerate excited state.

In the Soret band region (300–400 nm), orbitally degenerate HOMO − 1 → LUMO transitions are expected. However, the observed spectral envelope is broad due to the overlap of several close-lying bands. It is difficult to determine the band centers of each major component of the observed spectrum. Detailed analysis using band deconvolution techniques has been reported for ZnPc [92].

With an enlargement of the π system, generally, the Q band shifts to the red, but with an extent diminishing with increasing size of the π-system (i.e. Nc and Ac) [86]. The strength of the Q band increases, and the MCD intensity per unit absorption (*A/D* value) decreases. The angular momentum of the metallo Pc ring is always less than the angular momentum of the pure Pc orbital. This is due to the delocalization of a central metal d-electron onto a surrounding ligand through the empty $e_g(\pi^*)$ orbital [93].

3.3.1.2. Metal-free Pcs Fig. 14 contains (from top to bottom) the absorption spectra of tetra-*tert*-butylated metal-free TAP, Pc, Nc, and Ac. The Q band shifts to the longer wavelength with annulations with benzene rings, but to a decreasing extent with increasing molecular size. The split Q bands of H_2TAP and H_2Pc are related to the D_{2h} symmetry. In contrast, the Q bands of H_2Nc and H_2Ac do not split. There is a linear relationship between the splitting energy and Q-band

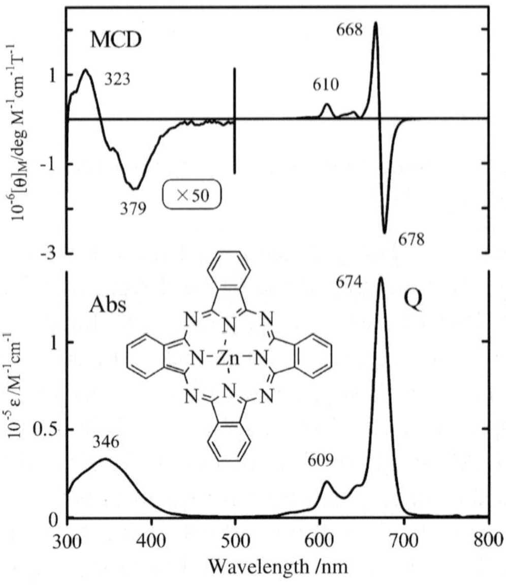

Fig. 13. Absorption (bottom) and MCD (top) spectra of ZnPc in pyridine.

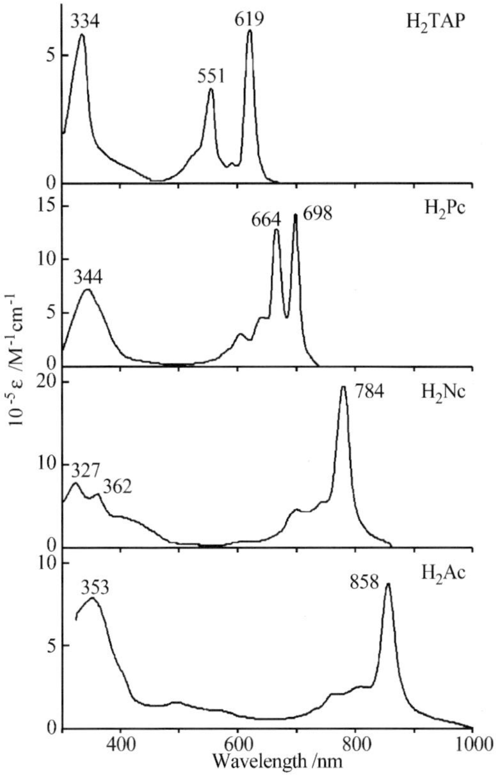

Fig. 14. Absorption spectra of tetra-*tert*-butylated metal-free TAP, Pc, Nc, and Ac (from top to bottom) in pyridine.

energy. The splitting is found to decrease for metal-free Pcs as the Q band shifts to longer wavelengths [94].

3.3.1.3. Low symmetry Pcs The geometrical perturbation on the π-system affects the splitting energy and intensity of the Q band significantly. In order to understand the symmetry-lowering effects, symmetry-adapted perturbation (SAP) theory based on the first-order perturbation theory, can be a useful approach [91, 95]. According to the SAP theory, perturbations at one or two opposite pyrrole sites of the Pc periphery cause Q-band splitting, while perturbations at adjacent pyrrole sites do not. Although the SAP theory is simple and intuitive, molecular orbital calculations should be performed for quantitative analysis.

Fig. 15 compares the absorption and MCD spectra of isomeric TABC and TAiBC complexes. The former has two fused benzene rings at opposite (*trans*) positions, while in the latter; the benzene rings are at the adjacent (*cis*) positions [88]. TABC has approximate D_{2h} symmetry, and shows two split Q bands at 842

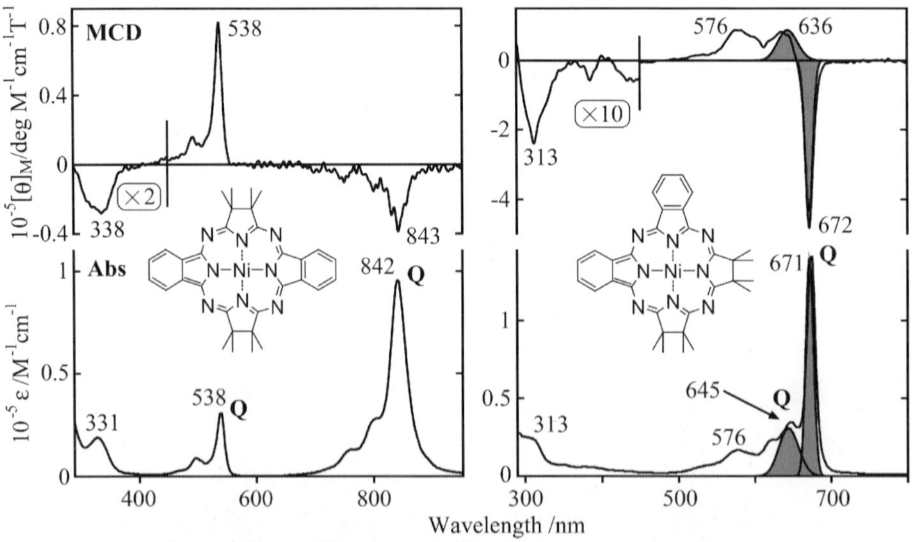

Fig. 15. Absorption (bottom) and MCD (top) spectra of TABC (left) and TAiBC (right) in chlorobenzene. The shaded region is the two deconvoluted Q band components of TAiBC.

and 538 nm. The Q bands can be unambiguously assigned on the basis of the MCD spectrum, i.e. the corresponding MCD signals are positive and negative Faraday *B* terms from longer to shorter wavelength [96]. In the case of TAiBC, sharp and unsplit Q bands appeared at 671 nm in accordance with the prediction of the SAP theory. However, band deconvolution analysis reveals that the Q bands are split slightly into two components (shaded two Gaussians in Fig. 15), which suggests that the first-order perturbation used for the SAP theory is insufficient in the case of the TAiBC Q band.

Co-planar dinuclear one-benzene-ring-shared Pc dimers represent another class of low-symmetry Pcs. This type of complex was first synthesized in 1987 [97], and their spectroscopic and electrochemical properties have since then been investigated [58]. A band deconvolution analysis was performed for Cu_2Pc_2 (Fig. 16), to interpret the absorption spectra of these dinuclear species [98].

Since the molecule has D_{2h} symmetry, the orbitally degenerate bands of monomeric D_{4h} metallo Pc split into separate *x*- and *y*-polarized bands. Since the resulting pairs of orbitals lie reasonably close in energy, transitions into these orbitals will mix under the influence of the applied magnetic field to give rise to pairs of intense, oppositely signed *B* terms within the MCD spectrum. Thus, in the Q-band region, two shaded Gaussians can be readily assigned to the Q transition, since the corresponding *B* terms show opposite signs. Assigning the Q bands would not be straightforward without the MCD spectrum. The absorption peak maximum at 773 nm appears to be one of the components of the symmetry-split Q bands, but this

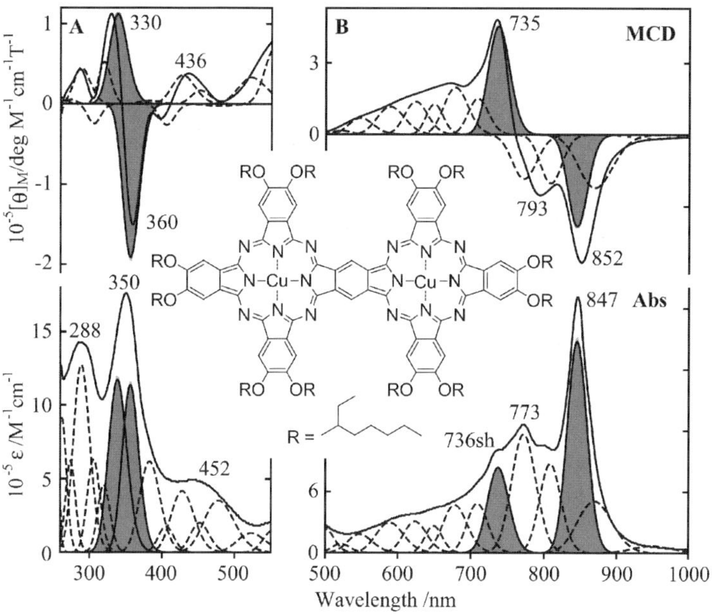

Fig. 16. Experimental (solid line) and deconvoluted Gaussian curves (broken line) of the (A) 260–550 and (B) 500–1,000 nm spectral regions of Cu_2Pc_2. The assigned Q and Soret band components are shown by the shaded Gaussians.

assignment is clearly not consistent with the MCD spectral data. The Soret bands were also assigned to two slightly split components in a similar manner (Fig. 16(A)).

3.3.2. Oxidized and reduced species

The Pc skeleton loses its aromatic nature upon ring oxidation or reduction. As a consequence, the resultant species shows a quite different absorption spectrum from that of the neutral species. Stillman's group [99] analyzed a variety of oxidized and reduced Pc species using the band deconvolution technique in order to assign the absorption bands. As an example, the absorption and MCD spectra of $[ZnPc(-3)]^-$ (which is generated by the photoreduction of ZnPc in the presence of hydrazine monohydrate in DMF) is shown in Fig. 17 (left) [100]. The generated radical anion has an unpaired electron in the LUMO of the parent ZnPc. If $[ZnPc(-3)]^-$ had D_{4h} symmetry, the molecule would have a degenerate ground state, and *C* terms would be expected in the MCD spectrum. However, the observation that the MCD spectrum is essentially temperature independent unambiguously defines the ground state as being orbitally non-degenerate. The most likely explanation is that the anticipated degenerate ground state splits under the influence of static Jahn–Teller effects. In the 10,000–13,000 cm^{-1} (1,000–750 nm) region, there is a sharp absorption band at 10,460 cm^{-1} (956 nm) with a corresponding MCD *B*

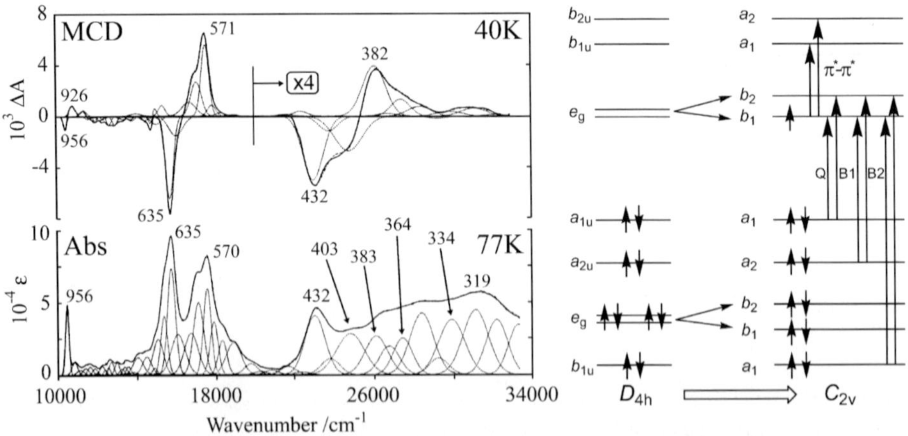

Fig. 17. Deconvolution results for the absorption and MCD spectra of $[ZnPc(-3)]^-$ in DMF (left). Note that peak maxima and minima are indicated in the nanometer units. Scheme of the energy levels in the neutral ZnPc and ring-reduced $[ZnPc(-3)]^-$ showing the origin of several possible transitions (right).

term, which is followed by two broad opposite-signed *B* terms. The *B* term at 957 nm arises from a transition into the lower partially filled orbital (Fig. 17 (right)).

The presence of the electron in the lower orbital means that there are two separate, allowed excited state configurations where an electron goes into the higher unfilled orbital. Namely, in this case, $a_1b_1\bar{b}_1$, $a_1b_1\bar{b}_2$, and $\bar{a}_1b_1b_2$ are spin-allowed configurations where bars indicates β spin of electrons. These configurations will interact with each other and could result in the two separate negative *B* terms seen to the blue of the 957 nm *B* term. In the 13,300–20,000 cm^{-1} (750–500 nm) region, two intense *B* terms were observed at 17,570 and 15,750 cm^{-1} (569 and 635 nm). Judging from their energy and intensity ratio, these have been assigned to $\pi^* \rightarrow \pi^*$ transitions. The result of the deconvolution analysis indicates that the bands at 23,040, 26,110, and 29,940 cm^{-1} (434, 383, and 334 nm) correspond to the B1 band, and 24,810, 27,470, and 31,150 cm^{-1} (403, 364, and 321 nm) correspond to the B2 band.

3.3.3. Solid-state spectra

Most practical applications for Pcs are based on their crystalline or amorphous phase. The solid-state absorption spectra differ significantly from those in solution, and depend on the nature of the crystal systems. Despite the importance of solid-state Pcs in many applications, the analysis of solid-state spectra was less than satisfactory until comparatively recently. Fig. 18 (top right) depicts the calculated solid-state absorption spectra of oxotitanium Pc (O = TiPc), in which exciton and charge transfer (CT) interactions, and molecular distortions are considered [101]. The calculated

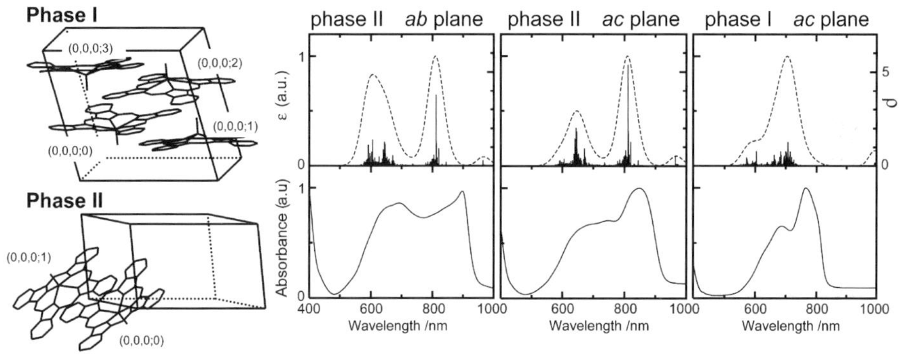

Fig. 18. Calculated (broken line) and experimental (solid line) absorption spectra when the incident light direction is perpendicular to the crystallographic *ab* plane of phase II (top left), the *ac* plane of phase II (top middle), and the *ac* plane of phase I (top right) with molecular structures in the unit cell of phase I (top left) and phase II (bottom left). Solid bars in the calculated spectra denote the product of oscillator strengths and coefficients of CT configurations, ρ.

spectrum is in close agreement with the experimental results (bottom right), in terms of the characteristics of phase dependence, dependence on the incident light direction, and the absorption intensity profiles. The calculation details can be obtained from the original paper [101]. The calculations helped to clarify that the large redshift and high CT character originates from large exciton interactions and large intermolecular resonance integrals, respectively. In addition, it has been revealed that the photoconductive properties correlate well with the intermolecular resonance integrals. These considerations are important for investigating certain Pc applications.

3.4. Electrochemistry

Redox properties have been reported for many Pcs [102]. In particular, the first oxidation and reduction potentials are important, since there is a close relationship between Q band energies and these potentials. When the Q band shifts to the red, the potential separation between the first oxidation and reduction potentials decreases, and *vice versa*. Reference [102] contains a comprehensive set of electrochemical data for the Pcs. In this section, several trends of redox properties are described in order to illustrate the relationship between structural characteristics and the electrochemical properties.

3.4.1. Mononuclear D_{4h} species

Generally speaking, radial extension of the phthalocyaninato π-system causes a marked shift in the oxidation potentials, while the shifts in the corresponding reduction potentials are significantly smaller. Fig. 19 shows how the shift in the oxidation and reduction potentials depends on the size of the π-system for a series of vanadyloxy (VO), copper (Cu), and cobalt (Co) macrocycles [86].

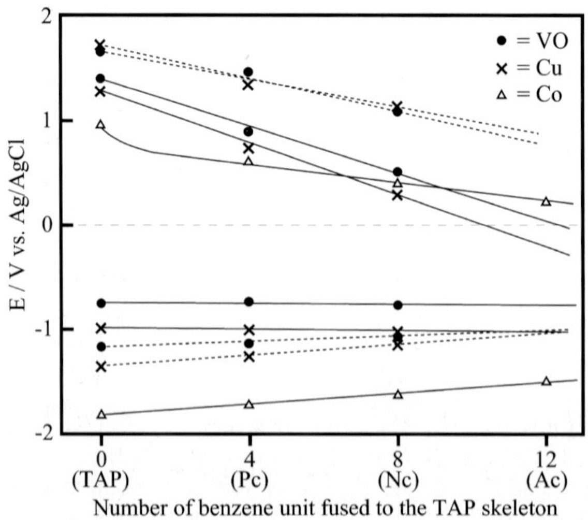

Fig. 19. Dependence of ligand oxidation and reduction potentials on the size of the π-system of macrocycles. The abscissa indicates numbers of benzene units fused to the TAP skeleton. Solid lines: first oxidation and reduction, broken lines: second oxidation and reduction. Note that the metal-centered redox processes are omitted.

Although the Co complex has a metal-centered first reduction, only the ligand-centered redox processes are shown within this figure. With increasing molecular size, the first ligand-oxidation potential shifts to negative voltage, while the first ligand-reduction potential is essentially invariant, except in the case of Co complexes, indicating that the LUMO changes only slightly while the HOMO destabilizes significantly. The second ligand-oxidation and reduction potentials shift negatively and positively, respectively, with increasing molecular size. In the case of Co complexes, the first reduction potential appears shifted to more negative potentials than in the case of other metal complexes. It is worth noting that the first oxidation potential of VO and Cu Acs can be extrapolated to negative potentials. This means that these complexes are unstable since oxidation of the Pc π-system occurs spontaneously.

3.4.2. Low-symmetry mononuclear species

As described above, when the symmetry of the π-system is lowered, the Q band region of the optical spectrum is modified substantially. In this section, the analogous electrochemical approach for describing the frontier orbitals of low-symmetry Pcs (Fig. 20) is outlined [103]. In Fig. 21, trends in the electrochemical data (A) are compared to the calculated energies of the four frontier orbitals (B) of the low-symmetry ZnPcs.

The first oxidation potential shifts to a more negative value in the order 0Zn > 1Zn > 2AdZn ≈ 2OpZn > 3Zn > 4Zn, while the first reduction potential

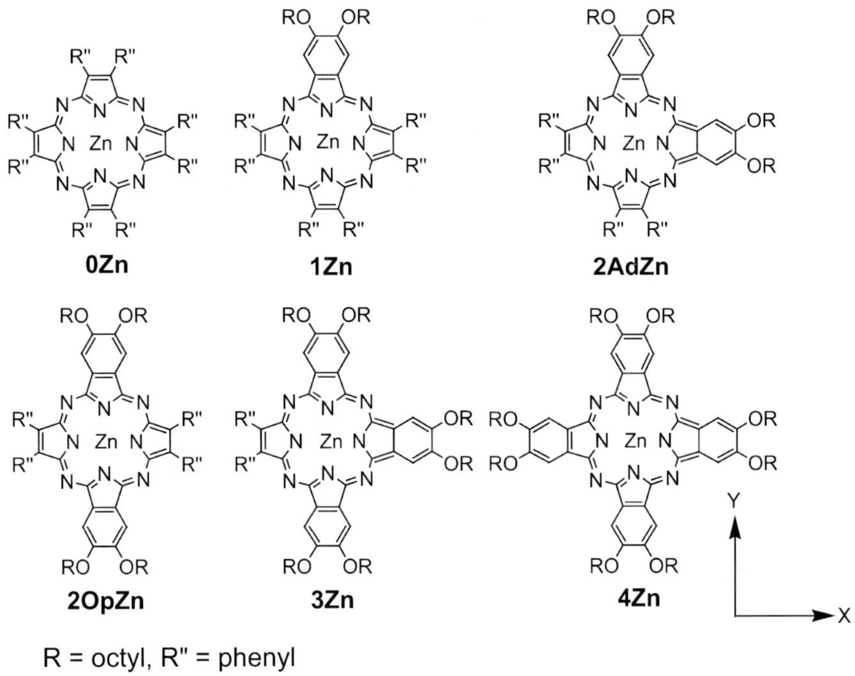

Fig. 20. Structures and the abbreviations used for low-symmetry ZnPcs in the text.

shifts to the negative in the order 0Zn ≈ 1Zn ≈ 2OpZn > 2AdZn ≈ 3Zn > 4Zn. The difference in the first oxidation couple between 2AdZn and 2OpZn is only 0.02 V, indicating that the potential shift depends only on the size of the π-system, and that the HOMO is destabilized by ligand expansion. In contrast to the HOMO, since the LUMO energy appears to have no dependence on the number of fused benzene rings, it is necessary to clarify the relationship between the LUMO energy and the π-system. MO calculations reveal that the HOMO becomes energetically destabilized as the π-system increases in size, which is consistent with the electrochemical results. In contrast, destabilization of the LUMO and LUMO + 1 depends on the symmetry of the π-system, and the variation in the LUMO energy resembles that seen for the first reduction potential in the electrochemical data.

The trends in the MO energies can be described in terms of the electron distribution of the MOs. In the case of the HOMO of 0Zn, the HOMO is destabilized by fused benzene rings, since the electron densities at the C_β positions of the pyrrole rings are large. In addition, the electron distributions at the C_β positions is independent of the pyrrole ring in the case of 0Zn. Therefore, the HOMO destabilization due to the fused benzene rings is independent of the pyrrole ring to which the benzo group is fused and exhibits a close relationship with the size of

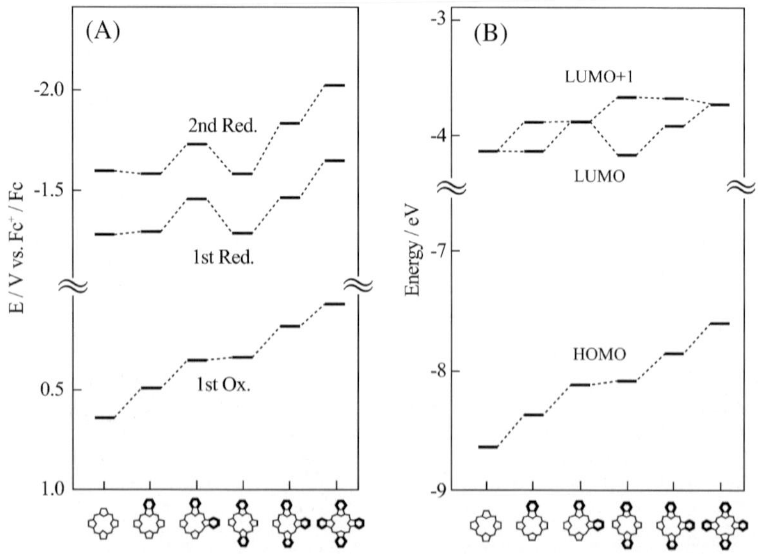

Fig. 21. Electrochemical data (A) and calculated MO energies (B) of low symmetry ZnPcs. The electrochemical data were recorded at a scan rate of 50 m V s^{-1} in *o*-dichlorobenzene containing 2.5×10^{-2} M pyridine and 0.1 M tetrabutylammonium perchlorate.

the π-system. The behavior of the e_{gx} and e_{gy} LUMOs is, however, more complex due to orbital degeneracy. In the e_{gx} (or e_{gy}) orbital, the electron densities at the C_β positions are larger in the pyrrole rings along one axis than on the other axis within the *x/y*-plane of the π-system (Fig. 22). The effect of additional peripheral fused benzene rings, therefore, differs along each axis. For example, in 1Zn, only the e_{gx} orbital is destabilized, while no change occurs in the energy of e_{gy}, and this results in a marked energy splitting between LUMO and LUMO + 1. The addition of a second benzene ring results in two isomers, 2OpZn and 2AdZn. In the former case, both benzene rings are fused on the *y*-axis. Therefore, only the LUMO + 1 energy of 1Zn is destabilized, while the LUMO energy is not influenced by the fused benzo rings (Fig. 22). The effect on 2AdZn is clearly very different, since the second benzene ring is fused on the *x*-axis. Compared with 1Zn, LUMO + 1, which originates from the e_{gx} orbital of 0Zn is hardly influenced, while the LUMO is destabilized by the adjacently fused benzene ring. Thus, the LUMO energy is very close to that of LUMO + 1 in the case of 1AdZn.

By using this model, the destabilization of the HOMO and LUMO energies of the low-symmetry Pcs can be readily rationalized.

3.4.3. Double-decker and triple-decker species

Some transition metals and most rare earth metals can form sandwich-type double- or triple-decker Pc multimers. The torsion angle between neighboring Pc

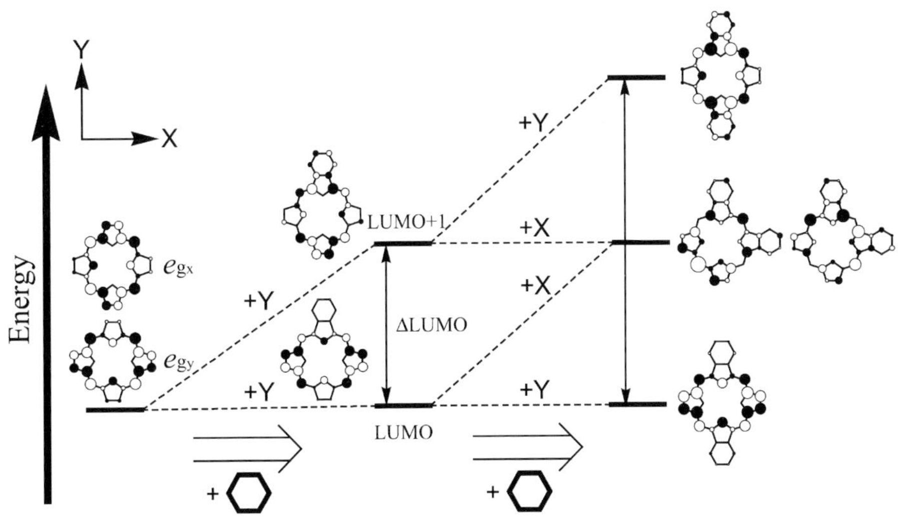

Fig. 22. Effects of the number and location of fused benzene rings on the LUMO and LUMO + 1 energies.

rings depends on the ionic radius of the central metal, i.e. the angle decreases from 45° for the small lutetium (Lu) ion to 6° for larger neodymium (Nd) Pc dimers [42]. A close relationship between the oxidation and reduction potentials and ionic radius of the central metal has also been observed. Fig. 23 depicts the results of electrochemical measurements on a series of $[NBu_4][Ln(III)Pc_2]$ ($[NBu_4]$ = tetra(*n*-butyl)ammonium; Ln = Y, Pr, Nd, Sm, Eu, Gd, Tb, Dy, Ho, Tm, Er, Yb, and Lu) [104]. The first and second oxidation potentials increase linearly with the ionic radius of the central metals, while the reduction potentials show no observable change.

The electrochemical characteristics of a series of heteroleptic *tris*(phthalocyaninato) complexes with two identical rare earth metals, (Pc)M(OOPc)M(OOPc) [M = Eu, Lu, Y, Pc = unsubstituted Pc, OOPc = 3,4,12,13,21,22,30,31-*octakis*(octyloxy)Pc], are shown in Fig. 24 [105]. In an analogous manner to the double-decker complexes, the first and second oxidation potentials depend on the size of the metal, while the fifth oxidation potential shows the opposite trend. By contrast, the reduction potentials are almost independent of the metal size. As a consequence, the difference between the first oxidation and reduction potentials decreases linearly with metal size in the range of 0.85–0.98 V. The optical properties are closely linked to this trend, i.e. the Q band shifts to the blue with increasing metal size. Since triple-decker Pcs show multiple redox couples in the observable spectroelectrochemical window, future applications in electrochromic display can be expected for these compounds.

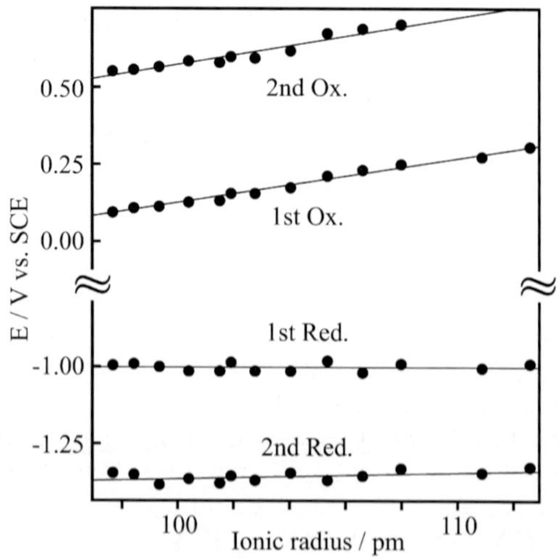

Fig. 23. Oxidation and reduction potentials of a series of double-decker $[NBu_4][Ln(III)Pc_2]$ in *o*-dichlorobenzene as a function of the ionic radius of M^{III}.

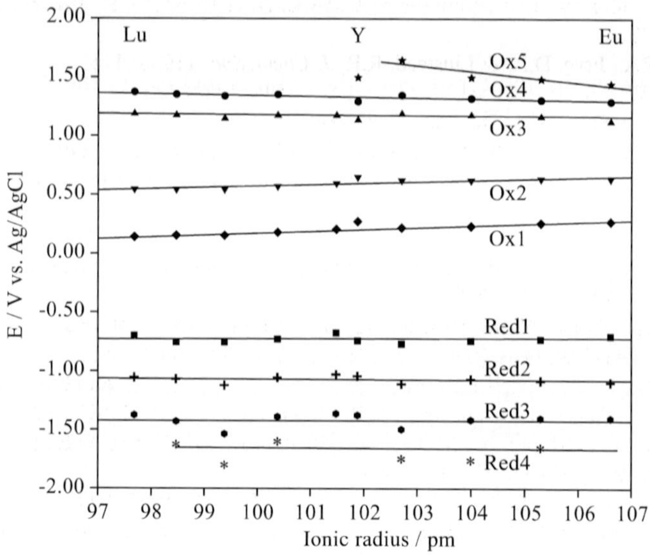

Fig. 24. Oxidation and reduction potentials of triple-decker (Pc)M(OOPc)M(OOPc) in CH_2Cl_2 as a function of the ionic radius of M^{III}.

4. CONCLUDING REMARKS

In this chapter, many representative synthetic methods of Pcs have been described with examples from the literature. In addition, recent development in the characterization of these species has been introduced. Progress in the field is rapid, and many newer methods and analyses are being reported both as academic reports and patents. For those who would like to consult a more comprehensive summary of Pc chemistry, Refs. [2–4] are readily available and have been read by many researchers active within the field.

REFERENCES

1. Broun, A. & Tcherniac, J., *Ber. Deut. Chem. Ges.*, **40** (1907) 2709.
2. Leznoff, C.C. & Lever, A.B.P., *Phthalocyanines – Properties and Applications*, VCH, Weinheim, New York, Vols. 1–4, (1989, 1992, 1993, 1996).
3. Shirai, H. & Kobayashi, N., *Phthalocyanines – Chemistry and Functions*, IPC, Tokyo (1997).
4. Kadish, K.M., Smith, K.M. & Guillard, R., *Porphyrin Handbook*, Academic Press, New York, Vols. 15–20 (2003).
5. Fukada, N., *Bull. Chem. Soc., Jpn.*, **34** (1961) 884.
6. Wohrle, D. & Wahl, B. *Tet. Lett.*, (1979) 227.
7. Von der pol, J.F., Neeleman, E., Miltenburg, J.C., Zwikker, J.W. & Nolte, R.J.M., *Macromolecules*, **23** (1990) 155.
8. Weber, J.H. & Busch, D.H., *Inorg. Chem.*, **4** (1965) 469.
9. Ruf, M., Laurence, A.M., Noll, B.C. & Pierpont, C.G., *Inorg. Chem.*, **37** (1998) 1992.
10. Shirai, H., Maruyama, A., Kobayashi, K., Hojo, N. & Urushido, K. *Makromol. Chem.*, **181** (1980) 575.
11. Barrett, P.A., Frye, D.A. & Linstead, R.P., *J. Chem. Soc.*, (1938) 1157.
12. Snow, A.W. & Jarvis, N.L., *J. Am. Chem. Soc.*, **106** (1984) 4706.
13. Tomoda, H., Saito, S. & Shiraishi, S., *Chem. Lett.* (1980) 1277.
14. Nemykin, V.N., Kobayashi, N., Mytsyk, V.N. & Volkov, S.V., *Chem. Lett.* (2000) 546.
15. Brach, P.J., Gramatica, S.J., Ossanna, O.A. & Weinberger, L., *J. Heterocycl. Chem.*, **7** (1970) 1403.
16. Uchida, H., Yoshiyama, H., Reddy, P.Y. & Nakamura, S., *Synlett.* (2003) 2083.
17. Kopylovich, M.N., Kukushkin, V.Y., Haukka, M., Luzyanin, K.V. & Pombeiro, A.J.L., *J. Am. Chem. Soc.*, **126** (2004) 15040.
18. Petit, M.A., Plichon, V. & Belkacemi, H., *New J. Chem.*, **13** (1989) 459.
19. Metz, J., Schneider, O. & Hanack, M., *Inorg. Chem.*, **23** (1984) 1065.
20. Uchida, H., Reddy, P.Y., Nakamura, S. & Toru, T., *J. Org. Chem.*, **68** (2003) 8736.
21. Poon, K.W., Yan, Y., Li, X. & Ng, D.K.P., *Organometallics,* **18** (1999) 3528.
22. Thompson, J.A., Murata, K., Miller, D.C., Stanton, J.L., Broderick, W.E., Hoffman, B.M. & Ibers, J.A. *Inorg. Chem.*, **32** (1993) 3546.
23. Yao, J., Yonehara, H. & Pak, C., *Bull. Chem. Soc. Jpn.*, **68** (1995) 1001.
24. Ogata, H., Higashi, H. & Kobayashi, N., *J. Porphyrins Phthalocyanines*, **7** (2003) 551.
25. Gacho, E.H., Naito, T., Inabe, T., Fukuda, T. & Kobayashi, N., *Chem. Lett.* (2001) 260.
26. Ohya, T., Takeda, J., Kobayashi, N. & Sato, M., *Inorg. Chem.*, **29** (1990) 3734.
27. Lowery, M.K., Starshak, S.J.A.J., Esposito, J.N., Krueger, P.C. & Kenney, M.E., *Inorg. Chem.*, **4** (1965) 128.
28. Sotiriou, C., Lee, W. & Giese, R.W., *J. Org. Chem.*, **55** (1990) 2159.
29. Kobayashi, N., *Chem. Commun.* (1998) 487.
30. Hu, M., Brasseur, N., Yildiz, S.Z., van Lir, J.E. & Leznoff, C.C., *J. Med. Chem.*, **41** (1998) 1789.
31. Kobayashi, N., Ashida, T. & Osa, T., *Chem. Lett.* (1992) 2031.
32. Fukuda, T. & Kobayashi, N., *Chem. Lett.* (2002) 866.
33. Nolan, K.J.M., Hu, M. & Leznoff, C.C., *Synlett.* (1997) 593.

34. Kobayashi, N., Fukuda, T., Ueno, K. & Ogino, H., *J. Am. Chem. Soc.*, **123** (2001) 10740.
35. Lee, H.–J., Brennessel, W.W., Lessing, J.A., Brucker, W.W., Young, V.G. & Gorum, S.M., *Chem. Commun.* (2003) 1576.
36. Janson, T.R., Kane, A.R., Sullivan, J.F., Knox, K. & Kenney, M.E., *J. Am. Chem. Soc.*, **91** (1969) 5210.
37. Ercolani, C., Gardini, M., Murray, K.S., Pennesi, G. & Rossi, G., *Inorg. Chem.*, **25** (1986) 3972.
38. Bottomly, L.A., Gorce, J.N., Goedken, V.L. & Ercolani, C., *Inorg. Chem.*, **24** (1985) 3737.
39. Rossi, G., Gardini, M., Pennesi, G., Ercolaini, C. & Goedken, V.L., *J. Chem. Soc., Dalton. Trans.* (1989) 193.
40. Ercolani, C., Gardini, M., Goedken, V., Pennesi, G, Rossi, G., Russo, U. & Zanonato, P., *Inorg. Chem.*, **28** (1989) 3097.
41. Kienast, A., Galich, L., Murray, K.S., Moubaraki, B., Razarev, G., Cashon, J.D. & Homborg, H., *J. Porphyrins Phthalocyanines*, **1** (1997) 141.
42. Koike, N., Uekusa, H., Ohasi, Y., Harnoode, C., Kitamura, F., Ohsaka, T. & Tokuda, K., *Inorg. Chem.*, **35** (1996) 5798.
43. Tojan, K.L., Kendall, J.L., Kepler, K.D. & Hatfield, W.E., *Inorg. Chim. Acta*, **198–200** (1992) 795.
44. Jiang, J., Liu, R.C., Mak, T.C.W., Chan, T.W.D. & Ng, D.K.P., *Polyhedron*, **16** (1997) 515.
45. Bennet, W.E., Broberg, D.E. & Baeninger, N.C., *Inorg. Chem.*, **12** (1973) 930.
46. Gorsch, M. & Homborg, H. *Zh., Anorg. All. Chem.*, **624** (1998) 634.
47. Huckstadt, H. & Homborg, H. *Zh., Anorg. All. Chem.*, **623** (1997) 292.
48. Huckstadt, H., Bruhn, C. & Homborg, H., *J. Porphyrins Phthalocyanines*, **1** (1997) 367.
49. Takahashi, K., Itoh, M., Tomita, Y., Nojima, K., Kasuga, K. & Isa, K. *Chem. Lett.* (1993) 1915.
50. Janczak, J. & Kubiak, R., *J. Chem. Soc., Dalton Trans.* (1993) 3809.
51. Kadish, K.M., Moninot, J., Hu, Y., Dubois, D., Ibnlgassi, A., Barbe, J.M. & Guilard, R., *J. Am. Chem. Soc.*, **115** (1993) 8153.
52. Dormond, A., Belkalem, B., Charpin, P., Lance, M., Vigner, D., Forcher, G. & Guilard, R., *Inorg. Chem.*, **25** (1986) 4785.
53. Linstead, R.P. & Robertson, J.M., *J. Chem. Soc.*, (1936) 1736.
54. Jiang, J., Mak, T.C.W. & Ng, D.K.P., *Chem. Ber.*, **129** (1996) 933.
55. Guyou, F., Pondaven, A., Guenot, P. & L'Her, M., *Inorg. Chem.*, **33** (1994) 4787.
56. Moussavi, M., De Cian, A., Fischer, J. & Weiss, R., *Inorg. Chem.*, **25** (1986) 2107.
57. Jiang, J., Lau, R.L.C., Chan, T.W.D., Mak, T.C.W. & Ng, D.K.P., *Inorg. Chim. Acta,* **255** (1997) 59.
58. Kobayashi, N., Lam, H., Nevin, W.A., Janda, P., Leznoff, C.C., Koyama, T., Monden, A. & Shirai, H., *J. Am. Chem. Soc.*, **116** (1994) 699.
59. Kobayashi, N., Higashi, Y. & Osa, T., *J. Chem. Soc., Chem. Commun.* (1994) 1785.
60. Budzikiewicz, H., Taraz, K., Munster, H., Wasgestian, F. & Heucher, R., *Monatsh. Chem.*, **116** (1985) 525.
61. Isa, K., Sasaki, K., Murano, I., Fukui, K. & Mizuno, K., *Nihon Kagaku Kaishi* (1985) 615.
62. Schildcrout, S.M., *J. Am. Chem. Soc.*, **105** (1983) 3852.
63. Muller, E.C., Kraft, R., Etzold, G., Drevs, H. & Taube, R., *J. Prakt. Chem.*, **320** (1978) 49.
64. Varmuza, K., Maresch, G. & Meller, A., *Monatsh. Chem.*, **105** (1974) 327.
65. Eley, D.D., Hazeldin, D.J. & Palmer, T.F., *J. Chem. Soc., Farad. Trans. 2* **69** (1973) 1808.
66. Bonderma, D., Cater, E.D. & Bennett, W.E., *J. Chem. Eng. Data,* **15** (1970) 396.
67. Starke, M. & Tummler, R., *Zh. Chem.*, **7** (1967) 433.
68. Hill, H.C. & Reed, R.I., *Tetrahedron,* **20** (1964) 1359.
69. Games, D.E., Jackson, A.H. & Taylor, K.T., *Org. Mass Spectrom.*, **9** (1974) 1245.
70. Freas, R.B. & Campana, J.E., *Inorg. Chem.*, **23** (1984) 4654.
71. Raynor, J.B., Robson, M. & Torrens-Burton, A.S.M., *J. Chem. Soc., Dalton Trans.* (1977) 2360.

72. Ukei, K., *Acta Crystallogr., Sect. B: Struct. Crystallogr. Crystl. Chem.*, **B29** (1973) 2290.
73. George, R.D., Chou, C.-W., Williams, P., Burrows, V.A. & McMillan, P.F., *Langmuir*, **12** (1996) 5736.
74. Nelson, R.W., Rainbow, M.J., Lohr, D.E. & Williams, P., *Science*, **246** (1989) 1585.
75. Usami, J., Masters Thesis, Tohoku University (1999).
76. Srinivasan, N., Haney C.A., Lindsey, J.S., Zhang, W. & Chait, B.T., *J. Porphyrins Phthalocyanines*, **3** (1999) 283.
77. Conneely, A., McClean, S., Smyth, W.F. & McMullan, G., *Rapid Commun. Mass Spectrom.*, **15** (2001) 2076.
78. Shankai, Z., Feng, Z., Weide, H., Zhongping, Y. & Hanhui, W., *Rapid Commun. Mass Spectrom.*, **9** (1995) 230.
79. Lau, R.L.C., Jiang, J., Ng, D.K.P. & Chan, T.-W.D., *J. Am. Soc. Mass Spectrom.*, **8** (1997) 161.
80. Kobayashi, T., *Spectrochim. Acta Part A*, **26** (1970) 1313.
81. Gong, X.D., Xiao, H.M. & Tian, H. *Int. J. Quantum. Chem.*, **86** (2002) 531.
82. Toman, P., Nešpůrek, S. & Yakushi, K. *J. Porphyrins Phthalocyanines*, **6** (2002) 556.
83. Tackley, D.R., Dent, G. & Smith, W.E., *Phys. Chem. Chem. Phys.*, **3** (2001) 1419.
84. Tackley, D.R., Dent, G. & Smith, W.E., *Phys. Chem. Chem. Phys.*, **2** (2000) 3949.
85. Braun, D. & Ceulemans, A., *J. Phys. Chem.*, **99** (1995) 11101.
86. Kobayashi, N., Nakajima, S.-I., Ogata, H. & Fukuda, T. *Chem. Eur. J.*, **10** (2004) 6294.
87. Wong, M.W., *Chem. Phys. Lett.*, **256** (1996) 391.
88. Fukuda, T., Makarova, E.A., Luk'yanets, E.A. & Kobayashi, N., *Chem. Eur. J.*, **10** (2004) 117.
89. Makarova, E.A., Fukuda, T., Luk'yanets, E.A. & Kobayashi, N., *Chem. Eur. J.*, **11** (2005) 1235.
90. Gouterman, M., *J. Chem. Phys.*, **30** (1959) 1139.
91. Gouterman, M., *J. Mol. Spectrom.*, **6** (1961) 138.
92. Nyokong, T., Gasyna, Z. & Stillman, M.J., *Inorg. Chem.*, **26** (1987) 1087.
93. Stillman, M.J. & Thomson, A.J., *J. Chem. Soc. Farad. Trans.*, **2** (1974) 805.
94. Kobayashi, N., Ogata, H., Nonaka, N. & Luk'yanets, E.A. *Chem. Eur. J.*, **9** (2003) 5123.
95. Kobayashi, N. & Konami, H., In *Phthalocyanine – Properties and Applications*; Leznoff, C.C., Lever, A.B.P., Eds., VCH, New York, Vol. 4 (1996) Chapter 13.
96. Note that the *B*-term with negative intensity is called the positive *B*-term and vice versa.
97. Leznoff, C.C., Lam, H., Marcuccio, S.M., Nevin, W.A., Janda, P., Kobayashi, N. & Lever, A.B.P., *J. Chem. Soc. Chem. Commun.* (1987) 699.
98. Kobayashi, N., Fukuda, T. & Lelièvre, D., *Inorg. Chem.*, **93** (2000) 3632.
99. Mack, J. & Stillman, M.J., In *The Porphyrin Handbook*; Kadish, K.M., Smith, K.M. & Guilard, R., Eds., Academic Press, New York, Vol. 16 (2003) Chapter 103.
100. Mack, J. & Stillman, M.J., *J. Am. Chem. Soc.*, **116** (1994) 1292.
101. Nakai, K., Ishii, K., Kobayashi, N., Yonehara, H. & Pac, C., *J. Phys. Chem., B* **107** (2003) 9749.
102. L'Her, M., Pondaven, A., In *The Porphyrin Handbook*, Kadish, K.M., Smith, K.M. & Guilard, R., Eds., Academic Press, New York, Vol. 16 (2003) Chapter 106.
103. Miwa, H., Ishii, K. & Kobayashi, N., *Chem. Eur. J.*, **10** (2004) 4422.
104. Konami, H., Hatano, H., Kobayashi, N. & Osa, T., *Chem. Phys. Lett.*, **165** (1990) 397.
105. Zhu, P., Pan, N., Ma, C., Sun, X., Arnold, D.P. & Jiang, J., *Eur. J. Inorg. Chem.*, (2004) 518.

Functional Dyes
Sung-Hoon Kim (Editor)

Chapter 2

Cyanine dyes for solar cells and optical data storage

He Tian and Fanshun Meng

Laboratory for Advanced Materials and Institute of Fine Chemicals,
East China University of Science and Technology, Shanghai, P. R. China

1. INTRODUCTION

The term "cyanine" was originally applied to one compound but subsequently extended to a group of dyes. The general structure of the most useful cyanine dyes is shown in Scheme 1, in which two nitrogen atoms are linked by a conjugated carbon chain with odd number to form an extended conjugated system over the nitrogen and carbon atoms and the nitrogen atom and part of the conjugated carbon chain form part of a heterocyclic system. The methine chain may lie at the *ortho*- or/and *para*-position of the two nitrogen atoms. A large number of resonance structures may be formulated, of which (**1**) and (**2**) are the most important, in which the electron density is lowest for the nitrogen atoms. The true structure of cyanine can be considered as the conjugate hybrid of the two structures (**1**) and (**2**), shown in Scheme 1. The aromatic heterocycles Ar_1 and Ar_2 can be selected from quinoline, benzoquinoline, benzimidazole, pyridine, benzothiazole, benzoxazole, indole, benzindole, etc. Cyanine dye is usually called monomethine, trimethine, pentamethine and heptamethine cyanine for $n = 0, 1, 2, 3$, respectively. Generally speaking, the maximum absorption wavelength is red shifted about 100 nm with the increase of two methine groups for every homogeneous cyanine. An unsymmetrical cyanine dye is a cyanine dye in which the ring structures joined by the methine bridge are not the same.

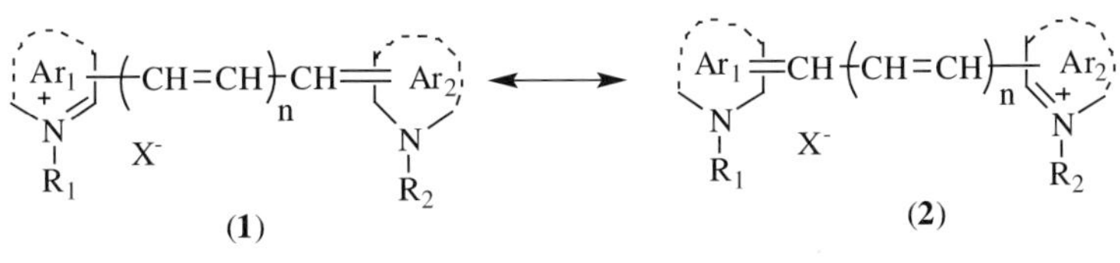

Scheme 1.

If one side of the methine chain is composed of nitrogen-containing heterocycle with a positive charge and the other side of the methine chain is connected to a substituted nitrogen through an aryl ring, the cyanine dye was called hemicyanine dye (for example, Scheme 2). The term styryl dyes are often used to refer to those hemicyanine dyes that contain only an ethylene group. In hemicyanine dye, the positive charge is localized in the heterocyclic nitrogen and forms large dipole moment in the ground state.

The first cyanine dye was discovered in 1856 by C.H. Greville Williams. Later in 1873, H.W. Vogel began to use cyanine dyes in photography [1]. Since then, research on cyanine dyes developed rapidly due to their extra sensitizing power on silver halide in the region of spectra from visible to near infrared (IR) in photography. In the early days, efforts mainly focused on the synthesis of new cyanine dyes, the elucidation of their chemical structures and the examination of their photographic properties. The synthesis of various cyanine dyes have been summarized previously by F.M. Hamer [1].

Cyanine dyes have relatively good stability, high molar absorption coefficients ($\sim 10^{-5}$ M^{-1} cm^{-1}), medium fluorescence intensity, narrow spectrum width and the ability to form H- or J-aggregates. The maximum absorption wavelength of cyanine can be tuned precisely from near-UV to near-IR by chemical structural modification. With these unique photophysical and photochemical properties, cyanine dyes are recently being used in many applications such as nonlinear optics, optical data storage, biomolecular labeling, dye laser, photorefractive materials and photodynamic therapy [2]. In this chapter, we will mainly introduce the applications of cyanine dyes in organic solar cells and in optical data storage.

2. THE APPLICATION OF CYANINE DYES IN ORGANIC SOLAR CELLS

Dye-sensitized solar cells (DSSCs) have attracted considerable attention since the report of highly efficient (the overall conversion efficiency was more than 10%) ruthenium complex-sensitized nanocrystalline TiO_2 solar cells by Grätzel and co-workers [3,4]. As an alternative to expensive heavy-metal-based polypyridyl complexes, organic dyes have also been studied as sensitizer in

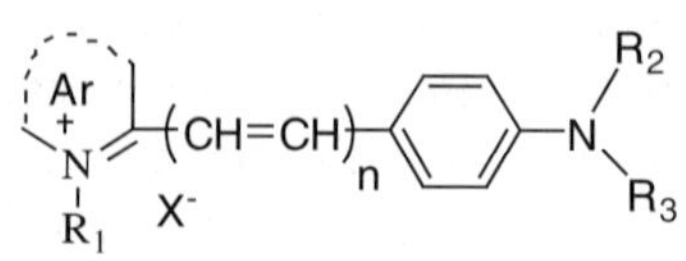

Scheme 2.

DSSCs because of their large absorption coefficients, easy preparation, low cost, saving of limited precious metal resources and easy handling for solar cell recycling without removal of metal. Among organic dye sensitizers, cyanine dyes are very important.

Cyanine dyes have been used in photography for more than 100 years to sensitize silver halide efficiently upon the absorption of light. So, it is natural to think of using cyanine dye to sensitize nanocrystalline TiO_2 in solar cells. The high extinction coefficients of cyanine dyes can make them absorb enough light and the absorption spectra of cyanine dyes can be tuned easily in the whole spectrum by tailoring their structures. Therefore, it is of great interest to study their sensitization for nanocrystalline TiO_2 (or other semiconductors) solar cells. This section will review the applications of cyanine dyes in DSSCs and solid-state organic solar cells.

2.1. Cyanine dyes for dye-sensitized nanocrystalline TiO_2 solar cells

Although cyanine dyes have high molar absorption coefficient, their sharp and narrow absorption band is one of the serious disadvantages. This disadvantage seems to inhibit cyanine dyes as solar cell sensitizers, which require strong absorption in the whole spectrum to harvest light. However, cyanine dyes form J- or/and H-aggregates when adsorbed on nanocrystalline TiO_2 film, and the absorption spectra are greatly broadened compared with that in the solution. All forms of cyanine dyes, J-, H-aggregates or monomers, can sensitize nanocrystalline TiO_2 solar cells with equal efficiency.

Spitler *et al.* [5,6] synthesized a series of cyanine dyes with carboxyl groups in the alkyl chain (Scheme 3) and studied their spectral sensitization of TiO_2 nanocrystalline electrodes with aggregated cyanine dyes. The spectrum of the TiO_2 attached cyanine **1c** is almost like that of the solution, showing a redshift of 10 nm for both the main peak and the blueshifted vibronic shoulder. The bulky methylbenzoic acid function seems to preclude the dye–dye interaction that produces aggregates. On the other hand, the other cyanine dyes can form different types of H-aggregates, and the spectra show both the aggregate and monomer absorption peaks with the H-aggregate peak predominant. But two distinct behaviors appear when TiO_2 surface coverage is less than unity. In one case, the initial dye adsorption is preferentially on monomer with subsequent attachment leading to growth in H-aggregation for cyanine dye **3a**, **3b** and **3c**. In another case, the dye attaches initially on an aggregated state and then gradually reverses its preferences as the monomer grows in the spectrum for cyanine dye **1a**, **1b** and **2**. In contrast, the spectrum of cyanine **3d** shows a sharper blue peak and a minimal red peak, due to the greater number of more extended H-aggregates and the monomeric contribution being effectively suppressed. The DSSC sensitized with cyanine **1b**, **2** and **3b** yields short-circuit currents comparable with that of N3 dye in the same conditions.

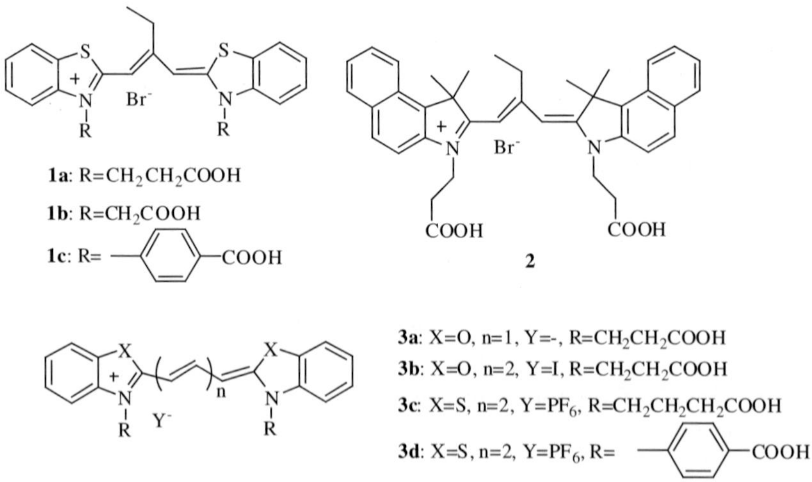

Scheme 3.

No J-aggregates were observed with the cyanine dyes bearing carboxyl groups in the alkyl chains. On the contrary, the cyanine dyes (Scheme 4) with carboxyl group in the aromatic rings can form both H- and small J-aggregates on the surface of TiO_2 electrode except cyanine **4a** [7]. In this case, the absorption spectra of the electrode are even broader, which is beneficial for collecting the light. Whether the cyanine dyes are attached to TiO_2 surface in any way (monomer, H- and J-aggregates), the photocurrent action spectra of the solar cell were found to match well with the absorption spectra. This means that all forms of cyanine dyes, aggregates or monomer, can sensitize nanocrystalline TiO_2 solar cells with equal efficiency. The IPCE (incident-photon-to-current conversion efficiency) of cyanine-dye-sensitized nanocrystalline TiO_2 solar cells increased with the decrease in the distance between the skeleton of the dye and the TiO_2 surface [8]. The maximum IPCE for cyanine **3c** was measured to be 70 [6] and 73% for cyanine **5** [9]. The maximum overall conversion efficiency of 3.9% was achieved for cyanine **5** sensitized nanocrystalline TiO_2 solar cell under 27 mW/cm^2 white light irradiation.

Matsui *et al.* [10] synthesized various near-IR absorbing heptamethine cyanine dyes derived from indole, benzoxazole, benzothiazole and quinoline with a cyclohexene linkage at the center. A 2-carboxyphenylthio-substituted indolium pentamethine cyanine dye **7** (Scheme 5) was used to study its application as a sensitizer for ZnO solar cells. The absorption band of cyanine **7** adsorbed on ZnO was slightly broader compared with that in DMSO solution. The maximum IPCE was observed to be 4.17% at 804 nm.

Although the absorption spectra of cyanine adsorbed TiO_2 electrode are expanded, each of these cyanine dyes absorbs only a fraction of the solar spectrum,

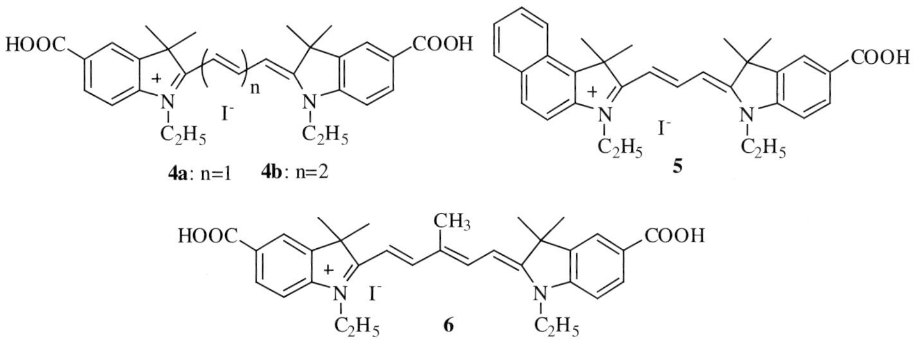

Scheme 4.

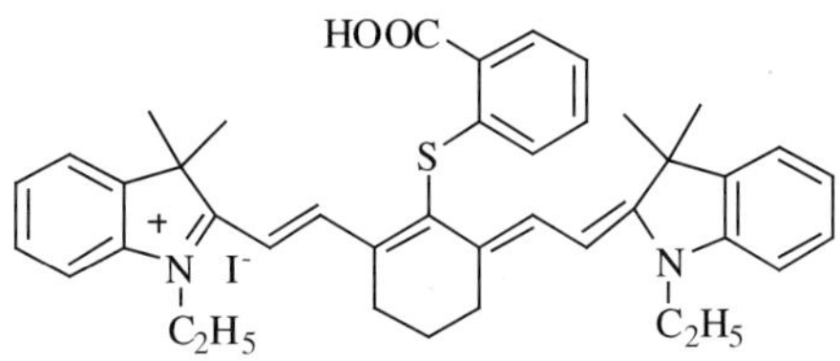

7

Scheme 5.

which is still far from the requirements for solar cell materials. The simultaneous adsorption of several dyes that have different absorption wavelengths on TiO_2 electrode is useful in order to harvest solar light over the entire spectral region. Spitler *et al.* [6] reported the possibility of sensitization of TiO_2 electrode with several cyanine dyes. The IPCE spectra of TiO_2 electrode sensitized by two cyanine dyes was in accord with the absorption spectra and the short-circuit current improved greatly compared with that of the photocurrent sensitized by single cyanine. For example, the short-circuit currents of **3a** and **1b** co-sensitized solar cells can be improved from 2.23 mA/cm^2 for **3a** and to 5.9 mA/cm^2 for **1b** 8.3 mA/cm^2. The use of three different cyanine dyes nearly spans the whole spectra from blue to near-IR and the photocurrent action spectra correspond well with the absorption spectra, meaning that the aggregate forms of the cyanine dye are efficient light harvesting elements. But the IPCE was measured to be only 10% in the whole wavelength region.

The mixture of pentamethine cyanine **4b** and trimethine cyanine **5** was employed to sensitize the solar cell over the entire visible spectrum [11]. The sensitizing properties of the mixed dyes with different ratio were systematically studied, and it was found that the 1:3 (v/v) ratio of **4b** and **5** cyanine mixture sensitized solar cell generated the highest photoelectric conversion efficiency of 3.4%. Co-sensitization was found to suppress the aggregation and affect the sensitization performance profoundly.

Sayama *et al.* [12] investigated the use of various kinds of cyanine dyes (Scheme 6) in DSSC to promote the short-circuit photocurrent and the solar light-to-power conversion efficiency. The sensitization ability of J-like aggregate of cyanine was not efficient, while the H-aggregate of **8a** showed efficient sensitization ability. The J-like aggregate of **8c** hardly showed sensitization. It was found that the short-circuit current and conversion efficiency of the nanocrystalline TiO_2 solar cell sensitized by three dyes **8a**, **8b** and **8c** was greater than those of each simple cyanine-sensitized solar cell. The IPCE spectra were broadened, but the IPCE value was located between the maximum IPCE values of **8a** and **8c** sensitized solar cells because of the possible filter effect and the energy and electron transfer between different cyanine dyes. The maximum conversion efficiency of 3.1% was achieved when three different cyanine dyes **8a**, **8b** and **9** were adsorbed simultaneously on TiO_2 electrode.

It is possible to improve the overall conversion efficiency of DSSC through the design of a cyanine to form different aggregates and the use of several dyes with different absorption spectra to co-sensitize the electrode. But the experiment on the combination of different dyes is very complicated because the performance of solar cells is greatly dependent on the immersion order of the dyes, the immersion time of the electrode in each dye solution, the aggregation of the dyes on the electrode and the energy or electron transfer between different cyanine dyes.

2.2. Hemicyanine dyes for dye-sensitized nanocrystalline TiO_2 solar cells

The hemicyanine dye is usually composed of a strong donor and an acceptor linked with a π-conjugation bridge. This molecular structure results in a large dipole moment, which facilitates efficient charge separation. Huang *et al.* [13] studied the photoelectric conversion properties of nanocrystalline TiO_2 electrode sensitized with hemicyanine dyes (Scheme 7). Hemicyanine **10** and **11** does not form any aggregate on TiO_2 film and the absorption peaks are only blueshifted

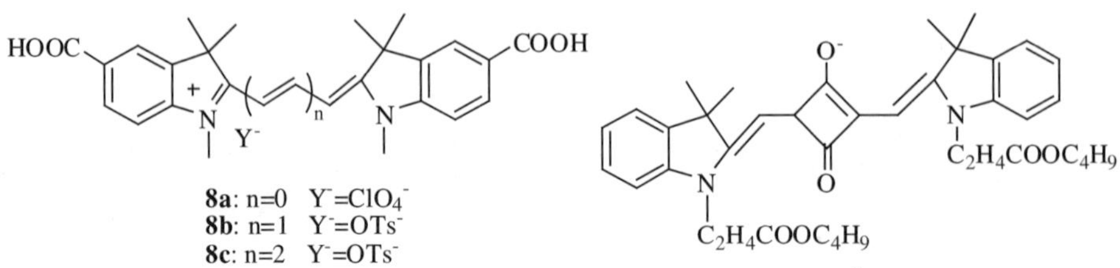

8a: n=0 Y^-=ClO_4^-
8b: n=1 Y^-=OTs^-
8c: n=2 Y^-=OTs^-

9

Scheme 6.

compared with their corresponding absorption peaks in methanol solution. These blueshifts could be attributed to the interaction between adsorbed dyes and the TiO_2 surface. Alkyl chain has great effect on the adsorbance, IPCE and overall conversion efficiency. A long alkyl chain is generally necessary for LB film formation, but for the dye-sensitized nanocrystalline solar cell it wastes surface and space of the electrode because of its large size and small positive effect on photocurrent generation. Therefore, the hemicyanine with methyl group result in a significant increase in adsorbance and IPCE. In addition, the quinoline ring is slightly larger than the pyridine ring. This difference in ring size can also affect the adsorption of the dye and consequently, the photoelectro-chemical performance. Correspondingly, the short-circuit currents for **10a** and **11a** sensitized solar cells are much higher than those of **11b** and **11c** based solar cells. The hemicyanine **11a** sensitized solar cell has the highest short-circuit current and overall conversion efficiency (about 2%) in the four dyes due to its higher IPCE values in a broad range. The steric effects on the amount of dyes adsorbed on the electrode and the conversion efficiency are also observed in other hemicyanine dyes.

Surface treatment of TiO_2 film is a key step for optimizing the photoelectrical properties of the solar cell. Huang *et al.* [14,15] studied the photocurrent enhancement of hemicyanine dyes (Scheme 8) containing sulfonic groups through treating TiO_2 films with hydrochloric acid. Dyes with sulfonic group are adsorbed onto the TiO_2 surface through electrostatic interaction between the sulfonic group and the surface Ti^{4+} ion. The surface without exposed Ti^{4+} ion may not be beneficial to dye adsorption, which is one of the key factors limiting the adsorbed

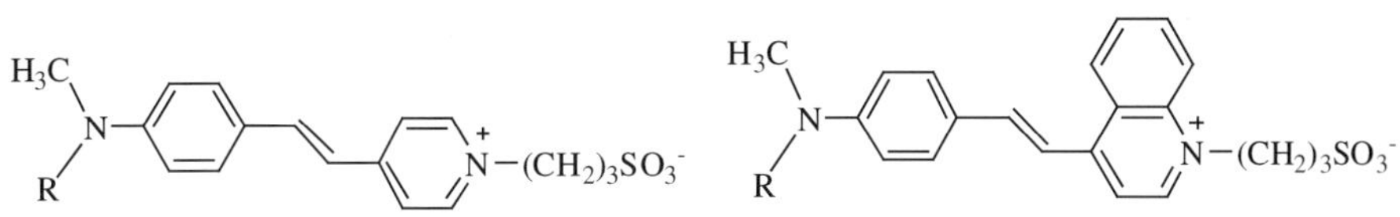

10a: R=CH_3
10b: R=$C_{16}H_{33}$

11a: R=CH_3
11b: R=$C_{16}H_{33}$

Scheme 7.

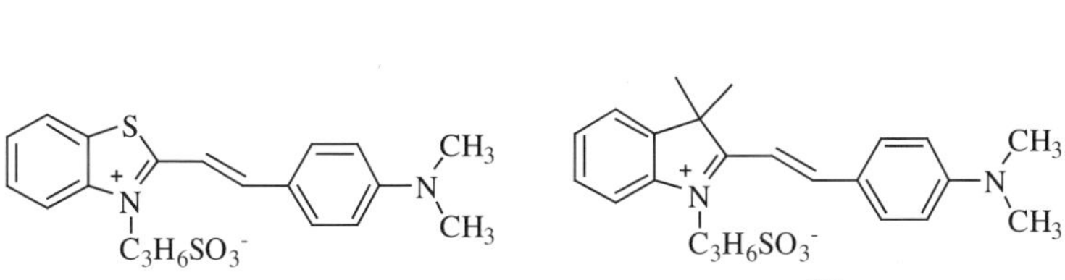

12 **13**

Scheme 8.

amount and conversion efficiency. Hydrochloric acid has been used to treat the TiO_2 film to increase the number of active sites on the TiO_2 surface. A large number of protons were adsorbed on the surface of TiO_2 film. These newly formed surface sites (H^+) are also active for dye adsorption besides the Ti^{4+} adsorbing site. As a consequence, dye molecules can be adsorbed in a more compact way, resulting in a remarkable increase in the amount of adsorbed dyes. The amount of dye adsorbed increased by 83 and 350% for **12** and **13**, respectively. The increased overall surface area of TiO_2 film caused by the smaller particles upon treating TiO_2 with HCl was also responsible for the increase in the adsorption amount of dyes. The amount of the adsorbed hemicyanine **12** is higher than that of the adsorbed hemicyanine **13** in both treated and untreated TiO_2 film because of the less-steric hindrance of hemicyanine **12**. The flat band potential of TiO_2 electrode was shifted more than 200 mV toward the positive direction after being treated with hydrochloric acid, which is attributed to the specific adsorption of H^+ ions at the TiO_2 surface. The positive shift of the flat band potential is favorable for electron injection from the excited state dye to the conduction band of TiO_2 and to reduce the charge recombination between dye cations and injected electrons. Both the hemicyanine dyes **12** and **13** show outstanding charge-transfer properties with nearly 100% IPCE at their maximum absorption wavelength. The short-circuit current was improved by 99 and 329% for hemicyanine **12** and **13**. The overall conversion efficiencies were hence improved remarkably as a result of short-circuit current increase: hemicyanine **12** increased from 3.1 to 5.1% and hemicyanine **13** increased from 1.3 to 4.8%.

For better adsorption on the nanocrystalline TiO_2 film, hemicyanine dye **14** (Scheme 9) bearing two carboxyl groups were synthesized [16]. The dye was covalently attached on TiO_2 nanoparticles, leading to satisfactory photoconversion efficiency with maximum IPCE value more than 45%. But the overall conversion efficiency is only 0.72% owing to the limited light-absorption ability of this hemicyanine.

Tian *et al.* [17] synthesized indolenine and benzindolenine-base hemicyanine dyes (Scheme 10) with two carboxyl groups and studied their optical and electrochemical properties. In comparison with hemicyanine **14**, the absorption spectra of hemicyanine **15** and **16** are redshifted as a result of the extended π-conjugation. When adsorbed on TiO_2 film, the absorption of hemicyanine **15** and **16**

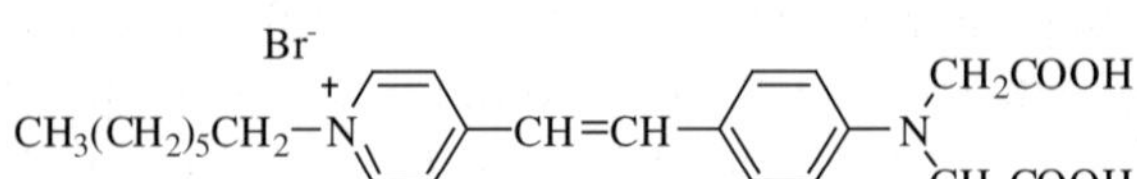

14

Scheme 9.

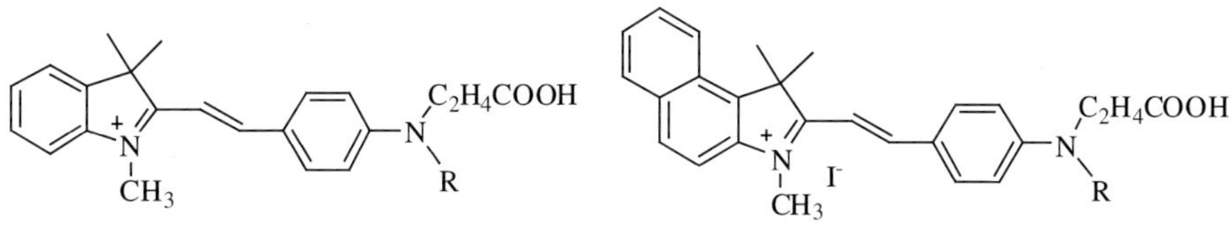

15a: R=C_2H_5
15b: R=C_2H_4COOH

16a: R=C_2H_5
16b: R=C_2H_4COOH

Scheme 10.

are broadened in the blue and red region of the spectra due to the formation of H- and J- aggregates. The absorption spectra are structureless and have high absorption in a broad wavelength range, which is favorable for the harvesting of solar energy. What is particularly noticeable is that the absorption spectra of hemicyanine (**15b** and **16b**) with two attaching groups on the sensitized electrode are broader than that of the corresponding hemicyanine (**15a** and **16a**) with only one attaching group. The adsorption amount of hemicyanine dyes with two carboxyl groups (**15b**, **16b**) on TiO_2 film is two or three times more than the dyes with only one carboxyl group (**15a**, **16a**). As a result, the DSSCs based on **15b** and **16b** have broader IPCE spectra with **16b** having the higher IPCE values ($> 60\%$) in the region from 460 to 640 nm. The short-circuit currents of **16b** sensitized solar cell reached a surprisingly high value of 21.4 mA/cm^2 under the irradiation of 90.0 mW/cm^2 white light. The overall conversion efficiencies of DSSCs based on **15a**, **15b**, **16a** and **16b** were 4.0, 4.6, 4.4 and 4.9%, respectively.

Compared to the adsorption amount of dyes with short-circuit current, hemicyanine dyes with carboxyl groups in the donor part have two different effects on the adsorption and photoelectric conversion properties of the dye. On the positive side, the adsorbed amount of dye molecules on the TiO_2 film is greatly increased, which is favorable for the conversion of light to electricity. On the negative side, the second carboxyl group impedes the charge separation of the molecules because it is attached to the electron-donor side and hence reduces the polarity of the molecules to some extent. Tian *et al.* [18] synthesized novel hemicyanine (Scheme 11) with multi-carboxyl group in the acceptor part and used them to sensitize TiO_2 electrode. Unfortunately, the adsorption amount of hemicyanine **18** and **19** on TiO_2 film is significantly lower than hemicyanine **17** because of their large steric hindrance, the resulting short-circuit current and overall conversion efficiency are lower than that of hemicyanine **17**. The overall conversion efficiency of **17** based solar cell is 2.12% under the illumination of 94.6 mW/cm^2 white light with the short-circuit current of 13.8 mA/cm^2 and open-circuit voltage of 356 mV.

Huang *et al.* [19] synthesized naphthothiazolium hemicyanine (Scheme 12) and systematically studied their absorption, emission, electrochemical and

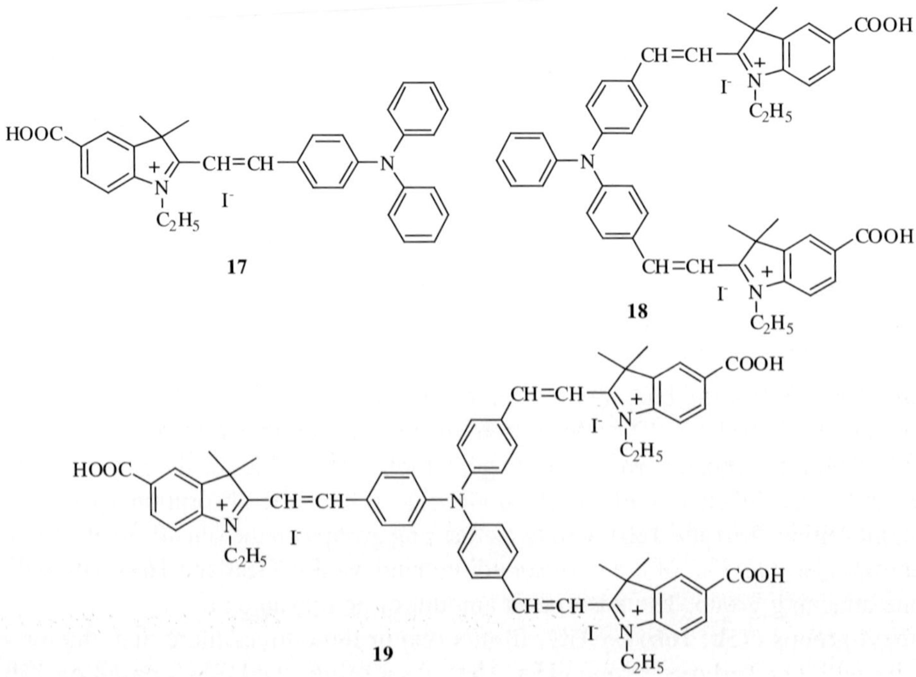

Scheme 11.

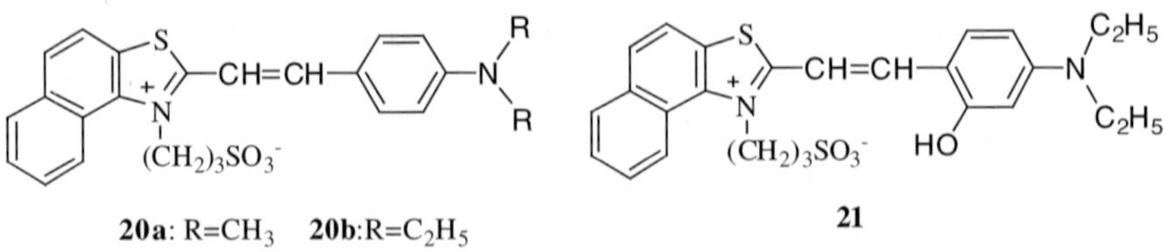

Scheme 12.

photoelectrochemical properties. The absorption peaks for the three dyes on TiO_2 films are all extremely broadened compared with their corresponding peaks in methanol solution, suggesting the formation of H- and J-aggregates. The amount of dye **20b** adsorbed on TiO_2 film is decreased by about 40% caused by the slightly large steric effect of the ethyl group, while the adsorption amount of dye **21** was much lower than those of dye **20**. Other factors may be responsible for such low adsorption except the steric effect. The contact area between dye **21** and TiO_2 nanoparticles may be increased by the cooperation effect of hydroxyl group and sulfonic group in the adsorption process, which subsequently leads to a decreased adsorption of dye per unit surface area of the TiO_2 film. By comparing the contribution of short-circuit current per dye molecules, it was shown that the short-circuit current generated by dye **21** is 4.7 times higher than **19**, revealing that the introduction of the additional hydroxyl group to the donor is favorable for the electron injection efficiency. The enhanced electron injection efficiency overcomes the inverse effect of decreased adsorption amount of dye **21** on TiO_2 film. Owing to this reason, the IPCE value of dye **21** are about 1.3 times higher than those of dye **20** in the whole spectra region from 400 to 700 nm. The overall conversion efficiency of **20a** and **20b** based solar cells were 4.1 and 4.0%, respectively, whereas the overall conversion efficiency of dye **21** based solar cell was as high as 6.3%. In an unsealed solar cell, dye **21** was proved to be able to generate constant conversion efficiency for more than 200 h under the illumination of 80 mW/cm^2 white light. The main reason for the gradual decrease of the efficiency after 200 h was found to be the evaporation of the solvent in the redox electrolyte solution.

Recently, Wang *et al.* [20] synthesized a series of new benzothiazolium hemicyanine dyes (Scheme 13) by introducing carboxyl, hydroxyl, or sulfonic acid anchoring groups onto the dyes' skeletons and studied their sensitization on nanocrystalline TiO_2 electrode. Carboxyl group is a stronger anchoring group than sulfonic acid and the combination of the carboxyl and hydroxyl groups will lead to a complexation reaction of the corresponding dyes with Ti^{4+} ion. Therefore, the adsorption amount of hemicyanine **22** and **23** on TiO_2 film is larger than that of hemicyanine **24** and **21** and the desorption of hemicyanine **22** and **23** from TiO_2 film in methanol needs much longer time than hemicyanine **24** and **21**. The presence of a hydroxyl group in the aniline moiety leads the adsorbed amount of **22a** and **23a** decreased by 7 and 16% with respect to **22b** and **23b** owing to the larger steric effect and the increased contact area of dye **22a**

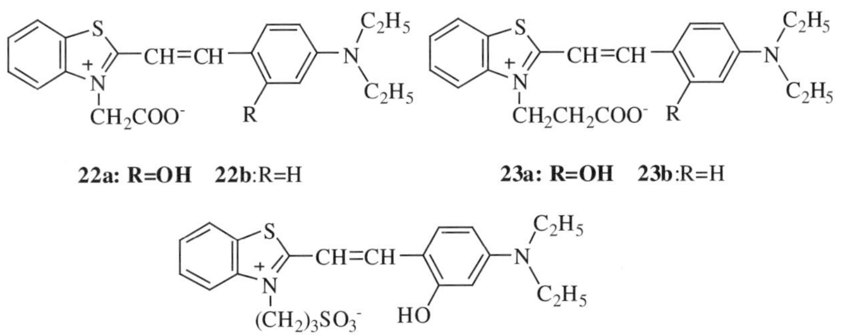

22a: R=OH **22b**:R=H **23a: R=OH** **23b**:R=H

24

Scheme 13.

and **23a**. The complexation of carboxyl and hydroxyl groups with Ti^{4+} ion enhances electronic coupling within the complexation interaction, which greatly improves the electron injection efficiency from dyes to TiO_2 electrode together with the cathodic shift of the excited state oxidation potential by the introduction of the hydroxyl group. The fluorescence quenching efficiencies of the dyes by colloidal TiO_2, the IPCE and overall conversion efficiencies of the DSSCs are all depend strongly on the anchoring group types and decrease in the order: carboxyl + hydroxyl > carboxyl > sulfonic acid + hydroxyl, indicating the importance of the dyes' anchoring groups for their sensitization effects in DSSCs. Accordingly, high IPCE values over a wide spectrum region with the maximum of 73.6% and overall conversion efficiency of 5.2% under AM1.5 Global simulate light (80 mW/cm^2) for dye **22a** based DSSC were achieved. Under similar conditions, the overall conversion efficiency of dye **21** based DSSC is only 2.6%.

2.3. Cyanine dyes for solid state solar cells

The use of a liquid electrolyte in the dye-sensitized nanocrystalline solar cell may limit device stability because the liquid may evaporate when the cell is imperfectly sealed, and more generally, the diffusion and reaction of water or oxygen molecules may worsen cell performance. The liquid electrolyte also makes the manufacture of multi-cell modules difficult because cells must be connected electrically yet separated chemically, preferably on a single substrate. One way to overcome some of these problems is to replace the liquid electrolyte with a solid conducting material. The other way is to develop pure solid-state organic photovoltaic devices.

The potential usage of cyanine dyes in thin film heterojunction photovoltaic devices has been investigated [21]. Heterojunction-multilayer devices were made from cyanine **25**, polymer **M2** (Scheme 14) and fullerene C_{60}. The cyanine and polymer were spin-coated from the solution and C_{60} was evaporated under high vacuum to make photovoltaic devices with different device architecture. The photocurrent spectrum for the device (**a**), ITO/cyanine **25**/C_{60}/Al, has an IPCE of more than 2% in the range 450–600 nm where cyanine **25** absorbs. In device (**b**), ITO/polymer **M2**/C_{60}/Al, the cyanine donor is replaced by polymer **M2** and IPCE values as high as 6% are obtained in a spectral domain where **M2** absorbs.

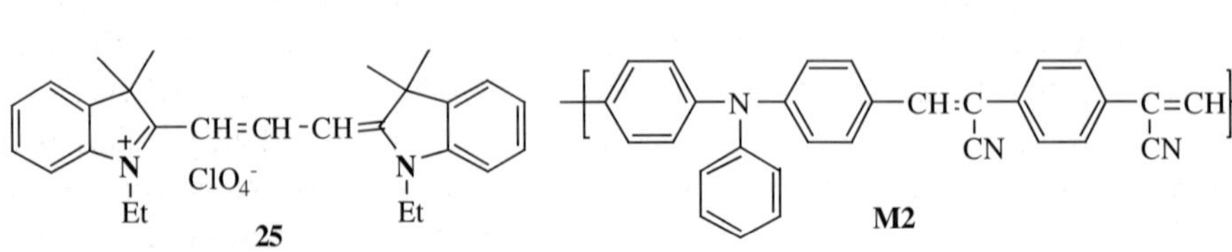

Scheme 14.

In device (**c**), ITO/polymer **M2**/cyanine **25**/Al, an important photocurrent in the **M2** absorption domain is indeed observed. However, no photocurrent matching the cyanine **25** absorption spectrum is observed. This phenomenon can be attributed to the presence of J-aggregate, which makes the highest occupied molecular orbital (HOMO) level of cyanine dye shift toward higher energies. The difference between the HOMO level of polymer and that of cyanine is not large enough to dissociate the excitons formed in the cyanine layer. While in device (**d**), ITO/polymer **M2**/cyanine **25**/C_{60}/Al, the photocurrent contribution from both **M2** and cyanine **25** occurred, this resulted in a fairly broad photocurrent spectrum. Excitons generated in polymer **M2** will dissociate at the **M2**/cyanine **25** interface, while excitons generated in cyanine **25** will dissociate at the cyanine **25**/C_{60} interface. In this device architecture, cyanine **25** acts both as an acceptor and as a donor at the same time and the maximum IPCE is over 10%. Another advantage of the use of cyanine as acceptor is that the device has high open-circuit voltage. The open-circuit voltage of device (**c**) is as high as 1.28 V and that of device (**d**) is 0.83 V, both are higher than device (**a**) and (**b**).

The astonishingly low open-circuit voltage of device (**a**) could be related to processes at the ITO/cyanine interface. In order to gain more insight into the interfacial mechanism, poly(styrene sulfonate) doped poly(3,4-ethylenedioxythiophene) (PEDOT:PSS) was used to coat ITO prior to the deposition of the cyanine layer and device (**e**), ITO/PEDOT/cyanine **25**/C_{60}/Al, was constructed [22]. By applying a positive bias to ITO, the perchlorate anions of cyanine **25** make a displacement toward the ITO surface and hole injection from ITO into the HOMO of cyanine **25** is significantly enhanced. As a result, a large hysteresis appears under the forward bias as soon as the voltage is raised above 4 V and the hysteresis is reversible even when the device is cycled at reverse bias. This behavior of device (**e**) is much less pronounced than that of device (**a**) as the PEDOT layer prevents the formation of a dipole layer on a short-length scale. The performance of device (**e**) with PEDOT layer is better than device (**a**) without PEDOT with respect to IPCE, short-circuit current and open-circuit voltage. By analyzing the spectral dependence of the photocurrent as a function of the applied voltage, it was found that photocurrent generation occurs at cyanine/ITO and cyanine/C_{60} interfaces in device (**a**). While in device (**e**), PEDOT inhibits electron injection from cyanine into ITO and therefore increases short-circuit current and open-circuit voltage.

Cyanine dye **26** (Scheme 15) having identical chromophores with cyanine **25** and a covalently counterion was used to investigate the counterion effects on the photoelectrical characteristics of heterojunction device (**f**), ITO/PEDOT/cyanine **26**/C_{60}/Al [23]. In device (**f**) where the counterion is covalently linked with cyanine, the hysteresis is not as obvious as that of device (**e**). No current generation in device (**f**) is observed in the blue to ultraviolet (UV) region of the spectra although C_{60} strongly absorbed at these wavelengths, while device (**e**) exhibits strong photocurrent response in the blue to UV region. This behavior is likely

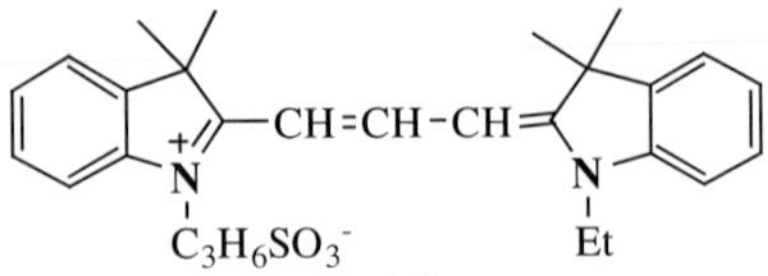

26

Scheme 15.

due to the mobile counterions in cyanine **25**. Mobile ions are susceptible to screen charges in the solid film and facilitate charge separation. So, cyanine dyes with mobile ions are beneficial to device performance.

One of the key problems for boosting the photovoltaic efficiency is the dissociation of the exciton into a positive- and negative-charge carrier. Excitons are formed upon photon absorption and are able to diffuse in the organic material until the decay or are dissociated by a charge-transfer process. In order to develop materials with high exciton dissociation efficiency, molecules were designed where C_{60} is covalently linked to the light-absorbing chromophore.

Tian *et al.* [24] synthesized novel cyanine-fullerene dyads (Scheme 16) and studied their optical, electrochemical and photovoltaic properties. The fluorescence of the cyanine-fullerene dyads was quenched efficiently as compared to the pristine cyanine, which was attributed to the photo-induced electron transfer from the cyanine to C_{60}. The redox potentials obtained by cyclic voltammetry provide evidence that this electron transfer is energetically favorable. The IPCE spectrum of the single-layer device, ITO/PEDOT:PSS/cyanine **27**/Al, shows a "in phase" correlation to the absorption spectrum of the cyanine, indicating that charge carriers are generated within the bulk of the film under excitation by light. When the irradiation intensity was increased from 0.31 to 310 mW/cm^2, the open-circuit voltages (V_{oc}) of the device increased from 0.18 to 0.33 V, but the short-circuit current increased in a sub-linear way. The fill factor and conversion efficiency of the single-layer devices is extremely low due to the high series resistance of the devices. C_{60} layer were inserted between the cathode and the cyanine-dyad layer to form double layer device, ITO/PEDOT:PSS/cyanine **27**/C_{60}/Al. Similar to single-layer device, the open-circuit voltage of the double-layer device increased from 0.18 to 0.33 V with increasing irradiation intensity. However, the short-circuit current and fill factor were much higher than those measured for single-layer devices. Under white light excitation of 310 mW/cm^2, the short-circuit current density reaches 1.2 mA/cm^2 and the conversion efficiency is 0.041%. At lower light intensity, the efficiency exceeds 0.1%. The better performance of the double layer device can be attributed both to the protective effect of the C_{60} layer during cathode deposition and the advantageous interface formation between C_{60} and aluminum.

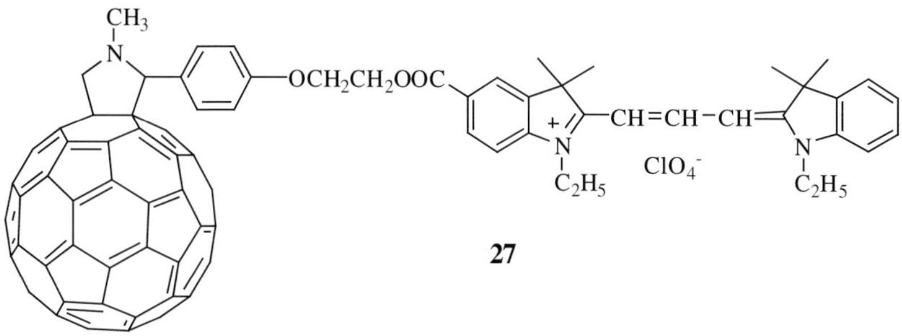

27

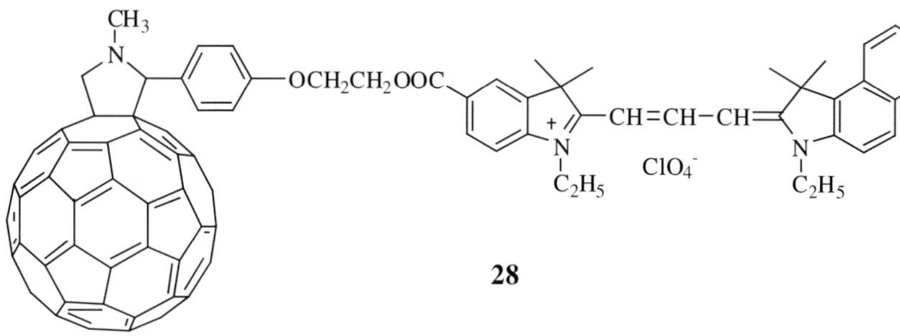

28

Scheme 16.

3. CYANINE DYES FOR APPLICATION IN OPTICAL DATA STORAGE

3.1. Cyanine dyes for CD-R

A recordable compact disk (CD-R) is an optical disk, which is capable of recording and reproducing information of characters, figures and sounds by irradiation of laser beam. It was invented in 1988 by the Japanese company Taiyo Yuden [25,26] and commercialized in 1994. The standard characteristics of CD-R were specified in the Orange Book II in 1990. Since then, CD-R has been accepted as a convenient medium for storage and retrieval of a great quantity of information and has been playing more and more important roles in data storage.

As shown in Fig. 1, CD-R is composed of a transparent substrate (e.g., polycarbonate PC), a recording layer made of organic dye (dye recording layer), a reflective layer made of metal (e.g., gold or silver) and a resin protective layer. The substrate contains a molded-in spiral pre-groove, which enables a tracking signal to be generated to guide the incident laser beam. The recording layer was formed on the substrate by spin-coating of a dye solution whose solvent is required not to dissolve the substrate, such as alcohols and aliphatic hydrocarbons.

The recording of data is carried out by irradiating the optical disk with a laser beam of the near-IR region (usually around 780 nm for CD-R) from the side of the transparent substrate. The irradiated part of the dye-recording layer absorbs

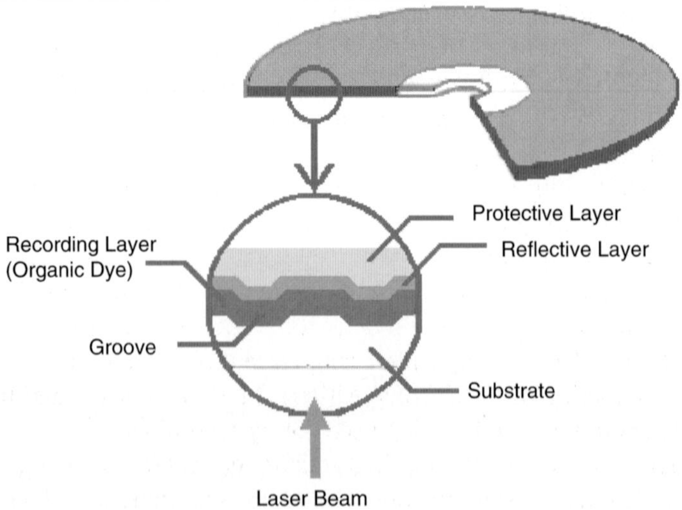

Fig.1. Configuration of CD-R and recordable digital versatile disk (DVD-R).

the light to generate heat and is thermally deformed (e.g., to form bumps at the PC/dye interface and pits at the dye/Au interface) to record the information [27]. The differences of the reflectivity and modulation between the recorded and unrecorded area are very big. The recorded information is reproduced by irradiating the optical disk with a laser beam having the same wavelength as the writing beam and by detecting the reflectance difference between the recorded area (the area where the dye-recording layer has been thermally deformed) and the non-recorded area (the area where the dye recording layer has not been deformed). The laser beam for the reading of data stored in the recording film has an output power less than that of the laser beam for the writing.

It is strongly desired for CD-R to follow the same standard as the widely used standards for CD, so that CD-R can be played by the commercially available CD player. To be compatible with CD, it is required that the reflectivity and modulation of the CD-R must be high enough (reflectivity $\geq$ 70% and modulation $\geq$ 60%) [27]. The reflectivity and modulation is determined by the complex refractive index $N=n-ik$ of a dye, where k is the extinction coefficient and n the real part of the refractive index. A high n of the dye is needed to obtain a high modulation and a low k is needed to obtain a high reflectivity. For recording a signal, the extinction coefficient k of the recording layer at the wavelength of the recording light and reproducing light is preferably 0.03–0.15. If $k < 0.03$, the absorption rate of the recording layer is lowered to make recording difficult with usual recording power. On the other hand, if k exceeds 0.15, the reflectivity will drop below 70% and fail to satisfy the standards thus making reproducing difficult. The refractive index n of the recording layer is preferably higher than 2.0. If $n < 2.0$, the reflectivity is lowered while the degree of signal modulation tends to be reduced to make reproducing signal difficult. Of course, both of n and k should be considered together for a given recording layer because these two factors affect the reflectivity and modulation.

Other than having good optical properties, the dyes used in CD-R should also have good thermal properties. To obtain a good quality record, it is preferred to employ a dye having a decomposition temperature in the range of 240–290°C. If the decomposition temperature of the dye is too low, the thermal stability of the dye layer becomes poor, and if it is too high, recording on the dye layer becomes difficult. The decomposition range (the difference in temperature from starting to terminating of the decomposition) should be narrow and the thermal conductivity should be low for a high sensitivity recording.

In order to effectively produce high-quality optical recording media in a relatively high yield by the spin-coating method, the recording dye should be satisfactorily soluble in easily volatile organic solvent, which do not dissolve the substrate, such as alcohols and aliphatic hydrocarbons. Besides, the dye should have good oxidative and hydrolytic stability, low toxicity and low cost.

The cyanine, azo and phthalocyanine dyes meet the above criteria best and are therefore used widely in CD-R. Cyanine dyes were the earliest developed ones and the standard characteristics of CD-R were established on the basis of cyanine-based optical recording disk. So, cyanine dyes played a very important role in the development of CD-R. We will review the developments and applications of cyanine dyes in CD-R.

In fact, when used as an optical recording material, cyanine dye has a number of advantages such as high absorption, small thermal conductivity and diffusivity with the possibility to form good pits configuration and high reflectivity in a laser beam wavelength. However, the cyanine-type dye has poor light resistance. When cyanine dyes return to ground state from the excited triplet state under laser beam irradiation, oxygen is activated to produce singlet oxygen even though the intersystem crossing yields for cyanine dyes are usually quite low. The singlet oxygen is so reactive that it is reacted with the methine chain of the cyanine dye to destroy the dye. Accordingly, the cyanine dye is degraded with respect to the absorbance, reflectance and the similar characteristics, making it difficult to perform good recording and reproducing operations. In order to prevent the degradation of cyanine dyes in optical recording medium by the singlet oxygen, singlet oxygen quencher was required to be added for providing a sufficient durability against the laser beam irradiation.

Since the writing and reproducing laser wavelength for CD-R is 780 nm, the cyanine used in this purposes is mainly pentamenthine cyanine dyes. The indoles- and benzindoles-type cyanine dyes (Scheme 17) were the most important cyanine dyes for CD-R because they tend to be more light stable than other cyanine dyes [28]. The structure of cyanine dye, such as the type of heterocycles (A and B), substitutents (R_1 and R_2) of the nitrogen atoms and the counterion (X), has a great effect on the optical and thermal properties and solubility, which are very important parameters for cyanine dyes used as recording medium.

Many manufacturers use proprietary chemical additives to make cyanine disks more stable. Pioneer Electronic Corporation claimed an optical recording layer containing a cyanine and a singlet oxygen quencher **S1** (Scheme 18). The cyanine is of the formula **29** in which R_1 and R_2 are methyl group and X is perchlorate [29], R_1 and R_2 are butyl and X is perchlorate [30], or R_1 is propyl, R_2 butyl and X perchlorate [31]. The optical disk made with these cyanine and singlet oxygen quencher have good durability against optical degradation, high reflection coefficient and modulation rate high enough to meet the CD standard. Taiyu Yuden Company claimed a class of aryl nitrogen compounds, including aryl nitroso compounds and 1-Pycryl-2,2-diarylhydrazyl free radicals, as light stabilizer capable of effectively rendering cyanine dyes light fast [32]. These compounds have high solubility in solvents and have good miscibility with cyanine dyes. The mixture of cyanine dye **29** and azo metal chelate compounds (e.g., Scheme 19) was also used as the recording layer for optical disk that is excellent in the light fastness and moisture resistance [33].

Media Chemical Corporation proposed to use leucoaryl amine (Scheme 20) as stabilizer (quencher) with cyanine dyes [34]. The amine stabilizer has improved solubility in desirable coating solvents and its presence increases the photostability of the dye layer, and stabilizes the dye solution such that the composition remains constant over extended periods of time, enabling the solution to be utilized much

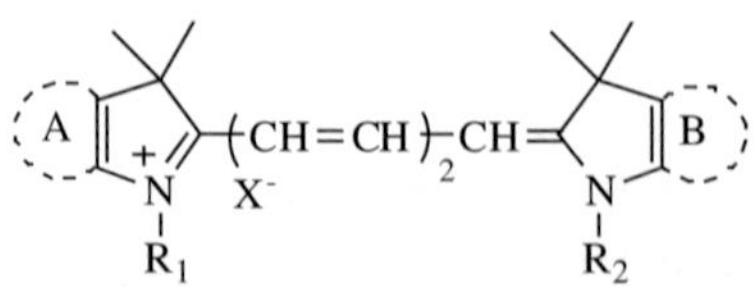

Scheme 17.

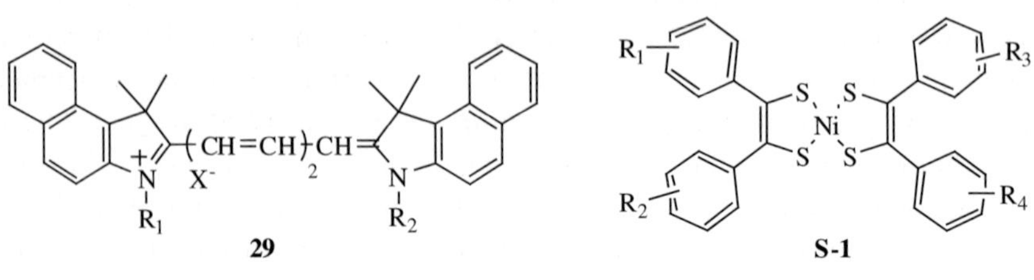

Scheme 18.

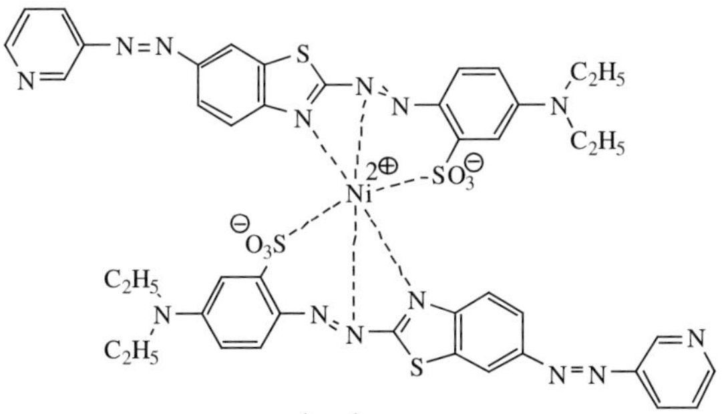

Azo-1

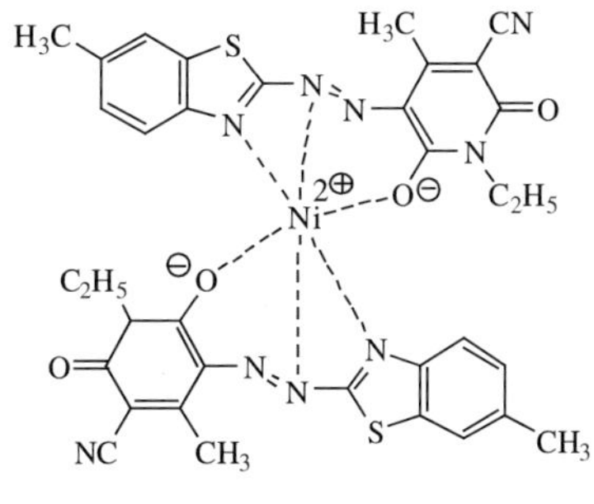

Azo-2

Scheme 19.

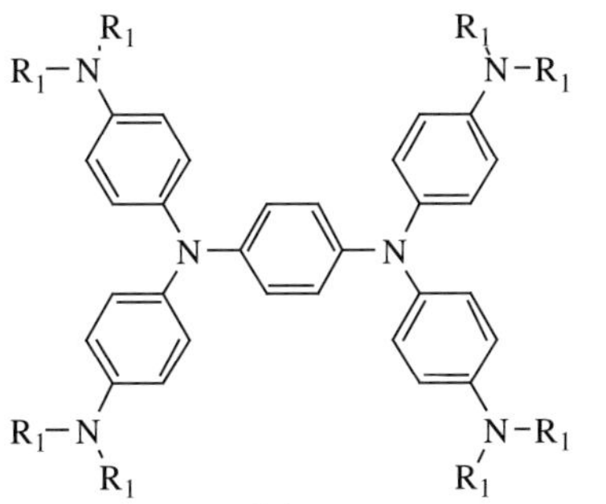

S-2

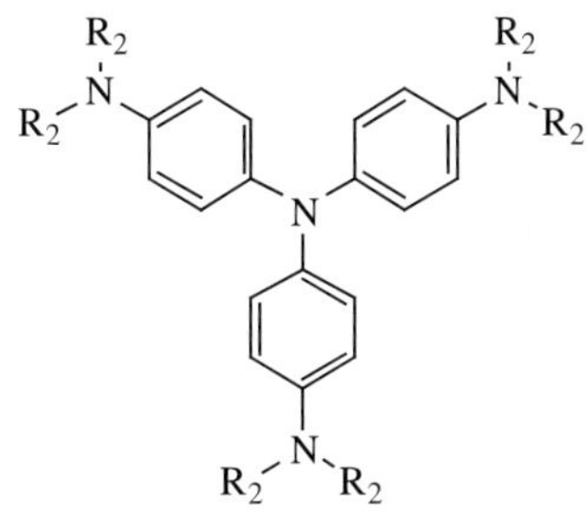

S-3

Scheme 20.

more effectively and economically. The leuco compound has no absorption in the long wavelength region, which can be used without altering the desirable absorption properties of the dye mixture at the recording wavelength. The produced optical disks have improved recording characteristics at a range of recording speeds.

An unsymmetrical cyanine dye **30** is used as a major component with certain compatible and stable non-cyanine dyes **S-4** or **S-5** (Scheme 21) [35]. The stable non-cyanine dyes **S-4** and **S-5** exhibit excellent solubility in preferred coating solvents and have relatively high absorption coefficients (typically >30,000 cm^{-1}) with absorption peaks in the range 900–1200 nm and relatively wide absorption bands. These spectral characteristics contribute to increasing absorbance with increasing wavelength in the critical recording range of wavelengths. These absorption characteristics counteract the decreased absorbance characteristic of cyanine dyes, thus producing a recording layer with specific absorption, stability and solubility parameters suitable to meet CD-R recording

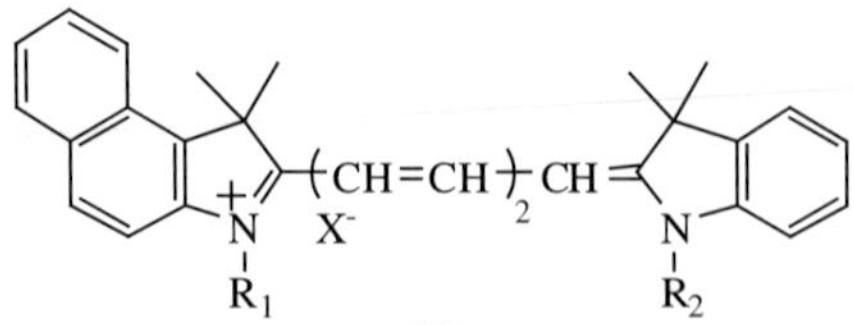

30

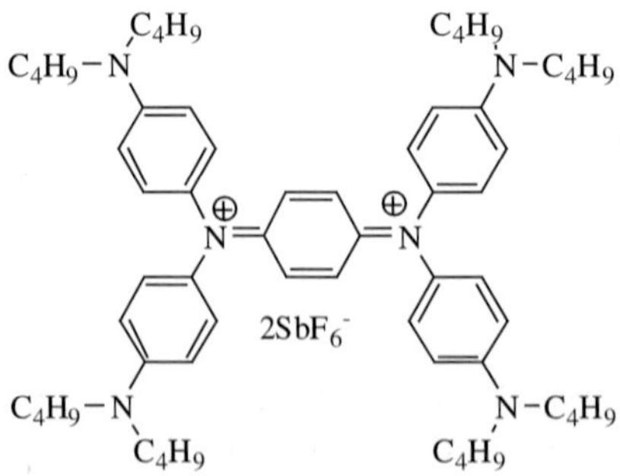

S-4

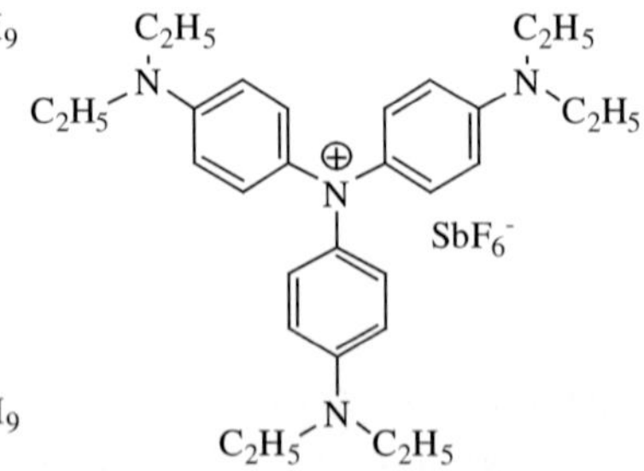

S-5

Scheme 21.

standards. The overlap of the absorption bands sufficiently at the recording wavelength facilitates energy transfer from the cyanine dye to the non-cyanine dye, stabilizing cyanine dyes. Furthermore, the disk has small variations in the parameters, which control the recording process such as dye layer film thickness, uniformity of the coating and substrate optical properties.

If a recording layer is formed from a mixture of a cation-type dye and a quencher, then four types of ions are present, a dye cation, a quencher anion and their counterions in a ratio corresponding to their mix ratio of the dye and the quencher. The presence of so many ionic combinations will result in a low melting point, a broad softening point and poor reliability of the recording materials. TDK Corporation proposed to use an ionic combination of a dye cation and transition metal complex anion (Scheme 22) as a singlet oxygen quencher for the purpose of preventing output lowering and improving light resistance [36,37]. In this combination, the dye cation and the quencher anion are present in a ratio of 1:1. This suggests that the use of an ionic combination, counterions of the dye and quencher being absent, has the advantage of less output lowering and higher light resistant than the mix system. In this respect, the ionic combinations are effective for reducing the reproduction deterioration of the heat mode and improving shelf stability as well as moisture resistance. If the quencher anions have a high coefficient of extinction *k*, the use of an ionic combination of the close contact type yields a recording layer having increased *k* and decreased reflectivity, failing to provide satisfactory reproduction, and therefore could not

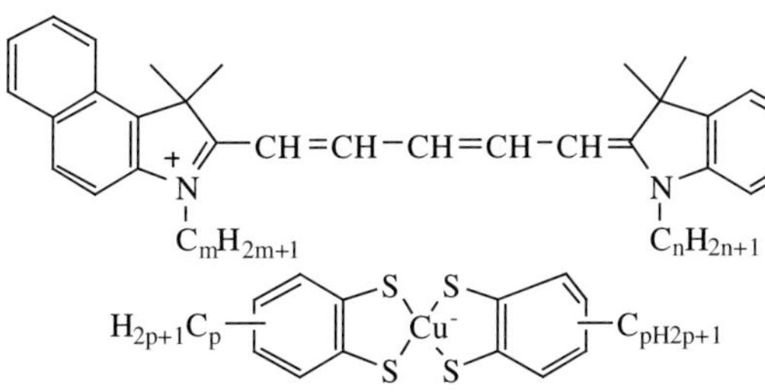

31

Scheme 22.

take advantage of the ionic combinations. A salt forming dye **31** (ionic bond compound) comprising an indolenine cyanine dye and a benzenedithiol metal complex anion are heat resistant, moisture resistant, light resistant as well as has minimal variation with a change in the wavelength of recording and reproducing light. The cyanine dye **31** have *n* and *k* which are less dependent on the light wavelength and exhibit a minimal change of *n* and *k* over the range 770–810 nm while having high resistance against light. The recording layer has the extinction coefficient *k* from 0.02 to 0.05 and refraction index *n* from 2.0 to 2.6 at the wavelength recording and reproducing light. The disk is resistant against light, heat and water to ensure satisfactory performance according to the CD standard.

Although cyanine dyes themselves are soluble in TFP (2,2,3,3-tetrafluoropropanol) at relatively high concentrations, most salt-forming dyes have substantially low solubility failing to reach concentrations sufficient to provide a necessary film thickness. TDK Corporation provides a new cyanine-quencher dye (Scheme 23), which is well soluble in the solvents having a high evaporation rate and are stable against light [38]. By attaching -SO_2R to the benzene ring of the benzenedithiol complex, the solubility of cyanine complex **32** in solvents having a high evaporation rate such as TFP (2,2,3,3-tetrafluoropropanol) becomes very high, and the coating time by spin-coating is reduced to half, leading to an improvement in productivity. Further, since the dye moiety playing an important role of governing optical constants is a cyanine dye, the degree of free design is very high and the restrictions on the available wavelength band are almost eliminated. For the adjustment of a wavelength and solubility, an unsymmetrical structure may be used. Also, adjusting the number of carbon atoms on the N-side chain in the range of 1 to 5 enables a fine adjustment of wavelength and a further improvement in solubility.

When the anion of cyanine dye consists of specific azo metal complexes, the recording media have improved properties [39]. The cyanine-azo complex (e.g., Scheme 24) has an improved compromise between high refractive index,

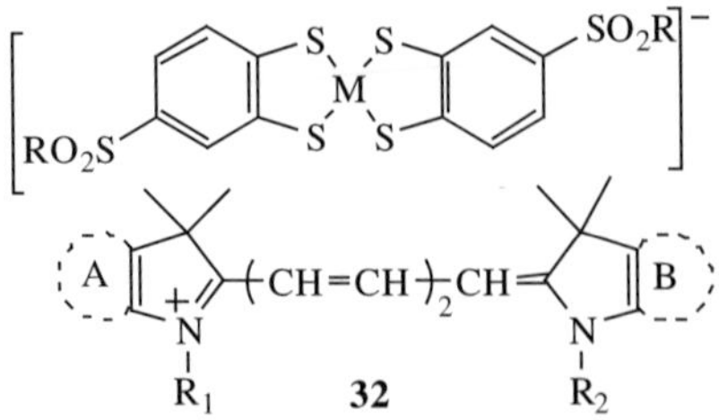

32

Scheme 23.

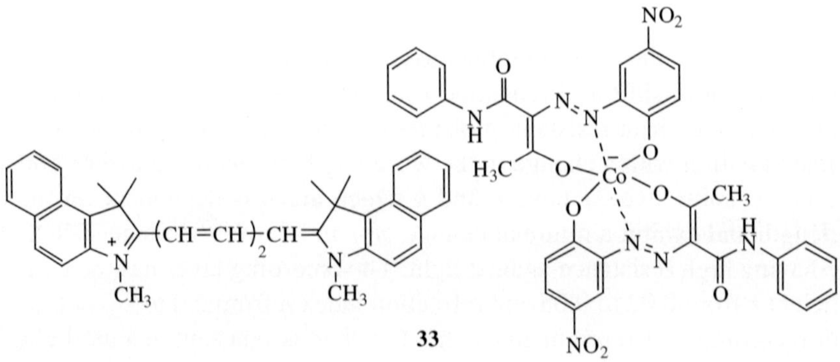

33

Scheme 24.

sensitivity and light stability. The dyes have a suitable amorphous morphology and an advantageous degradation temperature. During the recording process using a laser beam, the refractive index changes very much between the irradiated and unirradiated areas and the recording may surprisingly be effected already at lower energy. Owing to the higher sensitivity and the favorable morphology, which is substantially retained without any change in the volume during recording, the pit formation can be better controlled. This also contributes to the increase in the data-packing density as a less redundant code can be used.

According to the CD standard, the recording and reproducing operation wavelength is 780 nm. However, the laser wavelength may vary with temperature and somewhat different semiconductor laser products and is generally in the range from 770 to 790 nm. Normally, indolenine cyanine dyes have an extinction coefficient *k*, which is greatly dependent on the wavelength that for the dye having a desired *k* value at 780 nm, and *k* would deviate off the desired range if the wavelength varies within ±10 nm. Pioneer Electronic Corporation proposed to mix benzindolenine cyanine dyes **29** (Scheme 18) and indolenine cyanine **34** (Scheme 25) together with a quencher compound for the purpose of improving the wavelength dependency of k [40]. The cyanine dye **34** is preferably contained in the light-absorbing layer in a greater amount than the amount of cyanine dye **29**. If cyanine **34** is contained in an amount that is equal to or smaller than cyanine dye **29**, it is impossible to achieve the necessary reflectance or another disadvantage will occur in that the thickness of the dye film cannot be selected at such a value so as to achieve a good output balance between the push–pull tracking error and the reflectance. The mixture of pentamethine cyanine **35a** and heptamethine **35b** (Scheme 26) with maximum absorption in the near-IR region (650 and 780 nm) was used as the optical recording media [41]. This cyanine dye mixture is thermally stable for laser irradiating, and soluble in organic solvent for spin-coating.

A mixture of metallized formazan dye (Scheme 27) and symmetric indodicarbocyanine dye **29** was used as the optical recording layer with good anti-fade properties. The refractive index n of the recording layer is not less than 1.8 and the imaginary part k is not greater than 0.2 at 780 nm [42]. Whereas, most of the metallized formazans have strong electron withdrawing group (typically NO_2), which results in a relatively large absorption at 780. As a result, those formazans have high-k values (about 0.15). This relatively large k value has a negative impact on reflectivity when used with other recording layer dyes also having high-k values. So, a metallized formazan dye having a value of 0.03 to 0.07 for

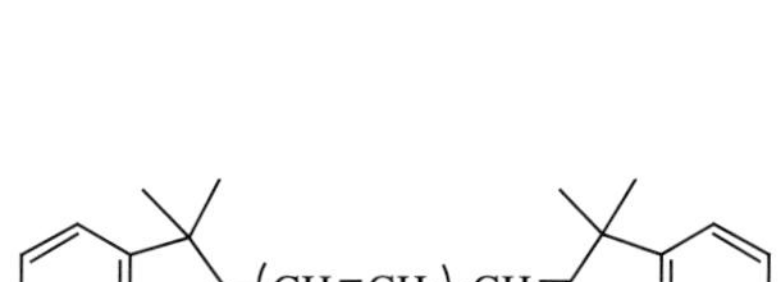

34

Scheme 25.

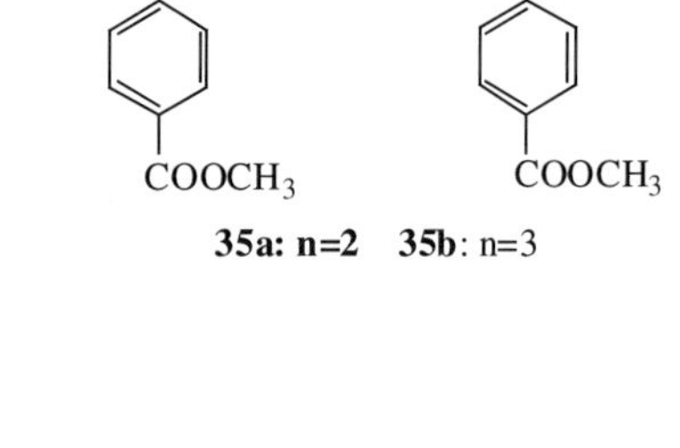

35a: n=2 **35b**: n=3

Scheme 26.

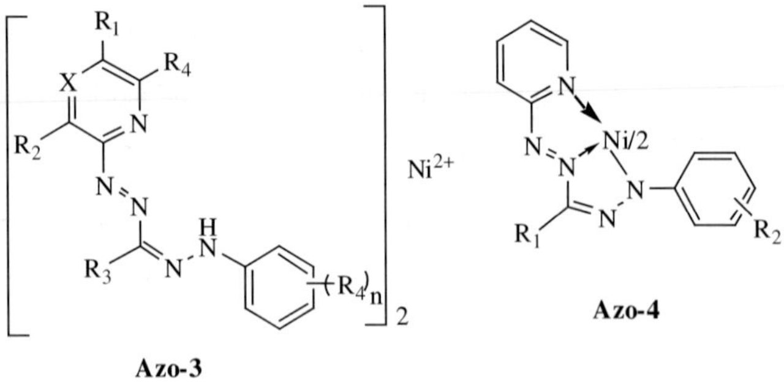

Scheme 27.

the imaginary part of the refractive index *k* was developed. The use of low-*k*-metallized formazan allows the use of greater concentrations of metallized formazans without adversely affecting the targeted *k* value of the recording layer [43]. On this basis, no *k* value (0.00–0.02) formazan dye is preferred to have improved incubation and light stability [44].

It is common to combine two symmetrical cyanine dyes (one with a high-*k* value and one with a low-*k* value) to meet the *k* requirement. However, Eastman Kodak Company patented a mixture of high-*k* symmetrical cyanine dye **29**, a low-*k* unsymmetrical cyanine dye **30** with low *k* [45] or no *k* [46] metallized formazan as the recording layer. The desired optical properties can be obtained by adjusting the component ratios of the cyanine dye. The unobvious advantage of the use of low *k* unsymmetrical cyanine is the unusually incubation stability that is not seen with the usually low-*k* symmetrical cyanine dyes.

Since the solubility of some known cyanine dye is relatively small (about 8 wt.% at the highest), the dye is liable to crystallize in the coating and drying procedure. Consequently, the recording layer thus formed has non-uniform thickness and the tracking characteristics at the inner part are inferior to those at the outer part. The solubility of cyanine dyes can be improved by changing the counteranion. The cyanine dyes with a counteranion selected from the group consisting of thiocyanate, lactate, hypophosphite, tetracyanonickelate, selenocyanate, trifluoromethanesulfonate, ferricyanide, 4-hydroxybutyrate, nitrite, 2-(3',5'-dimethyl-2'-hydroxyphenyl)-2H-benztriazole-4-sulfonate, 1,2-napthoquinone-2-diazide-5-sulfonate and picrate has improved solubility and greater solution stability, which makes possible the use of a coating solvent such as ethyl lactate having excellent film-forming characteristics, reduced toxicity and high solution stability, as well as excellent compatibility with a polycarbonate substrate[47].

Fuji Photo Film Co., LTD has developed cyanine dyes **36** (Scheme 28) consisting organic polyvalent anion [48]. The solubility of the cyanine dye **36** in 2,2,3,3-tetrafluopropanol is more than 13 wt.% at 25°C, therefore the cyanine is

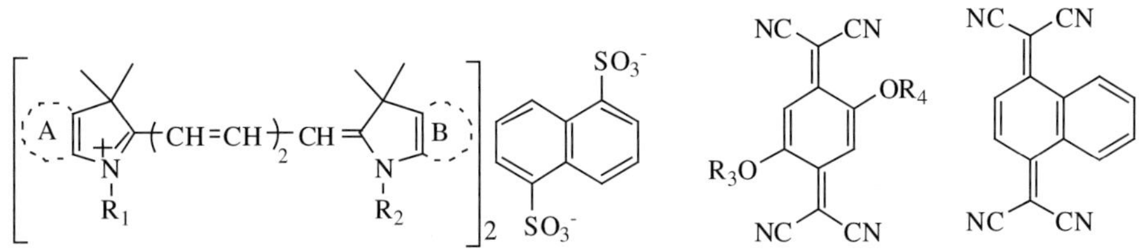

36 A1 A2

Scheme 28.

hardly crystallized and locally deposited in the coating procedure and a thin uniform recording layer can be easily formed by the spin-coating method without specially adjusting the coating conditions. The combination of the cyanine consisting polyvalent anion with a specific anti-fading agent with a formula of **A1** or **A2**, the light-resistance of the optical information recording medium can be remarkably improved [49].

Conventional cyanine dyes have perchlorate ion as the counterion and the reflectance of the disk is liable to be lowered when using Ag or Ag-containing alloy as the light-reflecting layer. Because Ag is more active and unstable than Au, ClO_4^- ion can react with Ag to form Ag compound at the interfacial zone between the recording dye layer and the light-reflecting layer. Using cyanine dye containing organic counterion as the recording layer material can solve the problem of using Ag as the reflecting layer without lowering the reflectance [50]. The cyanine dyes with fluorine-containing anions such as hexafluoro phosphoric acid as counterion have superior solubility and the disk made of this kind of cyanine can use Ag as the reflective layer [51]. An organic dye having PF_6^- and $CF_6SO_3^-$ as anion arranged in an indole-type cyanine dye is stable to heat and light and has excellent resistance as a recording layer by itself. An organic dye having SCN^- as anion arranged in the indole-type cyanine has very excellent writing sensitivity and a high C/N value [52].

In optical recording media, pits are formed when a dye melts and decomposes after absorbing a laser beam. Many cyanine dyes have a melting point and a decomposition point, which separate each other. Because of their significantly different temperatures, pits are formed slowly on a recording layer when irradiated by a laser beam and the heat of fusion and the decomposition conducts to the irradiated point and deforms the adjacent pits, which have been already formed. This makes it difficult to promptly form desired pits on the limited recording surface of optical recording media. Unsymmetrical indolenine pentamethine cyanine dyes **37** and **38** (Scheme 29) with electron withdrawing group (R_3 represents nitro, cyano, trifluoromethyl, trifluoromethoxy, carboxylic acid ester, carboxylic acid amide, alkylaminosulfonyl, or alkylsulfonyl) absorb visible

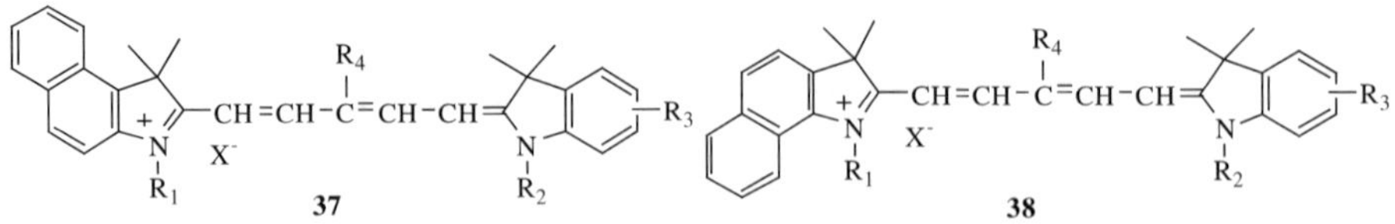

Scheme 29.

lights with wavelengths of around 780 nm when formed in a thin layer [53]. Most of them have only decomposition points or those that are undistinguishable from their melting points, where the decomposition points are not lower than 240°C. The cyanine dyes have a relatively high thermo-stability and promptly decompose at around their decomposition points (the decomposition range is usually within 10°C or narrower). Accordingly, the cyanine dyes exert excellent recording features in optical recording media using laser beams with wavelengths around 780 nm as a writing light, particularly high-speed-recordable-type CD-Rs using laser beams with wavelengths of 775–795 nm as a writing light.

3.2. Cyanine dyes for DVD-R

Analogous to CD-R, DVD-R is also composed of a PC substrate, a dye-recording layer, a metal reflective layer and a resin protective layer. The operation principle of DVD-R is the same as CD-R except the use of different laser wavelengths. Compared with CD-R, several changes were made in DVD-R as shown in Fig. 2: the recording and reproducing wavelength is 650 nm instead of 780 nm; numerical aperture of the read head is 0.6 instead of 0.5; the track pitch is 0.74 μm instead of 1.6 μm and the minimum mark length is 0.4 μm instead of 0.83 μm. These changes increase the data storage capacity significantly and the recording density of DVD-R is five to eight times higher than that of CD-R.

The performance of DVD-R depends mostly on the properties of dye in the recording layer. Selecting a proper dye possessing excellent recording and reading sensitivity with respect to shorter wavelength light for the recording layer of DVD-R is very important. Some criteria of dyes for CD-R such as chemical properties, thermal properties, solubility and solution properties can also be applied to DVD-R. Since the recording and reproducing wavelength in DVD-R differs from that of the CD-R, the optical properties needed in DVD-R differ from those of CD-R. Compounds having an absorption peak wavelength (λ_{max}) between 550 and 620 nm are suitable to be used as recording materials of DVD-Rs. The key to realization of DVD-R is the development of a new dye compound. Pentamethine cyanine dye has been well known for its successful application in CD-R. So, when searching for a suitable dye material for DVD-R, cyanine dye is one of the candidates to be considered. But due to the shorter laser beam wavelength used in DVD-R, indolenine and benzindolenine-based trimethine cyanine

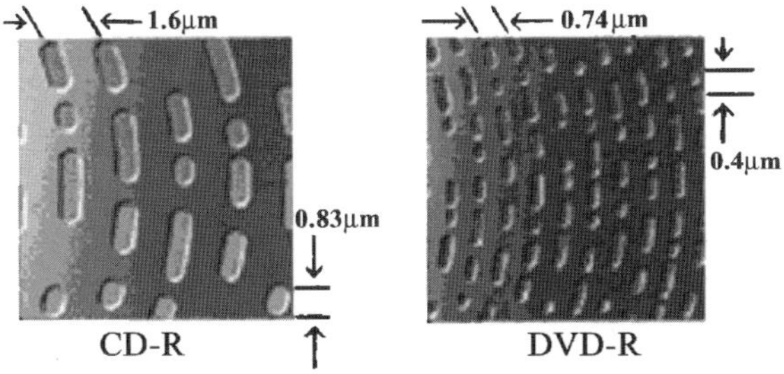

Fig.2. The difference between CD-R and DVD-R.

is used instead of pentamethine cyanine in order to get a suitable reflection and modulation of the DVD-R. One or more trimethine-based cyanine dyes (Scheme 30) having a different optical parameter were used to optimize the optical properties of the recording layer for the recording and reproducing data stored in DVD-R [54].

Trimethine cyanine dyes with an unsaturated substituent bonded to the nitrogen atom have high sensitivity to short-wavelength laser beam having a wavelength of 500–700 m and have good chemical, photochemical, physical and thermal stability. The optical recording medium made of these cyanine dyes meets the DVD standards, has low jitter component and is also superior in long-term reliability [55].

Indolenine cyanine dye (Scheme 31) having a side chain with a pull electrons effect on nitrogen is thermally stable and has a good solubility in organic solvent. Dyes mixture of cyanine **39a** and cyanine **39b** can be used as high-density optical recording layer stable for laser writing [56]. An optical recording disks made of dyes mixture of cyanine **39b** and cyanine dye **40** has a high recording sensitivity and signal-to-noise ratio value, where cyanine **39b** is used as a photosensitizing dye to raise the reflectivity of the recording disk [57].

Unsymmetrical cyanine **41** (Scheme 32) has a maximum absorption wavelength from 550–590 nm and molar absorptivity greater than 1×10^5 cm^{-1} mol^{-1} L [58]. This trimethine cyanine dye is very suitable to be used as a recording material for high-density optical recording medium, especially suitable for those using a laser beam with the wavelength of 620–660 nm, and can provide better recording properties for high-density optical recording applications.

Owing to the structural feature that the two benzindolenine rings bound to both ends of the trimethine chain are symmetric, the recording layer made of cyanine **42** (Scheme 33) has a high refractive index n (n = 2) and a suitable extinction coefficient k (k = 0.02) with respect to a laser beam having a wavelength of 650 nm [59]. As a result, the writing sensitivity increases and the recording power can be reduced, e.g., to below 11 mW. The optical properties,

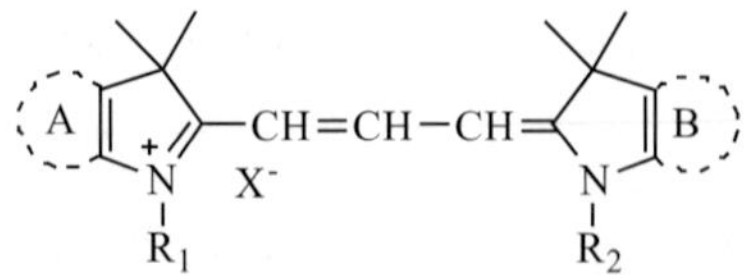

Scheme 30.

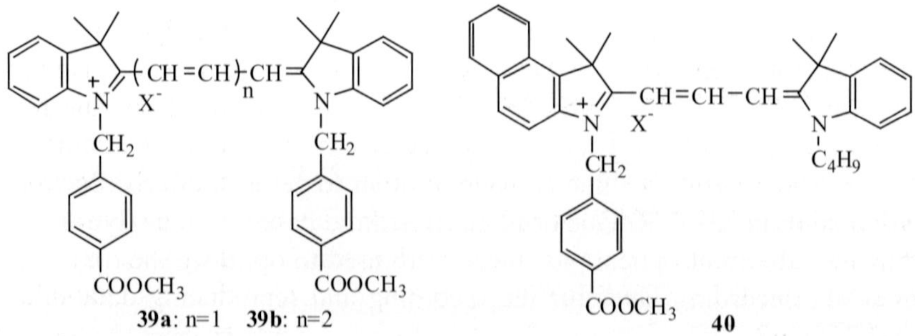

39a: n=1 **39b**: n=2

40

Scheme 31.

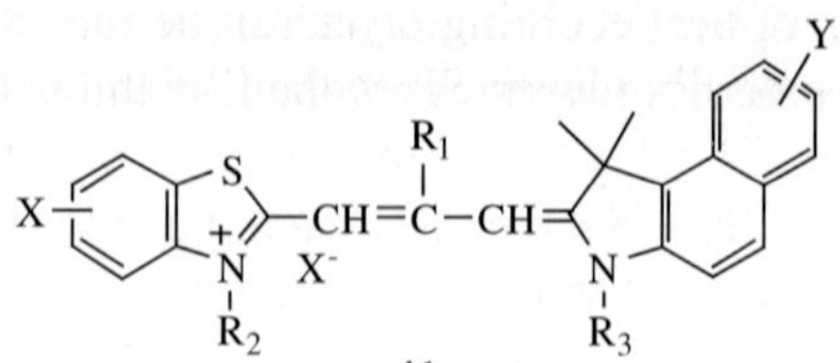

41

Scheme 32.

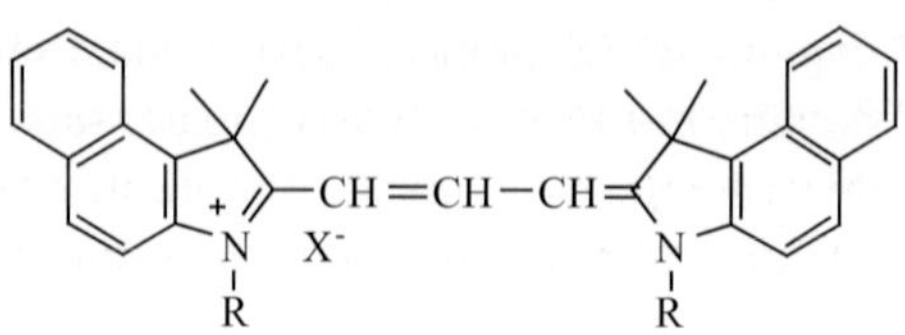

42

Scheme 33.

the solubility and the decomposition temperature of cyanine dye **42** may be adjusted and fine-tuned by varying the substituents R and X. For instance, if R is a straight or branched C3-7 alkyl group when X is PF_6 or SbF_6, and R is a straight or branched C1-5 alkyl group when X is ClO_4, BF_4 or ClO_3, the dye has good solubility and thermal stability. The recording layer having a good uniformity, a sufficient degree of modulation and a low jitter can be obtained with this kind of dyes.

A trimethine-based cyanine dye having a benzene ring or a naphthalene ring without a substituent group is accompanied with the problems that the peaks consisting of the main peak and the shoulder peak fail to become sharp. As a result, the recording sensitivity or the absorbance per unit film thickness of a recording layer at the occasion of forming pits cannot be sufficiently increased. This leads not only to the problems that the thickness of the dye layer and the recording power are required to be increased, or the recording speed is required to be decreased, but also to the problems that a so-called heat interference (wherein the deformation of the pits may be caused due to the accumulation of heat at the space between the pits) tends to be brought about at the occasion of recording, thus giving a bad influence to the characteristics of the recording layer such as modulation amplitude or jitter. A trimethine-based cyanine dye having nitro group attached to a benzene ring or to a naphthalene ring, for example cyanine **43** (Scheme 34), is capable of minimizing the shoulder peak relative to the main peak, thus enabling the spectrum consisting of the main peak and shoulder peak to become sharp [60]. The absorbance per unit film thickness of the recording layer can be increased and hence the recording sensitivity can be improved and the film thickness of the recording layer can be made thinner as compared with the conventional recording layer. Since the film thickness is made relatively thin, the accumulation of heat between pits at the moment of recording can be suppressed, thereby making it possible to inhibit the deformation of the configuration of the pits due to the accumulation of heat and to improve the jitter characteristics. If the recording sensitivity and the characteristics regarding the modulation amplitude, jitter, etc. can be improved, it becomes possible to reduce the power for recording and enhancing the recording speed.

When used in high-density optical recording media such as DVD-Rs, it is difficult for loosely decomposing organic dye compounds to form minute pits on a restricted recording surface at a relatively high density. Hayashibara Biochem Lab found that specific non-symmetric cyanine dyes have only relatively high decomposition points or decomposition points undistinguishable from their melting points, and promptly decompose at around their decomposition points. These

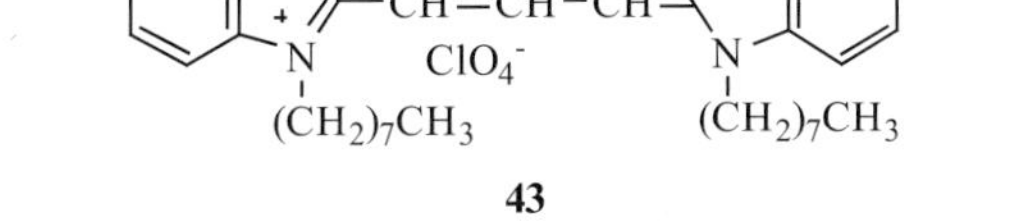

43

Scheme 34.

characteristic features indicate that high-density optical recording media, which have a relatively small jitter, insubstantial reading error and satisfactory stability against exposure to environmental light such as reading and natural light, can be obtained by using these cyanine dyes as light absorbents. The cyanine of the formulae **44** and **45** (Scheme 35) have absorption maxima in the visible region, and substantially absorb a visible light of a wavelength around 650 nm in a thin-layer form, wherein R_1 denotes a nitro group or a sulfonamide group and the counterion can be appropriately selected in view of the solubility of the cyanine dyes in organic solvents and the stability of the cyanine dyes [61]. The cyanine dyes promptly form minute pits stably on the recording surfaces and at a relatively high density when irradiated with a laser beam at a wavelength around 650 nm in optical recording media. Trimethine cyanine dyes of the formulae **46** (Scheme 36) with benzindolium or pyrazinimidazolium heterocycle do not only have the above properties but also have a relatively high decomposition points and thermal resistance with a decomposition point of over 270°C [62]. Although dyes with lower decomposition points can be generally used to write information using relatively low-power laser beams, however, as a disadvantage, when exposed to laser beams for a relatively long period of time on reading, optical recording media processed with such dyes may easily accumulate heat and deform parts around pits on recording surfaces, resulting in undesirable jitters and reading errors. These cyanine dyes provide excellent recording characteristics when used in high-density optical recording media.

CMC Magnetics Corporation also claimed a kind of unsymmetrical trimethine-cyanine dye (Scheme 37) with electron-withdrawing substituted group

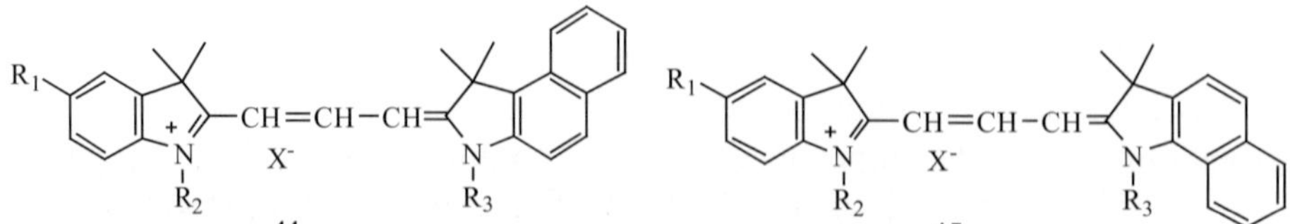

44 **45**

Scheme 35.

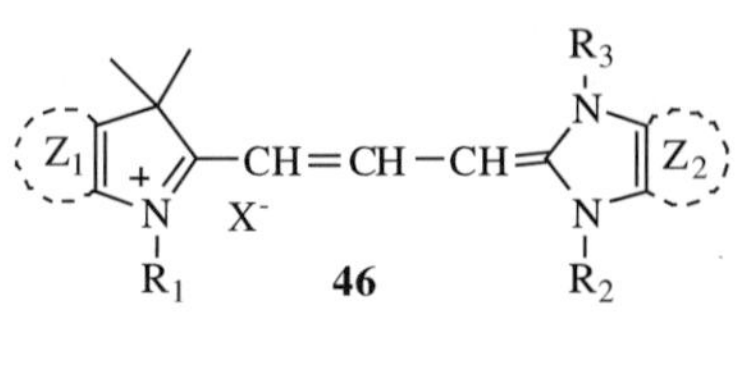

46

Scheme 36.

(EWG) to make high density optical information recording medium with improved write/read characteristics at the writing wavelength of 650 nm [63]. The increase in the recording sensitivity could be reached by increasing the absorption of the organic layer at the wavelength of writing laser. According to this opinion, new cyanine mixture composition with specific cyanine additives, which has a little absorption at the wavelength of write/read laser beam, could enhance the recording sensitivity. The additives can be unsymmetrical trimethine-cyanine dye (Scheme 38) containing electron-donating substituted group (EDG) [64,65]. The recording layer comprising the common trimethine-cyanine

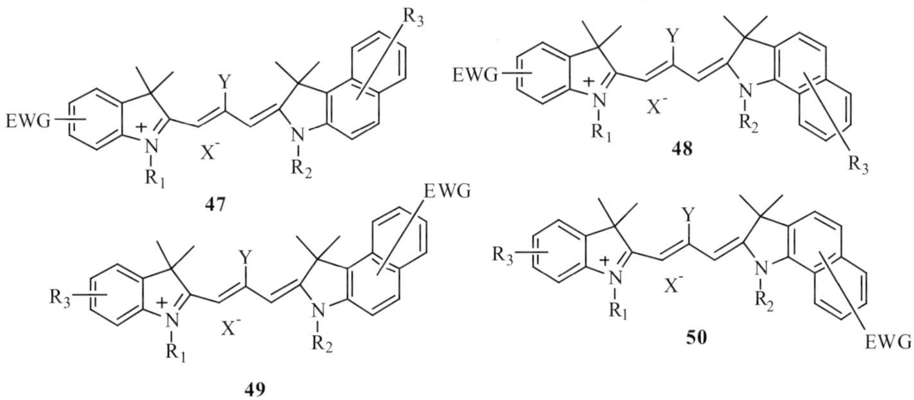

Scheme 37.

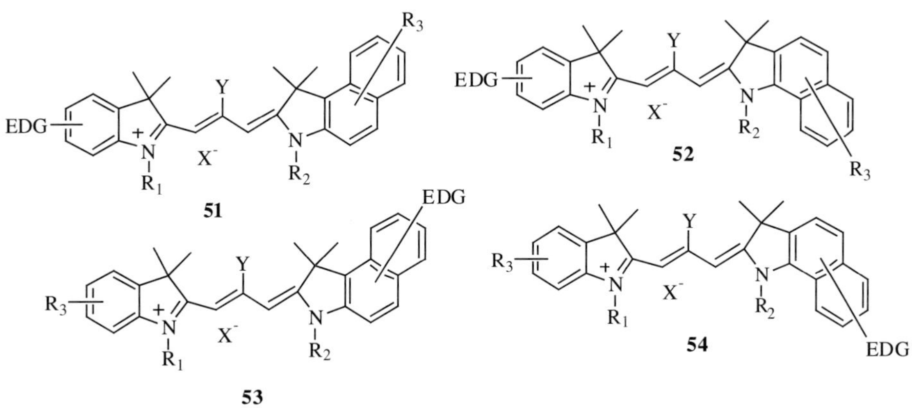

Scheme 38.

dye or trimethine-dye containing electron-withdrawing group and trimethine-cyanine dye containing EDG has the advantage of increasing the recording sensitivity, so that the writing power of laser can be reduced while recording information in DVD-R.

Eastman Kodak Company claimed a DVD-R recording element containing a mixture of metallized carbamoylazo dye (Scheme 39) and a cyanine dye. The recording layer has a refractive index of real part $n > 1.8$ and an imaginary part $k < 0.2$ at the wavelength from 400 to 660 nm. The recording disks exhibit good recording sensitivity and excellent light and dark stability. The metallized carbamoylazo dyes have an azo group linking a substituted 3-hydroxypyridine nucleus to a phenyl nucleus wherein the phenyl nucleus has a carbamoyl substituent ortho to the azo group [66], the phenyl nucleus has a thioether substituent ortho to the azo group [67] and the phenyl nucleus has an ether substituent ortho to the azo group and the phenyl nucleus is free of strong electron-withdrawing groups (such as nitro and alkylsulfonyl) [68].

Addition of neutral TCNQ derivative to cyanine dye can increase the photostability of the dye, but TCNQ is not quite soluble in the common organic solvents so that the doping concentration is limited, therefore the stability effect is also limited. The cyanine-TCNQ complex dye (Scheme 40) is produced by combining TCNQ molecules with cyanine dye molecules to form a stable dye complex, wherein the TCNQ molecules chemically bonds with the cyanine dye molecules to form a stable charge-transfer complex [69]. The cyanine-TCNQ complex dye has a better solubility in a variety of organic solvents so that a stable homogeneous solution with uniform distribution of cyanine-TCNQ complex dye can be obtained. Thus, a high quality coating film using this solution can be achieved for making a high-quality data-storage media. The cyanine-TCNQ complex dye **55** has a maximum absorption both in UV light region with the wavelength ranging from 200 to 400 nm and in near IR light region with the

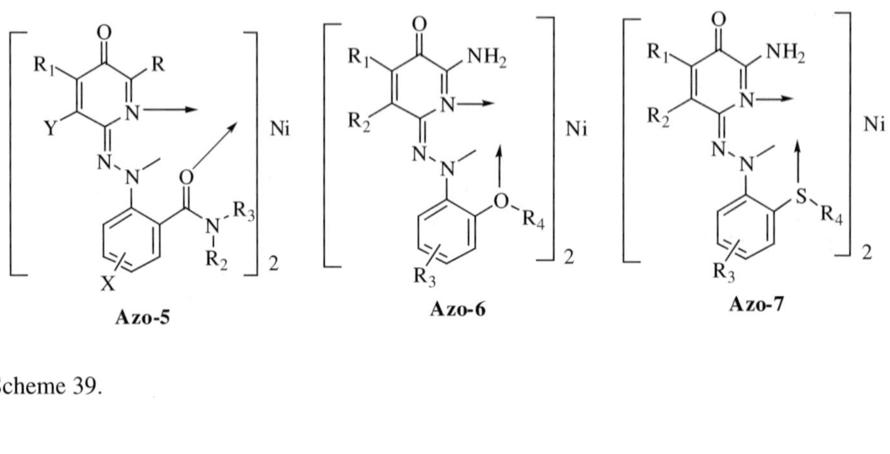

Scheme 39.

wavelength ranging from 800 to 1000 nm. Thus, the short wave light smaller than 400 nm can be eliminated, which would otherwise tend to break the chemical bonds. In addition, since the cyanine TCNQ charge-transfer complex possesses a higher-oxidation potential than that of a general cyanine halide such as a cyanine ClO_4 or cyanine PF_6 complex, and TCNQ possesses a better inhibition to singlet oxygen so that the main structure of cyanine dye can be effectively protected from singlet oxygen attack. The cyanine-TCNQ complex dye possesses an excellent photostability and the need for the addition of a photostabilizer can be eliminated. The dye-recording layer containing the cyanine-TCNQ complex dye (Scheme 41) is excellent in recording and reading performance and in storage properties such as light resistance and durability [70].

Unsymmetrical trimethine cyanine dyes play a very important role as recording materials in DVD-R owing to their suitable optical and thermo properties. But, the synthesis of unsymmetrical cyanine dyes by conventional method may have the problem like low yield and is difficult to purify. Tian *et al.* [72,73] developed a simple process to synthesize unsymmetrical trimethine cyanine dyes. As shown in Scheme 42, the Fisher's base was obtained through treating benzindolium iodide with sodium hydroxide and was then converted into the aldehyde by Vilsmeier reaction; finally the aldehyde reacted with another indolium salt to give the unsymmetrical trimethine cyanine dyes. With this method, the unsymmetrical cyanine dye can be obtained in high purity (>99%) and in high yield (>85%) by simple recrystallization from the solvent. The

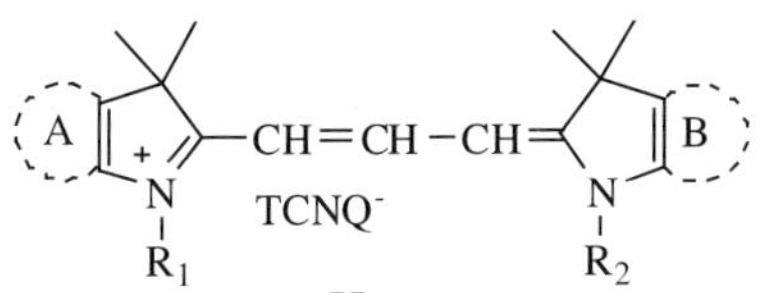

55

Scheme 40.

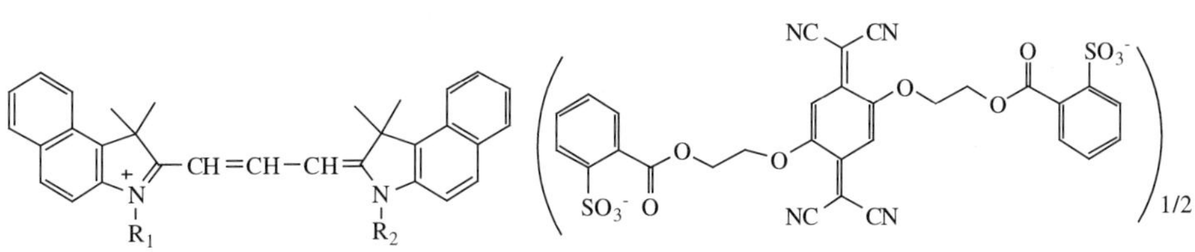

56

Scheme 41.

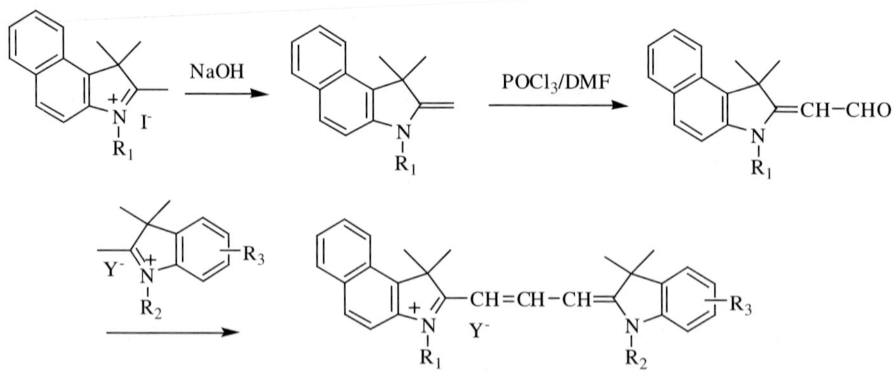

Scheme 42.

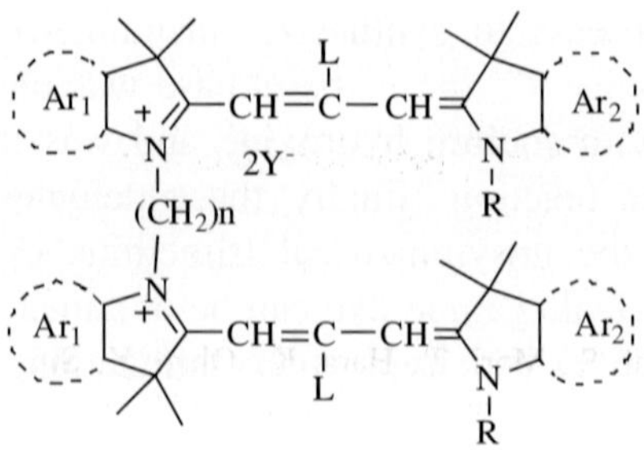

57

Scheme 43.

unsymmetrical cyanine dyes synthesized by this method are suitable to be used in DVD-R.

Recently, Tian *et al.* [74] claimed a novel trimethine cyanine dimmer (Scheme 43) to be used in high-speed DVD-R. This cyanine dimmer has increased molar absorption coefficient compared with common cyanine dyes, which makes it possible to absorb enough laser beam in less time to reach the decomposition time. Therefore, it is possible to use this novel trimethine cyanine dimmer to improve the writing sensitivity and speed of DVD-Rs.

REFERENCES

1. Hamer, F.M., *The Cyanine Dyes and Related Compounds*, Wiley, New York (1964).
2. Mishra, A., Behera, R.K., Behera, P.K., Mishra, B.K. & Behera, G.B., Cyanines during the 1990s: A review, *Chem. Rev.*, **100** (2000)1973.

3. O'Regan, B. & Grätzel, M., A low-cost, high-efficiency solar cell based on dye-sensitized colloidal TiO_2 films, *Nature*, **353** (1991) 737.
4. Nazeeruddin, M.K., Kay, A., Rodicio, I., Humphry-Baker, R., Muller, E., Liska, P., Vlachopoulos, N. & Grätzel, M., Conversion of light to electricity by cis-X2bis(2,2'-bipyridyl-4,4'-dicarboxylate) ruthenium(II) charge-transfer sensitizers(X=Cl-, Br-, I-, CN, and SCN-) on nanocrystalline TiO_2 electrode, *J. Am. Chem. Soc.*, **115** (1993) 6382.
5. Ehret, Stuhl, L. & Spitler, M.T., Variation of carboxylate-functionalized cyanine dyes to produce efficient spectral sensitization of nanocrystalline solar cells, *Electrochim. Acta*, **45** (2000) 4553.
6. Ehret, Stuhl, L. & Spitler, M.T., Spectral sensitization of TiO_2 nanocrystalline elecrode with aggregated canine dyes, *J. Phys. Chem. B*, **105** (2001) 9960.
7. Meng, F.S., Ren, Y.J., Gao, E.Q., Cai, S.M., Chen, K.C. & Tian, H., High efficient cyanine dyes used for nanocrystalline TiO_2 electrode, *SPIE* **4465** (2001) 143.
8. Sayama, K., Hara, K., Ohga, Y., Shinpou, A., Suga, S. & Arakawa, H., Significant effects of the distance between the cyanine dye skeleton and the semiconductor surface on the photoelectrochemical properties of the dye-sensitized porous semiconductor electrodes, *New J. Chem.*, **25** (2001) 200.
9. Ren, Y.J., Meng, F.S., Tian, H. & Cai, S.M., Highly efficient photosensitization of mesoporous TiO_2 electrode with a cyanine dye, *Chinese Chem. Lett.*, **13** (2002) 379.
10. Matsui, M., Hashimoto, Y., Funabiki, K., Jin, J.-Y., Yoshida, T. & Minoura, H., Application of near-infrared absorbing heptamethine cyanine dyes as sensitizers for zinc oxide solar cell, *Synth. Met.* **148** (2005) 147.
11. Guo, M., Diao, P., Ren, Y.J., Meng, F.S., Tian, H. & Cai, S.M., Photoelectrochemical studies of nanocrystalline TiO_2 co-sensitized by novel cyanine dyes, *Sol. Energy Mater. Sol. Cells,* **88** (2005), 23.
12. Sayama, K., Tsukagoshi, S., Mori, T., Hara, K., Ohga, Y., Shinpou, A., Abe, Y., Suga, S.& Arakawa, H., Efficient sensitization of nanocrystalline TiO_2 films with cyanine and merocyanine organic dyes, *Sol. Energy Mater. Sol. Cells*, **80** (2003) 47.
13. Wang, Z.S., Li, F.Y, Huang, C.H., Wang, L., Wei, M. & Jin, L.P., Li, N.Q., Photoelectric conversion properties of nanocrystalline TiO_2 electrode sensitized with hemicyanine derivatives, *J. Phys. Chem. B*, **104** (2000) 9676.
14. Wang, Z.S., Li, F.Y. & Huang, C.H., Highly efficient sensitization of nanocrystalline TiO_2 films with styryl benzothiazolium propylsulfonate, *Chem. Comm.* (2000) 2063.
15. Wang, Z.S., Li, F.Y. & Huang, C.H., Photocurrent enhancement of hemicyanine dyes containing RSO_3^- group through treating TiO_2 films with hydrochloric acid, *J. Phys. Chem. B*, **105** (2001) 9210.
16. Stathatos, E. & Lianos, P., Synthesis of a Hemicyanine Dye bearing two carboxylic groups and its use as a photosensitizer in dye-sensitized photoelectrochemical cells, *Chem. Mater.*, **13** (2001) 3888.
17. Yao, Q.H., Meng, F.S., Li, F.Y., Tian, H. & Huang, C.H., Photoelectric conversion properties of four novel carboxylated hemicyanine dyes on TiO_2 electrode, *J. Mater. Chem.*, **13** (2003) 1048.
18. Meng, F.S., Yao, Q.H., Shen, J.G., Li, F.L., Huang, C.H., Chen, K.C. & Tian, H., Novel cyanine dyes with multi-carboxyl groups and their sensitization on nanocrystalline TiO_2 electrode, *Synth. Met.*, **137** (2003) 1543.
19. Yao, Q.H., Shan, L., Li, F.Y., Yin, D.D. & Huang, C.H., An expanded conjugation photonsensitizer with two different adsorbing groups for solar cells, *New J. Chem.*, **27** (2003) 1277.
20. Chen, Y.S., Li, C., Zeng, Z.H., Wang, W.B., Wang, X.S. & Zhang, B.W., Efficient electron injection due to a special adsorbing group's combination of carboxyl and hydroxyl: Dye-sensitized solar cells based on new hemicyanine dyes, *J. Mater. Chem.*, **15** (2005) 1654.

21. Meng, F.S., Chen, K.C., Tian, H., Zuppiroli, L. & Nüesch, F., Cyanine dye acting both as donor and acceptor in heterojunction photovoltaic devices, *Appl. Phys. Lett.*, **82** (2003) 3788.
22. Nüesch, F., Tornare, G., Zuppiroli, L., Meng, F.S, Chen, K.C. & Tian, H., Interface modification to optimize charge separation in cyanine heterojunction photovoltaic devices, *Sol. Energy Mater. Sol. Cells*, **87** (2005) 817.
23. Nüesch, F., Faes, A., Zuppiroli, L., Meng, F.S., Chen, K.C. & Tian, H., Counterion effects in cyanine heterojunction photovoltaic devices, *J. Mater. Sci.*, **40** (2005) 1353.
24. Meng, F.S., Hua, J.L., Chen, K.C., Tian, H., Zuppiroli, L. & Nüesch, F., Synthesis of novel cyanine-fullerene dyads for photovoltaic devices, *J. Mater. Chem.*, **15** (2005) 979.
25. Hamada, E., Shin, Y., & Ishiguro, T., That's CD-R, *SPIE*, **1499** (1989) 45.
26. Hamada, E., Shin, Y., & Ishiguro, T., CD-compatible write-once disc with high reflectivity, *SPIE* **1078** (1989) 80.
27. Holtslag, A.H.M., MaCord, E.F. & Werumeus Buning, G.H., Recording mechanism of overcoated metallized dye layers on polycarbonate substrates, *Jpn. J. Appl. Phys.*, **31** (1992) 484.
28. Hamada, E., Arai, Y., Takagisi, Y. & Ishiguro, T., Optical information recording medium and recording method, Taiyu Yuden Co., Ltd., US5213955 (1993).
29. Yanagisawa, S., Sakai, T., Chuman, T., Araki, Y. & Matsui, F., Optical recording medium, Pioneer Electronic Corporation, US5204220 (1993).
30. Yanagisawa, S., Sakai, T., Chuman, T., Araki, Y. & Matsui, F., Optical recording medium, Pioneer Electronic Corporation, US5336584 (1994).
31. Yanagisawa, S., Sakai, T., Chuman, T., Araki, Y. & Matsui, F., Optical recording medium, Pioneer Electronic Corporation, US5328802 (1994).
32. Ootaguro, K., Hamada, E., Takagisi, Y. & Fujii, T., Optical recording medium containing a light stabilizer comprised of aryl nitrogen compound, Taiyu Yuden Co., Ltd., US5318882 (1994).
33. Maeda, S., Imamura, S., Mitsuhashi, K. & Tsukahara, T., Dye-incorporated composition, Mitsubishi Kasei Corporation, US5330542 (1994).
34. Hurditch, R., Stabilized dye compositions for optical recording media, Media Chemical Corp., US6077584 (2000).
35. Hurditch, R., Multi-component dye compositions for optical recording media, Carl M. Rodia & Associates, US5817388 (1998).
36. Namba, K., Inoue, T., Kitagawa, S. & Shinkai, M., Optical recording medium, TDK Corporation, US5154958 (1992).
37. Namba, K., Shinkai, M., Inoue, T. & Kitagawa, S., Optical recording medium, TDK Corporation, US5574715 (1996).
38. Namba, K., Kitagawa, S., Shinkai, M., Suzuki, M., Kimura, S. & Hirako, K., Photo-stabilized cyanine dyes and optical recording media, TDK Corporation and Sumitomo Seika Chemicals Co., Ltd., US6071672 (2000).
39. Wolleb, H. & Schmidhalter, B., Complex polymethine dyes and their use, Ciba Specialty Chemicals Corporation, US5958650 (1999).
40. Yanagisawa, S., Sakai, T., Tanaka, S., Chuman, T., Araki, Y. & Matsui, F., Optical recording medium, Pioneer Electronic Corporation, US5391413 (1995).
41. Hu, A.T., Lee, H.-J., Huang, J.-L., Chang, J.-C., Ye, S.-J., Huang, D.-R., Chiang, D.-Y., Liao, W.-Y., Jeng, T.-R. & Chen, J.-S., Cyanine dyes mixture as optical recording media, National Tsing Hua University and Industrial Technology Research Institute, US5900348 (1999).
42. Cunningham, M.P. & Evans, S., Optical recording elements having recording layers containing mixtures of formazan and cyanine dyes, Eastman Kodak Company, US5547728 (1996).
43. Chapman, D.D., Cunningham, M.P. & Goswami, R., Optical recording elements having recording layers containing mixtures of low k metallized formazan and cyanine dyes, Eastman Kodak Company, US5922429 (1999).

44. Chapman, D.D., Cunningham, M.P. & Goswami, R. & Fleming, J.C., Optical recording elements having recording layers containing mixtures of no k metallized formazan and cyanine dyes, Eastman Kodak Company, US5922504 (1999).
45. Chapman, D.D., Cunningham, M.P. & Goswami, R., Mixtures of low k metallized formazan dyes with symmetrical and unsymmetrical cyanine dyes, Eastman Kodak Company, US5731054 (1998).
46. Chapman, D.D., Cunningham, M.P. & Goswami, R., Optical recording layers containing no k metallized formazan dyes mixed with symmetrical and unsymmetrical cyanine dyes, Eastman Kodak Company, US5773193 (1998).
47. Hurditch, R. & Griffiths, J., Dye compositions for optical recording media having selected anions, Media Chemical Corp., US5952073 (1999).
48. Hiroshi, K. & Kanagawa, O.-S., Optical information recording medium, Fuji Photo Film Co., LTD, EP892397 (1999).
49. Ishida, T., Shibata, M., Wariishi, K. & Morishima, S., Optical information recording medium, Fuji Photo Film Co., Ltd., US5998094 (1999).
50. Usami, Y. & Shibata, M., Optical Information recording medium, Fuji Photo film Co., Ltd., US6258509 (2001).
51. Ayako, H., Toshio, K. & Shigeo, Y., Cyanine dye, Hayashibara Biochem Lab, EP1092753 (2001).
52. Osamu, S., Hisamitsu, K., Masashi, S., Ryo, N., Shinichiro, I., Hitoshi, W. & Hideo, F., Optical Data Recording Medium, Hitachi Maxell Ltd., EP352068 (1990).
53. Ayako, H., Dai, M., Toshio, K. & Shigeo, Y., Cyanine dye, Hayashibara Biochem Lab, EP1090961, (2001).
54. Tomizawa, Y., Fujii, T. & Hamada, E., Optical information recording medium, Taiyo Yuden Co., Ltd., US5976658 (1999).
55. Kanno, T., Optical recording medium, Fuji Electronic Co., Ltd., US6063467 (2000).
56. Liao, W.-Y., Huang, C.-L., Huang, D.-R., Chiang, D.-Y., Hu, A.T., Lee, H.-J., Ye, S.-J. & Kao, Y.-J., Preparation of cyanine dye for high-density optical recording disk, Industrial Technology Research Institute and National Tsing Hua University, US5958087 (1999).
57. Liao, W.-Y., Hu, A. T., Huang, C.-L., Yang, H.-W., Huang, D.-R., Hu, D.-Y, Lee, M.-C. & Lee, H.-J., Preparation of cyanine dye for high-density optical recording disk, Industrial Technology Research Institute and National Tsing Hua University, US6589706 (2003).
58. Chen, H.-J., Huang, C.-S., Hu, M.-J., Lin, Y.-S., Wang, S.-Y. & Yeh, C.-J., Asymmetrical trimethine cyanine dyes for high capacity optical recording medium, Ritek Corporation, US6306478 (2001).
59. Fujii, T., Okitsu, I., Negishi, R. & Hamada, E., Optical information recording medium, Taiyo Yuden Co., Ltd., EP1107243 (2001).
60. Tajima, T., Fujii, T., Tomizawa, Y. & Hamada, E., Optical information recording medium, Taiyo Yuden Co., Ltd., US6291045 (2001).
61. Kasada, C., Hata, Y., Kawata, T. & Yasui, S., Cyanine dyes, Hayashibara Biochem Lab, US6413607, (2002).
62. Kasada, C., Orita, K., Yano, K., Kawata, T. & Yasui, S., Cyanine dyes, Hayashibara Biochem Lab, US6525181 (2003).
63. Yeh, S.-M., Chang, K.-M., Chiu, W.-P. & Tsai, L.-C., Optical information recording medium, CMC Magnetics Corporation, US6667087 (2003).
64. Yeh, S.-M., Chang, K.-M., Chiu, W.-P., Huang, C.-M. & Tsai, L.-C., Optical recording medium, CMC Magnetics Corporation, US6716509 (2004).
65. Yeh, S.-M., Chang, K.-M., Chiu, W.-P., & Huang, C.-M. Optical recording medium, CMC Magnetics Corporation, US6835433 (2004).

66. Chapaman, D. D., Goswami, R. & Kovacs, C. A., Optical recording elements containing mixtures of metallized carbamoylazo and cyanine dyes, Eastman Kodak Company, US5821029 (1998).
67. Chapaman, D.D. & Kovacs, C.A., Optical recording elements containing mixtures of metallized azo thioether and cyanine dyes, Eastman Kodak Company, EP961266 (1999).
68. Chapaman, D.D., Carroll-Lee, A.L. & Kovacs, C.A., Optical recording elements containing mixture of metallized azo ether and cyanine dyes, Eastman Kodak Company, US6582881 (2003).
69. Liao, W.-Y., Lee, M.-C., Huang, C.-L., Yan, C.-F., Jeng, T.-R., Hu, A. T. & Lee, C.-C., Cyanine-TCNQ complex dye data storage media and manufacturing method thereof, Industrial Technology Research Institute and National Tsing Hua University, US6821708 (2004).
70. Morishima, S., Cyanine dye component having tetracyanoquinodimethane compound as counter anion and optical recording medium containing the same, Fuji Photo Film Co., Ltd., US6770347 (2004).
71. Yano, T., Shigeno, K. & Okada, M., Cyanine compound, optical recording materials, and optical recording medium, Asahi Denka Co., Ltd., EP1505125 (2005).
72. Meng, F.S., Yang, S.J., Tian, H., Su, J.H. & Chen, K.C., Indole derivatives and their preparation, Chinese Patent No. ZL 01105570.7, (CN1311184A, 2001).
73. Meng, F.S., Su, J.H., Yang, S.J., Tian, H. & Chen, K.C., Preparation of unsymmetrical cyanine dyes for DVD-R discs, Chinese Patent No. ZL 01105622.3 (CN1312249A, 2001).
74. Su, J.H., Meng, F.S., Tian, H., Li, C., Wang, H.L. & Chen, K.C., Preparation of novel trimethine cyanine dimmer for high speed DVD-R discs, Chinese Patent CN1563201A (2004).

Functional Dyes
Sung-Hoon Kim (Editor)

Chapter 3

Photochromic naphthopyrans

John D. Hepworth[a] and B. Mark Heron[b]

[a]Faculty of Science, University of Central Lancashire, Preston, UK
[b]Department of Colour and Polymer Chemistry, University of Leeds, Leeds, UK

1. INTRODUCTION

The ring-chain tautomerism of 2*H*-pyrans (Scheme 1) is markedly influenced by substituents. Although 2*H*-pyran itself is still to be synthesised, substituents at C-2 stabilise the system and the pyran and its acyclic tautomer co-exist. Fused 2*H*-pyrans show similar behaviour; but the pyran is now usually the dominant species. The ratio of tautomers in the equilibrium mixture is affected not only by the electronic and steric factors associated with the substituent, but also by temperature and solvent [1–6]. The tautomers have different geometries and exhibit different absorption spectra and other physical and chemical properties.

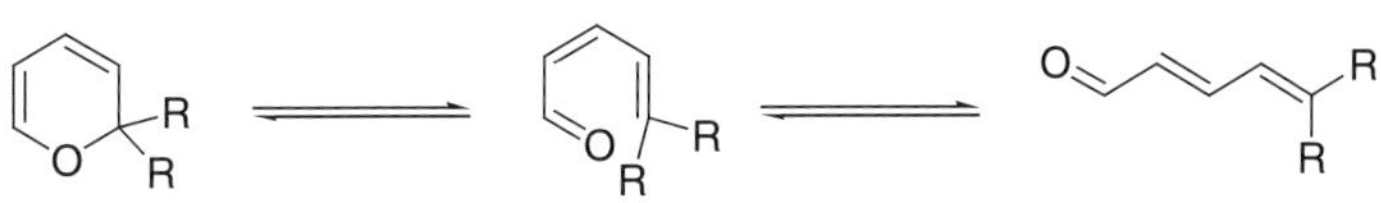

Scheme 1.

Benzannulation has a dramatic influence on the stability of the pyran ring and many 2*H*-[1]-benzopyrans occur naturally in both plant [7, 8] and marine hosts [9]. However, the racemisation of chiral benzopyrans under the influence of sunlight probably occurs through electrocyclic opening of the pyran ring and subsequent ring closure of the acyclic tautomer [10]. The isolation of many naturally occurring benzopyrans as racemates can be attributed to a similar sequence [11]. Benzopyrans are also readily cleaved at the O–C2 bond by nucleophiles, a process that is facilitated by the electronegativity of the heteroatom and that is exploited in the synthesis of heterocycles, aromatic molecules and acyclic materials from pyrylium salts (Scheme 2) [12, 13]. It is this same bond cleavage that plays a pivotal role in the photochromism of naphthopyrans, spiropyrans and spirooxazines [14].

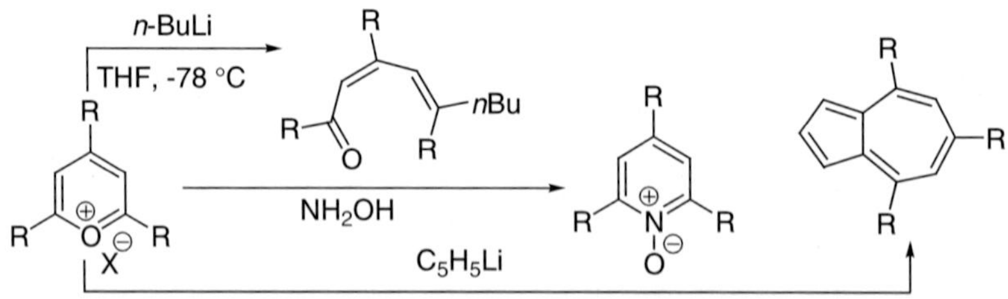

Scheme 2.

Photochromism is defined simply as the light-induced reversible transformation of a chemical entity into an isomeric species that has different absorption characteristics. It is a phenomenon that has been well documented and has been the subject of a number of reviews [14–21]. In the case of benzopyrans **1**, the heterocycle is the stable colourless ground state that upon UV-excitation generates a ring-opened species that absorbs at longer wavelength, towards and possibly in the visible region. On cessation of irradiation, the unstable acyclic species reverts over time to the original cyclic state (Scheme 3).

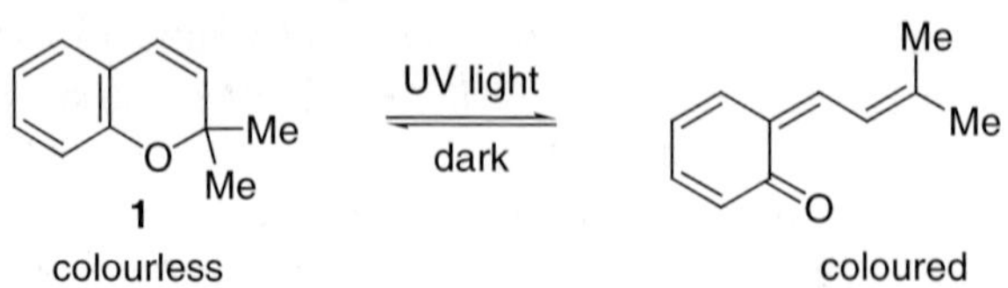

Scheme 3.

The weak photochromic behaviour of 2*H*-[1]benzopyrans **1** associated with the electrocyclic ring-opening process, first noted by Becker [22], is enhanced on annulation of an additional benzene ring, with the ring-opened tautomer exhibiting both a more intense colour and having an increased lifetime. These desirable features are further improved by *gem* diaryl substitution adjacent to the heteroatom, and such naphthopyrans are currently the system of choice for imparting photochromic properties to a variety of polymeric host materials.

Of the three isomeric naphthopyrans **2–4**, the linear isomer 2*H*-naphtho[2,3-*b*]pyran **4** displays no significant photochromic response at ambient temperature, a feature that may be rationalised by considering the extensive π-system reorganisation that must accompany an electrocyclic ring opening, and which would disrupt the aromaticity of both the rings of the naphthalene unit.

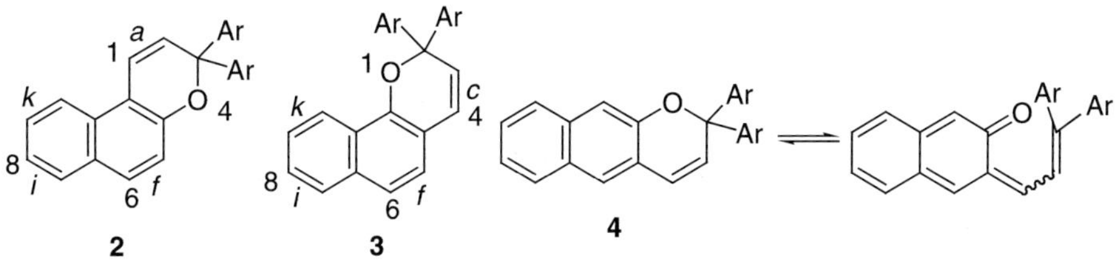

The angular isomers, **2** and **3**, have received much attention since they display good photochromic properties in solution under ambient conditions [23, 24]. Their behaviour on irradiation with UV light is considered in detail later, but the major differences are apparent from the spectra in Fig. 1. The [1,2-*b*] isomer not only absorbs more strongly but also exhibits two absorption bands in the visible region. Structural diversity has been achieved by the fusion of aromatic and heterocyclic moieties onto the isomeric naphthopyrans. This chapter discusses the consequences of the

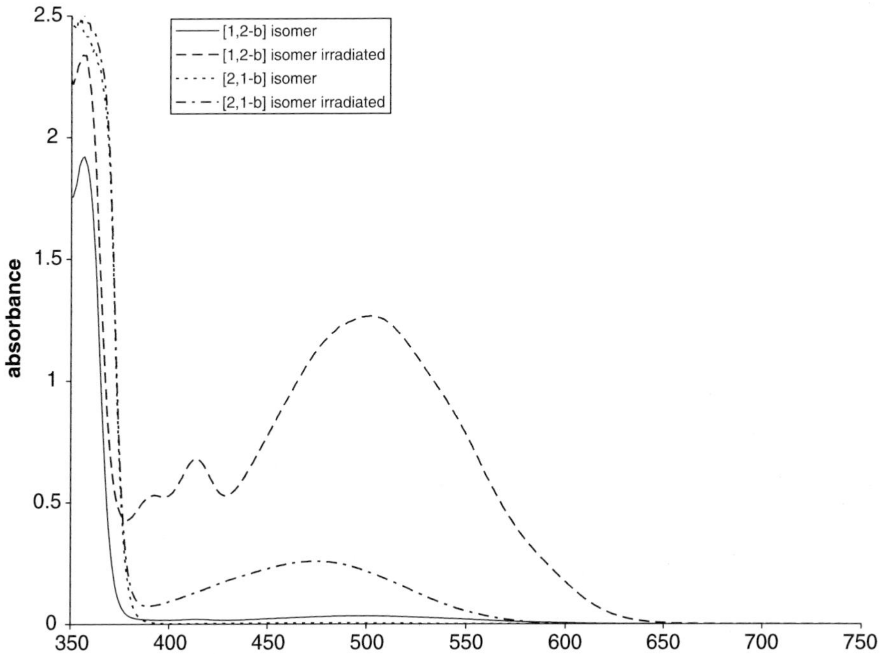

Fig. 1. UV–visible spectra of 2,2-bis(4-methoxyphenyl)-2*H*-naphtho[1,2-*b*]pyran and 3,3-bis(4-methoxyphenyl)-3*H*-naphtho[2,1-*b*]pyran in toluene.

reversible opening of the pyran ring in such compounds under the influence of UV-irradiation and draws together the information reported in the scientific and patent literature concerning the synthesis and photochromic properties of these molecules.

2. SYNTHESIS OF THE DIARYL SUBSTITUTED PYRAN RING

Routes to 2*H*-[1]benzopyrans have been reviewed [2, 25–27], and in many cases these methods are readily adaptable to naphthopyran synthesis. However, the requirement for a *gem* diaryl unit in the pyran ring of both benzo- and naphthopyrans imposes restrictions on the synthetic approach.

The classical reaction of aryl Grignard reagents with coumarins suffers from moderate yields and by-product formation; this is especially the case when applied to naphthopyranones (benzocoumarins) **5** [28–30].

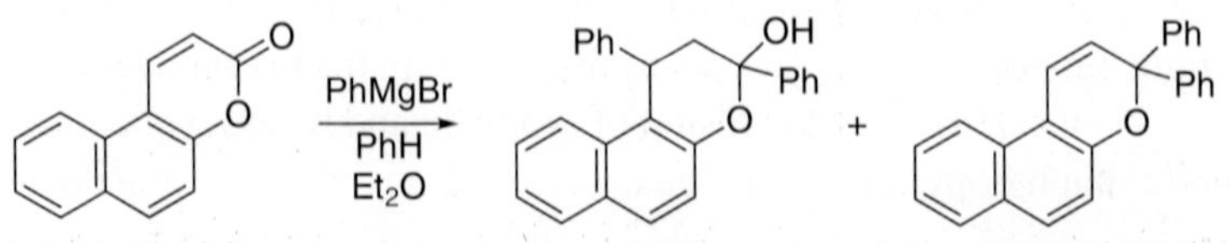

Similarly, the widely used route [31] to 2,2-dialkyl- and 2-alkyl-2-arylbenzopyrans by reduction and dehydration of dihydrobenzopyran-4-ones **6**, readily available from 2′-hydroxyacetophenones and ketones [32], is not appropriate for the diaryl derivatives because of low yields even when *t*-butoxide is used as the condensing reagent [33].

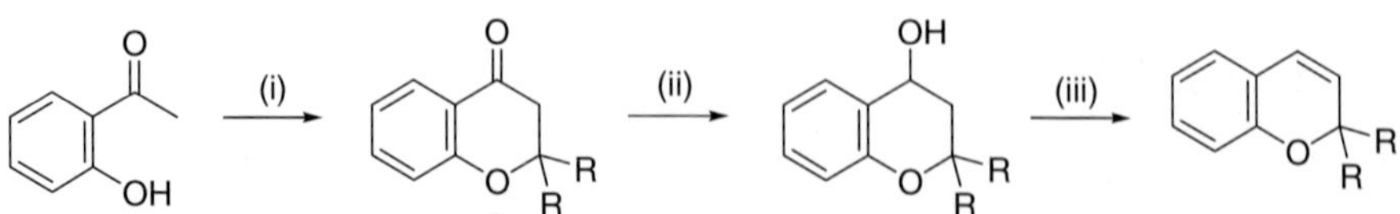

Reagents: (i) $R_2C{=}O$, PhMe, pyrrolidine, reflux; (ii) $NaBH_4$, EtOH, reflux; (iii) 4-TsOH, PhMe, reflux

The compatibility of substituents with the organolithium reagent is the only limitation to the formation of benzopyrans by reaction of α,β-unsaturated aldehydes with dilithiated *o*-bromophenols [34]. This methodology has been adapted for the synthesis of a 2,2-diaryl-2*H*-naphtho[1,2-*b*]pyran (Scheme 4) [35]. Equally successful is the reaction of a metallated heterocycle with 2-hydroxy-1-naphthaldehyde to give naphthopyrans (e.g. **7**, Scheme 5) [36].

The reaction of titanium phenolates, derived from phenols and titanium(IV) ethoxide, with β-phenylcinnamaldehyde **8** [37] can be successful where other strategies fail. The extra effort involved in the synthesis of the cinnamaldehyde,

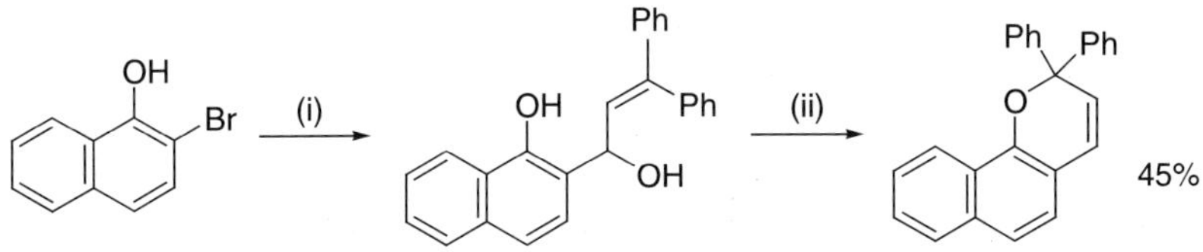

Reagents: (i) 2 *n*-BuLi, RT, Et_2O then β-phenylcinnamaldehyde; (ii) 4-TsOH, PhMe, 60 °C

Scheme 4.

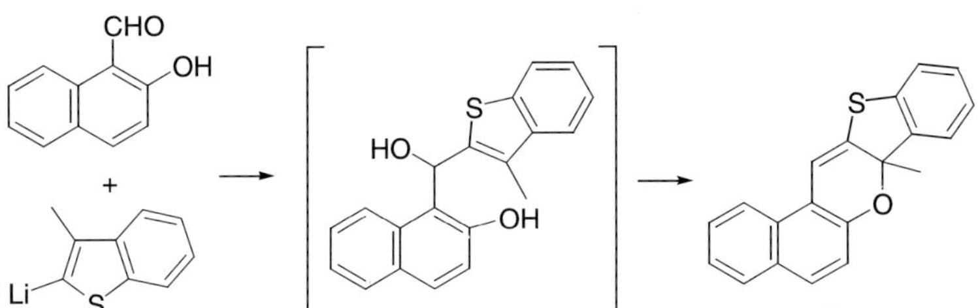

7

Scheme 5.

of which relatively few structurally diverse aryl substituted examples are readily available, may be justified, as for example in their reaction with electron-deficient hydroxy-substituted heterocycles (Scheme 6) [38, 39].

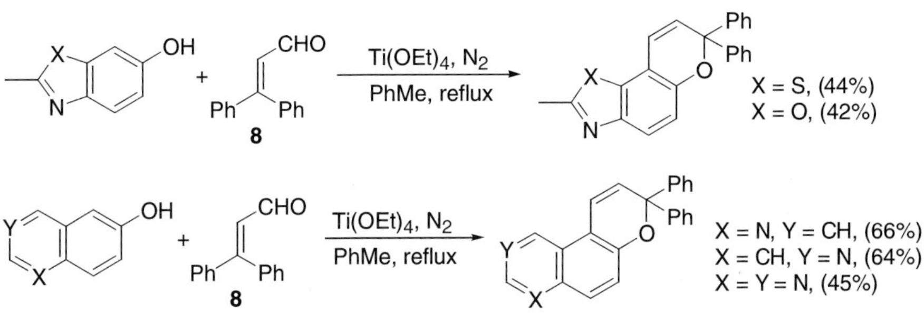

Scheme 6.

The most expeditious route to diaryl substituted naphthopyrans that offers good flexibility is based upon the thermal rearrangement of naphthyl propargyl ethers **9** [40], derived from the alkylation of a naphthol with a haloalkyne, to substituted naphthopyrans **10** (Scheme 7) reported by Iwai and Ide [41]. Catalysis by

Cu(I) or (II) has been noted for the synthesis of aryl dimethylpropargyl ethers [42] and zeolites facilitate the reaction of naphthols with 2-phenylbut-3-yn-2-ol [43].

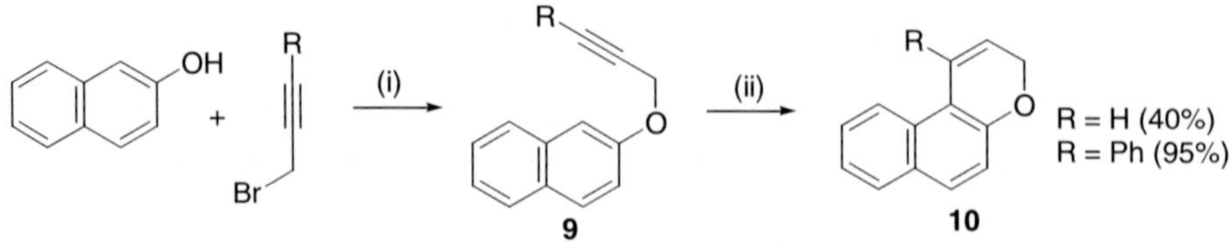

Scheme 7.

In a substantially modified version of this protocol that yields diarylnaphthopyrans in a single step and in good yield [44], 1,1-diarylprop-2-yn-1-ols are heated with a naphthol in toluene containing an acidic catalyst that promotes the *in situ* formation of the naphthyl propargyl ether (Scheme 8). This protocol is suitable for hydroxy-substituted heterocyclic systems [45–47] and has been adapted for the solid-state synthesis of naphthopyrans [48]. However, it should be noted that interception of the intermediate carbocation by a nucleophilic C-site in the naphthol may result in the formation of propenylidenenaphthalenones **11** along with, or to the exclusion of, the naphthopyran [49, 50]. α,β-Unsaturated aldehydes, derived from either a Rupe [51] or Meyer–Schuster rearrangement [52] of the prop-2-yn-1-ol, are also frequently encountered by-products [49, 53]. Although aryl 1,1-dimethylpropargyl ethers are readily available [42], attempts to synthesise diarylpropargyl naphthyl ethers from naphthols and diarylpropynols under Mitsunobu conditions [54] failed, producing instead the naphthopyran in low yield. In the presence of 4-nitrophenol, yields were improved offering a mild and facile route to the photochromes [49]. A further development of the propynol route incorporates $(MeO)_3CH$ as a dehydrating agent [55].

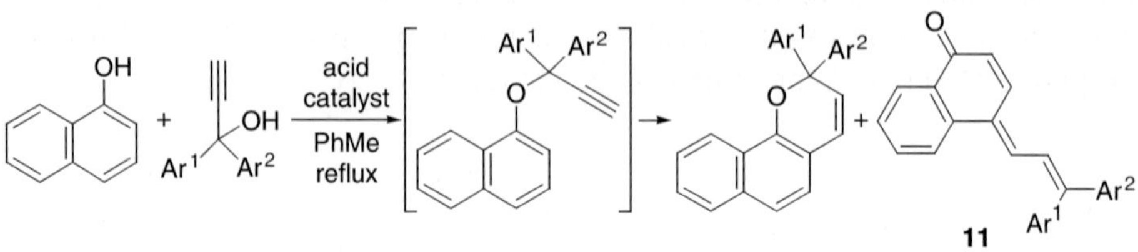

Scheme 8.

1,1-Diarylprop-2-yn-1-ols condense with enolisable ketones under acidic catalysis to afford merocyanine dyes. Dehydrogenation with concomitant electrocyclisation of dye **12** affords the naphthopyran [56].

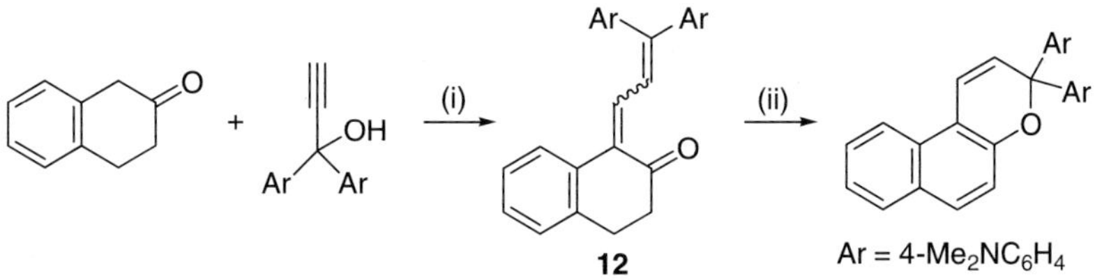

Reagents: (i) 4-TsOH, PhMe, reflux (46%); (ii) *p*-chloranil, PhMe, reflux (36%)

A major aspect of the value of this route to naphthopyrans is the availability of the precursors. Thus, 1,1-diarylprop-2-yn-1-ols are accessible from the reaction between benzophenones and lithium acetylide [57] or, better, its trimethylsilyl derivative (Scheme 9) that avoids the formation of 1,1,4,4-tetraarylprop-2-yn-1,4-diols [56, 58].

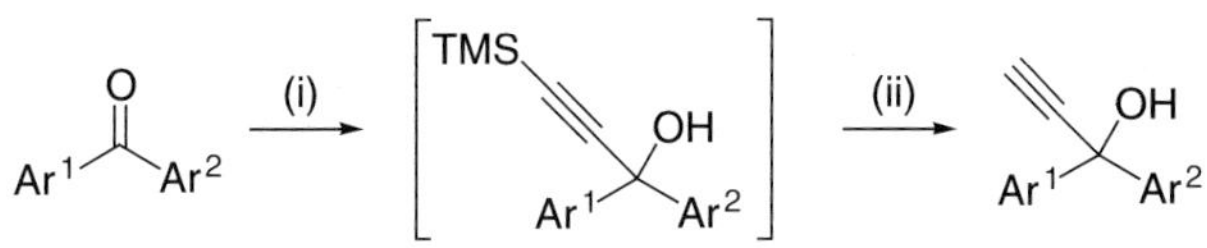

Reagents: (i) *n*-BuLi,TMS-acetylene, THF, 0 °C - RT, N_2;
(ii) either KOH, MeOH, THF, RT or TBAF, THF, RT

Scheme 9.

Many 1- and 2- naphthols are commercially available and a wide range of substituted derivatives are easily obtained by standard methodologies. In the case of less accessible 3-substituted 1-naphthols, Stobbe condensation is particularly valuable and has been much used in the production of photochromic naphthopyrans (Scheme 10) [59]. The half ester formed from the reaction of dimethyl succinate with either a benzaldehyde, acetophenone or a benzophenone is cyclised to the naphthoate and thence hydrolysed to the methyl 4-hydroxynaphthalene-2-carboxylate. A wide variety of substituted naphthopyrans [60–64] and phenanthropyrans [65] have been obtained *via* this protocol.

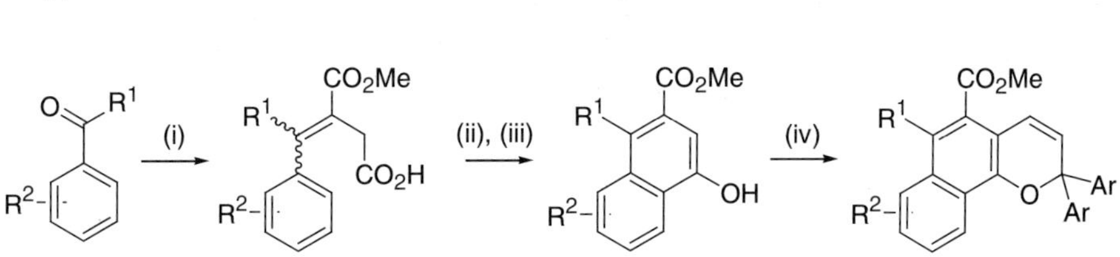

Reagents: (i) dimethyl succinate, NaH, PhMe, RT; (ii) NaOAc, Ac_2O, reflux; (iii) MeOH, c. HCl;
(iv) 1,1-diarylprop-2-yn-1-ol, acidic catalyst, PhMe, heat

Scheme 10.

1-Amino-3-naphthols have proved to be particularly important substrates for the formation of intense colouring naphthopyrans (Section 4.1). Their efficient synthesis from 2-naphthol by chlorination, amination and hydrogenation sequence was first described by Pilkington PLC in 1993 [66]. Other routes to these useful naphthols have been described by PPG Industries Inc. involving the hydrolysis of a naphthoquinone hydrazide derived from 1,2-naphthoquinone-4-sulfonic acid K salt [67] and by The Perkin–Elmer Corporation through the transition metal mediated amination of triflates derived from 1,3-dihydroxynaphthalene (Scheme 11) [68].

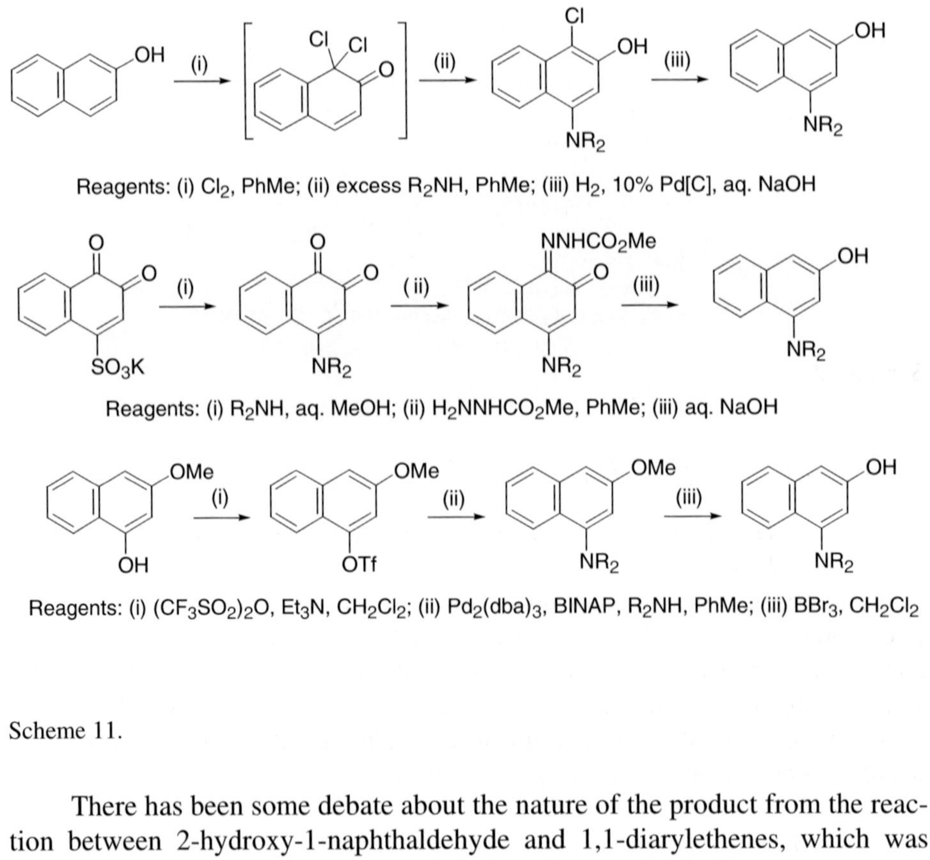

Scheme 11.

There has been some debate about the nature of the product from the reaction between 2-hydroxy-1-naphthaldehyde and 1,1-diarylethenes, which was originally considered to give a naphtho[2,1-*b*]pyran [69]. The structure was revised to two different 1:2 adducts, cyclopenta[*b*]naphtho[1,2-*d*]furans [28, 70] and finally to a substituted dihydronaphthopyran **13** (X = H, Ar = Ph) [71, 72]. A careful re-examination of this procedure has shown that naphthopyrans can indeed be prepared in moderate yield in this way from both 2-hydroxy-1-naphthaldehyde and the corresponding 2-naphthaldehyde (Scheme 12) and that the dihydronaphthopyran is often an accompanying product [73]. This synthesis proceeds through the capture of a quinonemethide by the diarylethene followed by an acid-catalysed dehydration that generates a carbocation. Deprotonation results in the formation of a naphthopyran, but interception by a second molecule of the diarylethene leads to the 1:2 adduct **13** (X = morpholino, Ar = 4-methoxyphenyl).

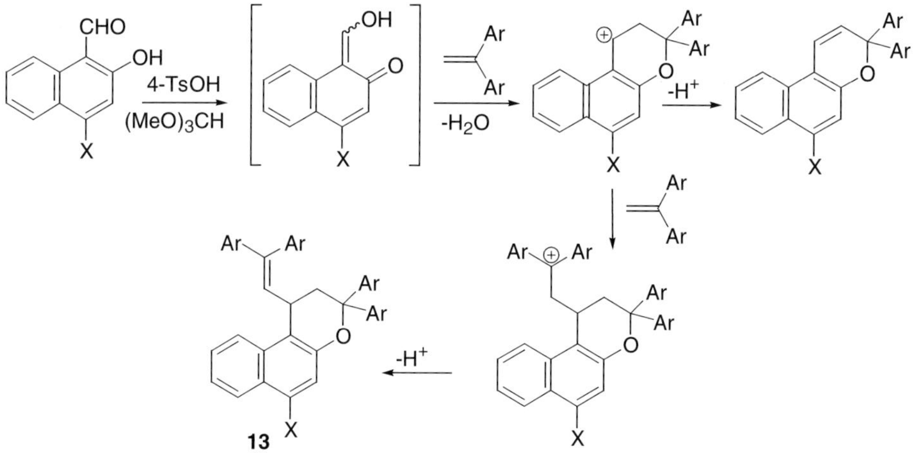

Scheme 12.

3. STRUCTURAL CHARACTERISATION

There are relatively few papers dedicated to the structural characterisation of photochromic naphthopyrans. However, there is a wealth of such data spread thinly throughout the scientific literature in papers reporting the synthesis and photochromic properties of new naphthopyrans. Of the various spectroscopic techniques employed to characterise naphthopyrans, ^{1}H NMR and UV–visible spectroscopy are the most predominant. UV–visible spectroscopy is fundamental to the study of the photochromism of naphthopyrans and is considered in detail in appropriate sections of this chapter.

3.1. NMR spectroscopy

The characteristic feature of the ^{1}H NMR spectra of 2,2-disubstituted benzopyrans is the pair of doublets ($J_{3,4}$ *ca.* 10 Hz) at *ca.* δ 5.5 and 6.5 associated with H-3 and H-4 and these are generally replicated in the corresponding naphthopyrans albeit usually further downfield. The latter proton always resonates downfield of H-3 in keeping with its benzylic nature [74, 75]. There are relatively few reports concerned with the ^{1}H NMR spectroscopy of simple diaryl substituted naphthopyrans and Fig. 2 serves to show the differences between the spectra of the [1,2-*b*] and [2,1-*b*] isomers [76].

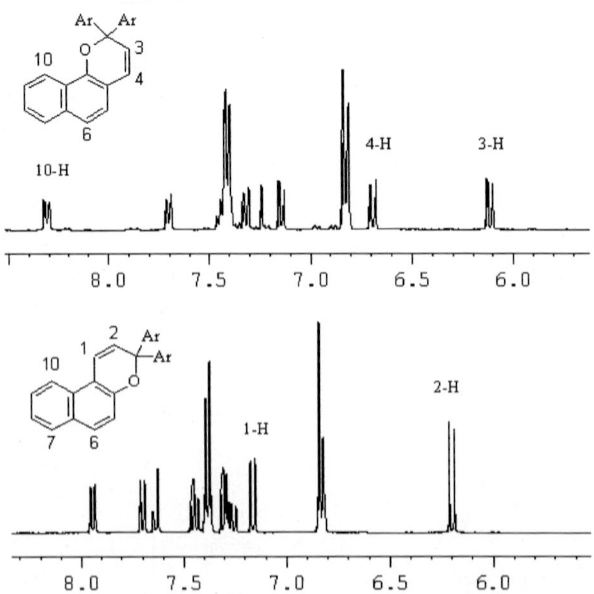

Fig. 2. Comparative ¹H NMR spectra of isomeric diaryl substituted naphthopyrans.

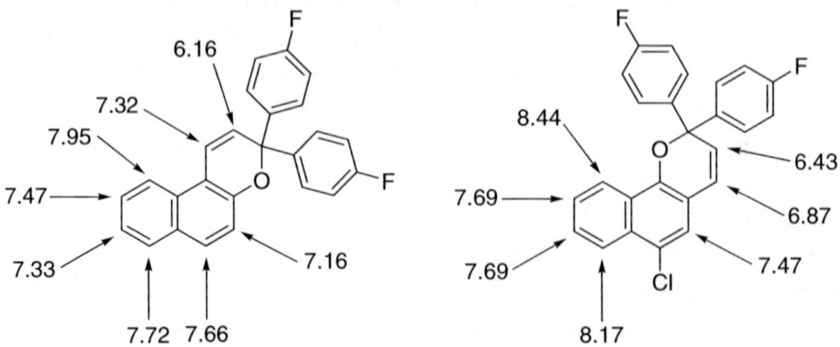

Fig. 3. Comparative ¹H NMR data for fluorine containing naphthopyrans.

Much of the reported work has focused on fluorine-containing naphthopyrans and the major chemical shifts that have been assigned are shown in Fig. 3. It can be seen that 2-H in the 3*H*-naphtho[2,1-*b*]pyran resonates at *ca.* δ 6.2 as a doublet with $J = 9.8$ Hz and the corresponding proton (3-H) at *ca.* δ 6.4 for the 2*H*-naphtho[1,2-*b*]pyran series [77, 78]. Greater differences in the chemical shift of the benzylic protons in the isomeric naphthopyrans are noted with 1-H resonating at *ca.* δ 7.3 some 160 Hz further downfield than the corresponding proton, 4-H, at *ca.* δ 6.9 in the naphtho[1,2-*b*]pyran isomer. ^{1}H–^{1}H COSY NMR experiments have proved useful in the assignment of these benzylic protons [53]. In both isomers, 10-H typically resonates furthest downfield, with that of the naphtho[1,2-*b*]pyran isomer, *peri* to the pyran ring oxygen atom, at *ca.* δ 8.4 [77, 78]. The influence of substituents on the chemical shift of the aromatic protons of the naphthopyrans have been documented and is predictable [39, 64, 79, 80].

The chemical shift of the pyran ring proton in the ^{1}H NMR spectra of the naphthopyrans **14** and **15** derived from 1,1,3-triaryl substituted prop-2-yn-1-ols is relatively unaffected by the adjacent aryl unit [49], though moderate shifts in the chemical shift of the pyran ring protons has been observed for isomeric pyranocarbazoles [81].

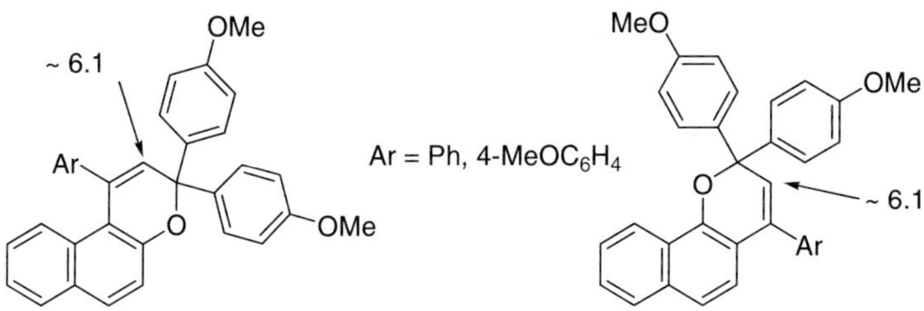

14 **15**

When a fluorine atom is incorporated in the *ortho* position of one of the *geminal* aryl rings, long range ^{1}H–^{19}F coupling to 2-H of 3–5 Hz is noted, resulting in a double doublet; more complex signals result for 2-H when additional fluorine atoms are incorporated into the other *ortho* positions (Fig. 4). ^{1}H {^{19}F} decoupling experiments have been used to verify this unusual interaction [82, 83]. ^{19}F Chemical shifts have been documented for these fluorine containing naphthopyrans (Fig. 4) [78, 82].

Key ^{13}C NMR chemical shifts for the isomeric naphthopyran systems are presented in Fig. 5. Perhaps of greatest importance as confirmation of the pyran unit is the chemical shift of the sp^3 hybridised ring carbon at *ca.* δ 82 [78], a chemical shift that is unaffected by replacement of the naphthalene rings by a carbazole unit [81] and which is not dissimilar to that for 2,2-dimethyl-2*H*-[1]benzopyrans (*ca.* δ 78) [84].

NMR spectroscopy is particularly useful as a tool to probe the structure and geometry of the coloured ring-opened species generated by the irradiation of the naphthopyrans. In an initial study employing a battery of NMR techniques, two relatively long-lived isomers characterised as the *trans–cis* (TC) and the *trans–trans* (TT) forms of an *s-trans* diene and a third short-lived isomer were detected upon irradiation of naphthopyran **16** [85]. The coupling constant for the *s-trans* unit was measured at 11.7 Hz, which compares favourably with those of merocyanine dyes that possess a similar geometry [49, 56, 86]. The presence of the F atoms in **16** enabled ^{19}F NMR spectroscopy to probe the evolution of the isomers

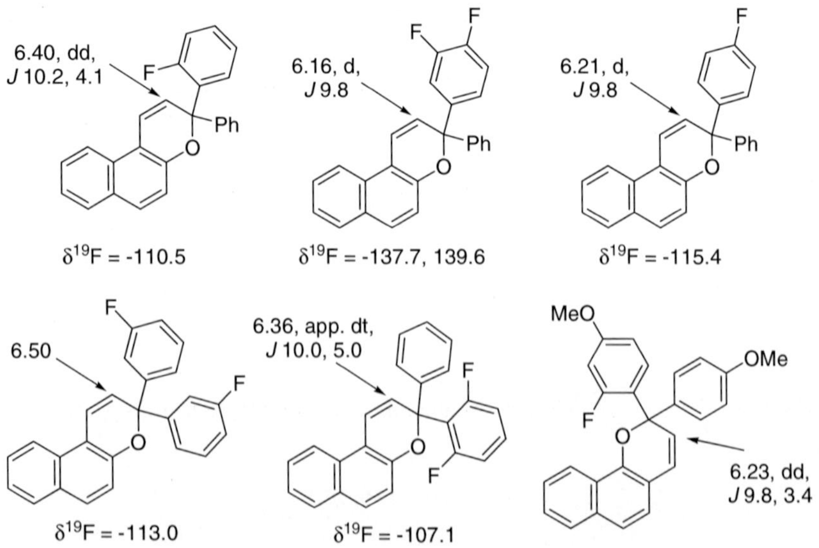

Fig. 4. Selected ¹H and ¹⁹F NMR data for some fluorine containing naphthopyrans.

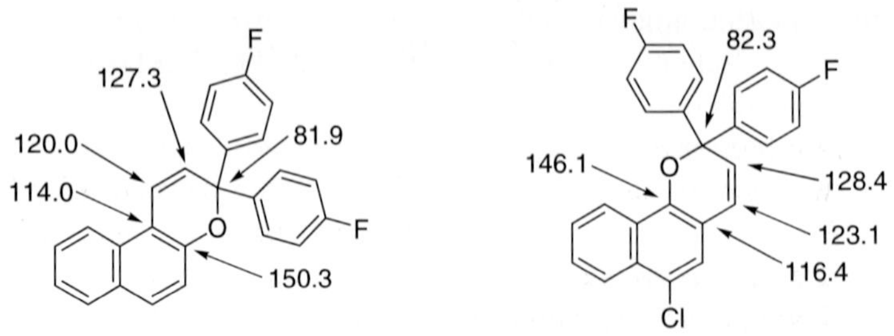

Fig. 5. ^{13}C NMR signals for the pyran ring in naphthopyrans.

with time. For the TC and TT isomers, two new ^{19}F signals were observed as a consequence of the asymmetric nature of the ring-opened species (Fig. 6). The use of ^{1}H and ^{19}F NMR spectroscopy to characterise the ring-opened products derived from various naphthopyrans has become more widespread [80, 87–89].

A further photogenerated intermediate has been characterised by NMR spectroscopy [90, 91]. The allene **17**, derived by a 1,5-hydrogen shift from either the TC or TT isomer, displays a signal at *ca.* δ_C 211 for the central C atom, a value that compares favourably with those (~δ_C 205) obtained for allenes derived from naphthopyrans by an anionic ring opening – electrophilic trapping protocol [92].

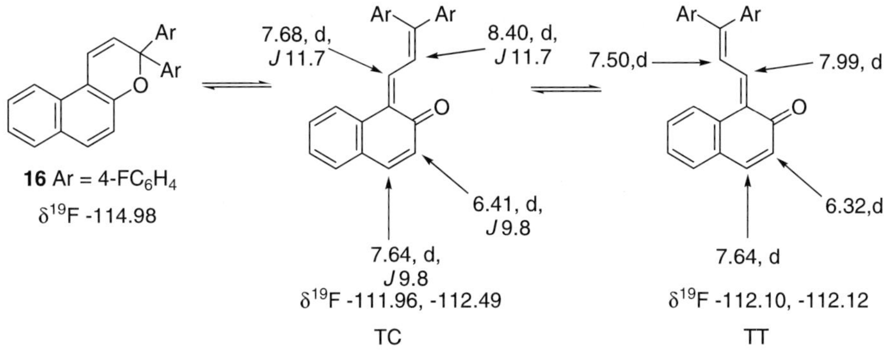

Fig. 6. Selected ^{1}H and ^{19}F NMR data for the ring-opened forms of naphthopyran **16**.

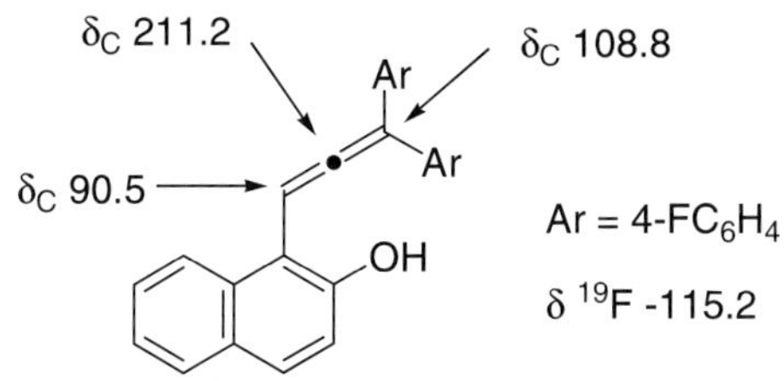

17

3.2. X-Ray crystallography

A review of the structural studies of photochromic molecules by X-ray diffraction includes six examples of diaryl substituted naphthopyrans [93]. The *gem* diaryl rings are orthogonal in 3,3-diphenylnaphtho[2,1-*b*]pyran and in several derivatives, although the orientations with respect to the pyran ring vary [94, 95]. The 2*H*-pyran ring is folded as a consequence of the sp^3 hybridised oxygen and the adjacent diaryl substituted tetrahedral carbon atom. The O–C(Ar)$_2$ bond is longer than a typical O–C ether bond (*ca.* 1.42 Å) [96], a feature that has been attributed to steric repulsions between the O atom and the proximal aryl groups, although the orientation of the aryl rings does not apparently influence the bond length [93]. Some authors have claimed that the O–C(Ar)$_2$ bond is elongated for *ortho* disubstituted compounds, e.g. **21** [97]. However, comparison of recent crystallographic data (Table 1) suggests that factors such as the electronic nature of the substituent, its location on the aryl ring, and the bond angles between the substituents on the (distorted) tetrahedral C atom, all impact upon the O–C bond length in a complex relationship.

Space filling representations based on crystallographic measurements of **22** and **23** have been used to demonstrate the influence of the steric interactions between a cyclic amine group and an adjacent function on the electronic donor

Table 1
Selected bond length data for some 3*H*-naptho [2, 1-*b*] pyrans

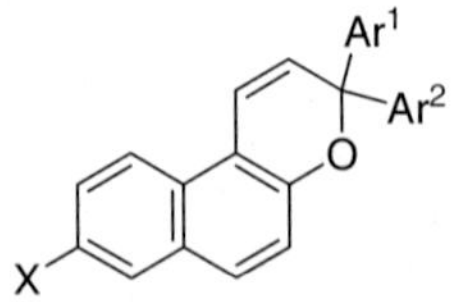

	X	Ar^1	Ar^2	$O–C(Ar)_2$ (Å)	$C–Ar^1$ (Å)	$C–Ar^2$ (Å)	$Ar^1–C–Ar^2$ (°)
18 [82]	H	2,6-diFC_6H_3	Ph	1.440	1.539	1.544	108.1
19 [82]	H	2-FC_6H_4	4-$MeOC_6H_4$	1.464	1.525	1.527	113.1
20 [98]	Br	4-$MeOC_6H_4$	4-$MeOC_6H_4$	1.473	1.521	1.530	112.3
21 [97]	H	2-$MeOC_6H_4$	2-$MeOC_6H_4$	1.462	—	—	93.3
22 [53]	H	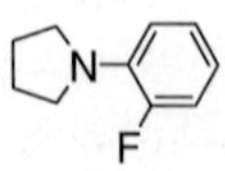	Ph	1.450	1.522	1.538	107.5
23 [53]	H	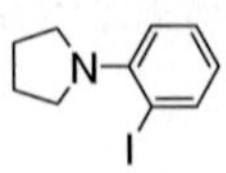	Ph	1.451	1.532	1.521	111.7

properties of the amine. Such interactions facilitate the fine tuning of λ_{max} of the photogenerated dye [53]. In the tricarbonylchromium complex with 3,3-diphenyl-3*H*-naphtho[2,1-*b*]pyran, the transition metal moiety and the axial phenyl ring are on opposite sides of the plane of the naphthopyran system [99].

Crystal structures of two 2*H*-naphtho[1,2-*b*]pyrans **24** [80] and **25** [100] have been reported. The dihedral angles ($Ar^1–C–Ar^2$) are 111.7 and 109.8° and the $O–C(Ar)_2$ bondlengths are 1.459 and 1.454 Å respectively, which compare favourably with the data for the naphtho[2,1-*b*]pyrans in Table 1.

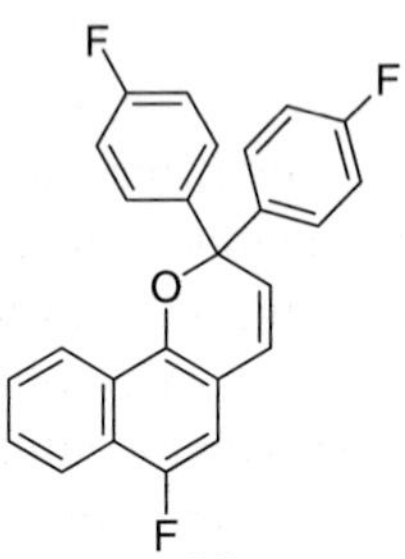

24

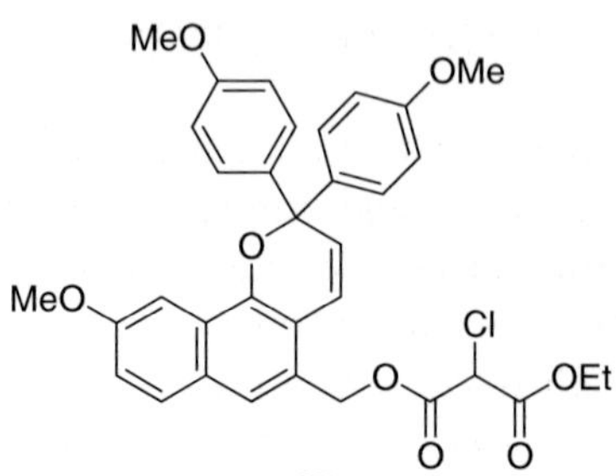

25

3.3. Infrared spectroscopy

Useful information gained from infrared spectroscopy is restricted to the confirmation of substituent functions and such data can be located in papers describing the synthesis of substituted diaryl naphthopyrans. The major absorptions of the diaryl naphthopyrans are very similar to those described for simple 2*H*-[1]benzopyrans [101] and relate to stretching vibrations for the C=C bond in the pyran ring, which appears at 1630–1650 cm^{-1} and the C–O bond at *ca.* 1250 cm^{-1}.

3.4. Mass spectrometry

The mass spectral fragmentation of simple 2*H*-pyrans and 2*H*-[1]-benzopyrans is well established and operates through loss of one of the C-2 substituents from the molecular ion to afford a (benzo)pyrylium ion [102]. The electron impact mass spectra of symmetrical diaryl substituted naphtho[2,1-*b*]- and [1,2-*b*]-pyrans indicate a similar pathway to the naphthopyrylium ion, which then undergoes complex fragmentation (Scheme 13). The fragmentation of the symmetrically substituted naphthodipyran **26** proceeds in a similar manner to give the monopyrylium cation; the loss of a second C-2 substituent from the remaining diaryl pyran unit to afford the naphthodipyrylium ion was not observed, but the base peak appeared at *m/z* 227. Unsymmetrical diaryl substituted naphthopyrans, e.g. **27,** fragment through loss of either one of the aryl groups (Scheme 14) [103].

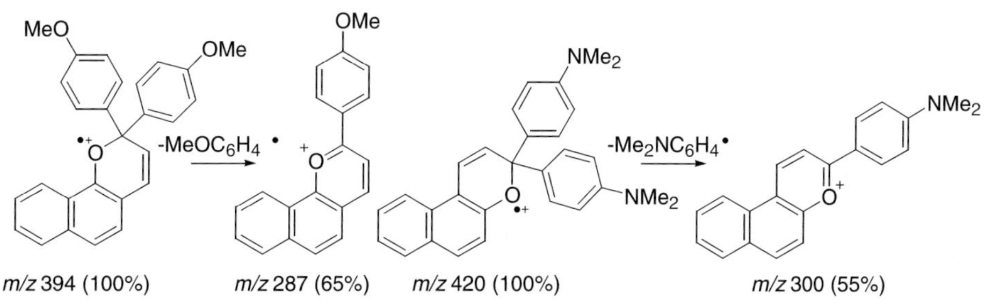

Scheme 13.

4. PHOTOCHROMIC PROPERTIES

The photochromic characteristics of a compound are usually measured in terms of λ_{max} of the ring-opened and ring-closed forms and the induced optical density of the coloured (ring opened) species at its λ_{max} (colourability) achieved after irradiation to constant value and at a specified temperature. The speed of the backward reaction (ring closure) is measured by recording the loss of colour with time, reporting the data as $t_{½}$, the time in seconds required for the sample to return to half the optical density of the equilibrium value [19]. The ideal combination of

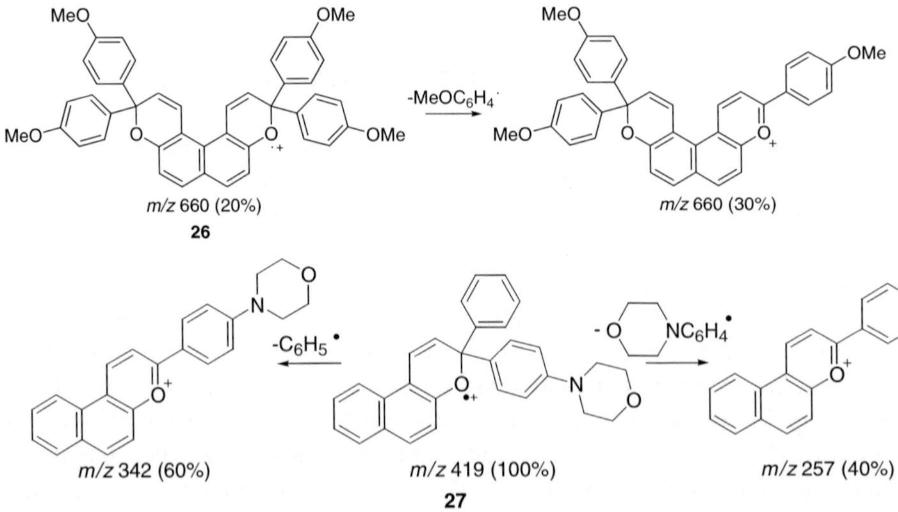

Scheme 14.

photochromic properties required for variable optical transmission devices is intense colour generation with a reasonably rapid fade rate (bleaching) at ambient temperatures. It is also important that the compound exhibits good fatigue resistance; the ring-opening – ring-closing cycle must be repeatable many times ($>10^6$) without loss of performance. The medium in which the photochrome is dissolved or dispersed can exert a significant effect on these properties. For example, some photochromic naphthopyrans exhibit solvatochromism [104, 105] and halochromism [106]. More significant is the influence of a polymer matrix, which in addition to causing minor shifts in λ_{max} generally hinders ring closure, thereby increasing $t_{1/2}$ [24]. The thermal fading of naphthopyrans is retarded by solvents of high viscosity, in which it is proposed that the substituted ethenyl group changes its position in the solvent sphere during cyclisation, while the naphthalene unit remains undisturbed [107, 108]. The sol–gel method has been used to entrap naphthopyrans in thin film ormosil coatings. Naphthopyrans substituted with methoxy groups showed slow bleaching of the coloured forms, attributed to a combination of stabilising interactions between the methoxy groups and the matrix and to the restricted mobility within the matrix pore. The bleaching process is accelerated by irradiation with visible light [109, 110].

The photochromic process for the naphthopyrans involves initial photolytic cleavage of the O–C bond that leads to the generation of two coloured ring-opened structures, a *trans–cis* (TC) and a *trans–trans* (TT) merocyanine, of which the latter is the more stable (Scheme 15) [85, 111–113]. The involvement of the TC and TT isomers in the solid-state photochromism of some 3*H*-naphtho[2,1-*b*]pyrans has been noted [114]. Pariser, Pople, Parr molecular orbital (PPPMO) calculations predict the absorption characteristics of 3*H*-naphtho[2,1-*b*]pyrans in better agreement with the experimental values when TT geometry is assumed [115]. Opening of the pyran ring and fusion of a cyclopropane ring across the original 5,6-bond ensues when a naphtho[1,2-*b*]pyran reacts with a rhodium carbenoid. X-ray crystallography of the product confirms the TT structure for the *s-trans* diene array [86]. The photochemical behaviour of a TT merocyanine has been described [116]. The *trans* → *cis* conversion is slower than the thermal ring closure of the TC form but is accelerated by irradiation with visible light. Consequently, after a fast initial fade, some colour remains for an appreciable time with certain photochromes. This feature is significant for 5-acylnaphtho[2,1-*b*]pyrans for which ^{19}F NMR spectroscopy indicates that the TC and TT isomers are formed on irradiation with UV light in approximately equal amounts together with an allenylnaphthol. Thermal cyclisation of the TC species is observed over time at low temperatures, but the TT form remains and is only cyclised on irradiation with visible light [89]. The reddish ring-opened species precipitates from CCl_4 solution when an 8-aminoxylnaphtho[2,1-*b*]pyran is irradiated, perhaps indicating that the radical possesses some dipolar attributes. Stabilisation of cationic species by dimethylaminophenyl groups is well established and may contribute to the stabilisation of the betaine **28**. Recyclisation to the pyran occurs on treatment with SiO_2 [117].

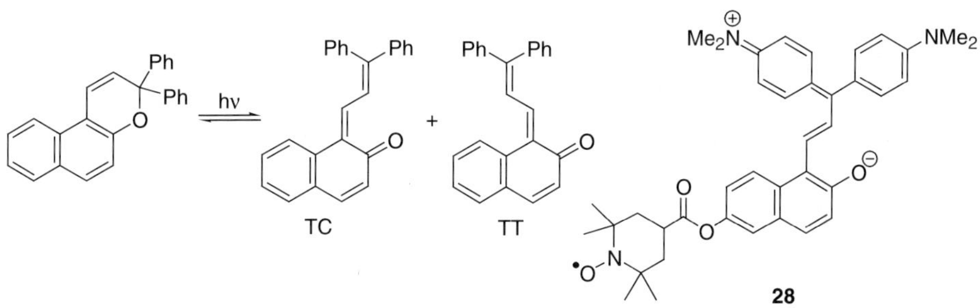

Scheme 15.

A study of the racemisation of chiral 2-aryl-2-methylnaphthopyrans and hetero-fused benzopyrans proceeding through thermal cleavage of the O–C2 bond has indicated that $\Delta G^{\neq}$ decreases with the electron donating power of a 4-substituent in the pendant phenyl ring in naphtho[1,2-*b*]pyrans. Presumably, the transition state for thermal ring opening is stabilised by the additional conjugation with the substituent. Similarly, fusion of an additional benzene ring, giving the phenanthropyran, has a stabilising influence such that $\Delta G^{\neq}$ is reduced. Conversely, fusion of either a benzene or a pyridine ring onto 2*H*-[1]benzopyran has little effect on $\Delta G^{\neq}$ irrespective of the site of fusion. There appears to be a correlation between $\Delta G^{\neq}$

and the calculated π-bond order for the fusion bond between the pyran and benzene rings; the more electron rich the bond, the lower is $\Delta G^{\neq}$ [118].

4.1. 3*H*-Naphtho[2,1-*b*]pyrans

The photochromic response of the angular 3*H*-naphtho[2,1-*b*]pyran isomer **2** is typically characterised by the production of a weak colour associated with the photochemically induced electrocyclic ring opening of the colourless pyran ring to a coloured quinoidal form on irradiation with UV light (Scheme 15). The low equilibrium concentration of the ring-opened yellow species under steady state conditions combined with its rapid cyclisation gives the overall impression of a weakly colouring molecule, e.g. for **2** Ar = Ph, λ_{max} = 432 nm with $t_{1/2}$ ~ 45 s [(diethyleneglycol bis(allyl carbonate)] [23].

Through judicious choice of substituents, the performance of the 3*H*-naphtho[2,1-*b*]pyran system can be significantly improved. The data in Table 2

Table 2
Influence of substitution in the 3, 3-diphenyl rings on photochromic properties of 3*H*-naptho [2, 1-*b*] pyrans

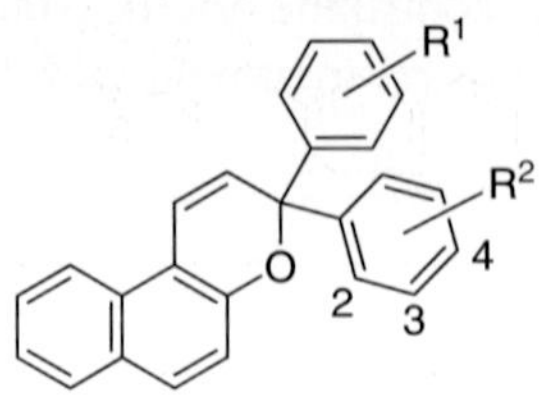

R^1	R^2	λ_{max} (nm)[a]
H	H	430
H	4-MeO	458
4-MeO	4-MeO	475
4-MeO	2-F	456
2-F	H	421
2-F	4-F	419
4-F	4-F	428
H	4-CF_3	422
4-MeO	4-NMe_2	512
4-NMe_2	4-NMe_2	544, 442

[a]Data recorded for PhMe solutions [24, 82].

illustrate the effect of substitution in the phenyl rings at the 3-position. Generally, electron-releasing groups at the *para* positions bring about a red shift of the absorption band and this is accompanied by an increase in the fade rate. Electron-withdrawing groups cause a blue shift and slow the fade rate to some extent.

The subtle variations that can be achieved in λ_{max} through careful selection of substituents are illustrated by the 4-aminophenyl derivatives shown in Table 3. The absorption characteristics correlate well with the electron donating abilities of the amino functions observed in a study of some triarylmethine dyes [119, 120] and with those derived from the basicities of enaminoketones [121] and from NMR studies of *N,N*-dialkylated anilines [122, 123].

Table 3
The effect of amino-substituents on photochromic properties of 3, 3-diphenyl-3*H*-naphtho [2, 1-*b*] pyrans

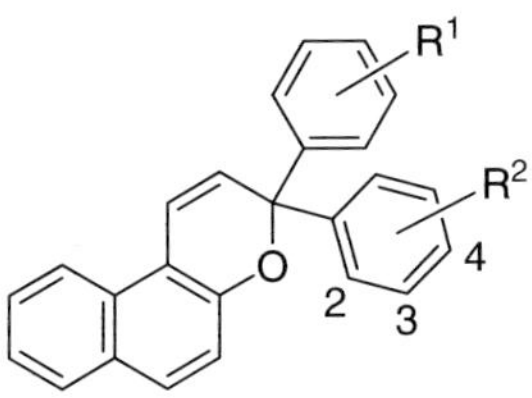

R^1	R^2	λ_{max} (nm)[a]	$t_{1/2}$ (s)
4-NMe_2	H	528	10.1
4-NEt_2	H	544	4.8
4-pyrrolidino	H	538	5.0
4-piperidino	H	514	6.7
4-homopiperidino	H	549	4.8
4-morpholino	H	502	8.6
4-NMe_2	4-MeO	532	<1
4-NMe_2	4-NMe_2	544, 442	<1
4-NEt_2	4-NEt_2	576, 458	<1
4-pyrrolidino	4-pyrrolidino	572, 456	<1
4-piperidino	4-piperidino	544, 436	<1
4-morpholino	4-morpholino	526, 412	<1

[a]Data recorded for PhMe solutions at 20°C [124].

Further shifts of λ_{max} to the red, the order of which again relate to the donor ability, follow the introduction of an identical 4-amino substituent into the second phenyl ring (Table 3). However, of greater significance is the development of a second weak absorption band (shoulder) in the visible region that prompts comparison with the spectrum of Malachite Green, 4,4′-bisdimethylaminotriphenylmethine, the colour of which is derived from the combination of yellow (λ_{max} 427.5 nm) and blue (λ_{max} 621 nm) components. Here, though, the importance is that the two bands overlap to an appreciable extent and thus give dull shades of red and purple. This feature is a characteristic of the symmetrical substitution in the 3-phenyl rings, since only one absorption band is shown when different substituents are introduced into the second phenyl ring (Table 3). The extremely fast fading ($t_{½}$ < 1 s) shown by these compounds is noteworthy.

Additional control of the photochromic parameters can be achieved for a given aminophenyl derivative through the introduction of further substituents. A substituent adjacent to the pyrrolidine function in the relatively fast-fading

Table 4
The influence of halogen substituents in the aryl rings on photochromic properties of 3-phenyl-3-(4-pyrrolidinophenyl)-3*H*-naphtho[2, 1-*b*]pyrans

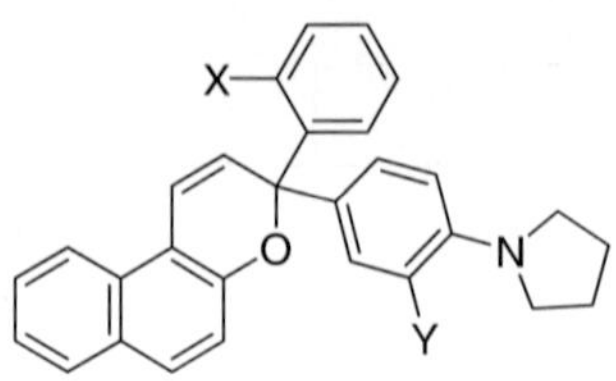

X	Y	λ_{max} (nm)[a]	$t_{½}$ (s)
F	H	554	40
Cl	H	554	741
Br	H	554	1024
I	H	553	1167
H	H	538	5
H	F	529	4
H	Cl	513	4
H	Br	493	4
H	I	494	5

[a]Data recorded for PhMe solutions at 20 °C [53, 83].

3-phenyl-3-(4-pyrrolidinophenyl)naphtho[2,1-*b*]pyran **29a** (λ_{max} 538 nm, $t_{1/2}$ 5 s) allows variation in λ_{max} between 493 and 529 nm. The bulkier the group Y, the larger the blue shifts that result while $t_{1/2}$ is essentially unchanged (Table 4). Clearly, the steric effect of the new substituent influences the electron donation by the pyrrolidine (Fig. 7) [53]. A 2-methyl group **29d** exerts a hypsochromic shift but also increases the colourability and generates a second absorption band so that the merocyanine is a dull red. On the other hand, a 2′-methyl substituent **29b** gives a strongly coloured, red shifted dye [83]. A 3-methyl substituent **29c** results in a hypsochromic shift in λ_{max} (Fig. 7).

Substitution at the *ortho* positions has a major effect on the photochromic properties, causing a pronounced increase in $t_{1/2}$ [44, 125]. The lifetime of the open form increases with increasing size of the *ortho* substituent. Although the size of the halogen atom has no effect on λ_{max} for photochromes derived from 3-aryl-3-(4-pyrrolidinophenyl)naphtho[2,1-*b*]pyrans, the half-lives dramatically increase through the series X = F → I (Table 4) [83].

Substitution around the periphery of the molecule has a variable effect. The data for the methoxy derivatives given in Table 5 indicate the importance of a donor substituent at the 6- or 8- positions for manipulating λ_{max} [126, 127]. Dramatic increases in colourability (initial optical density IOD upon cessation of

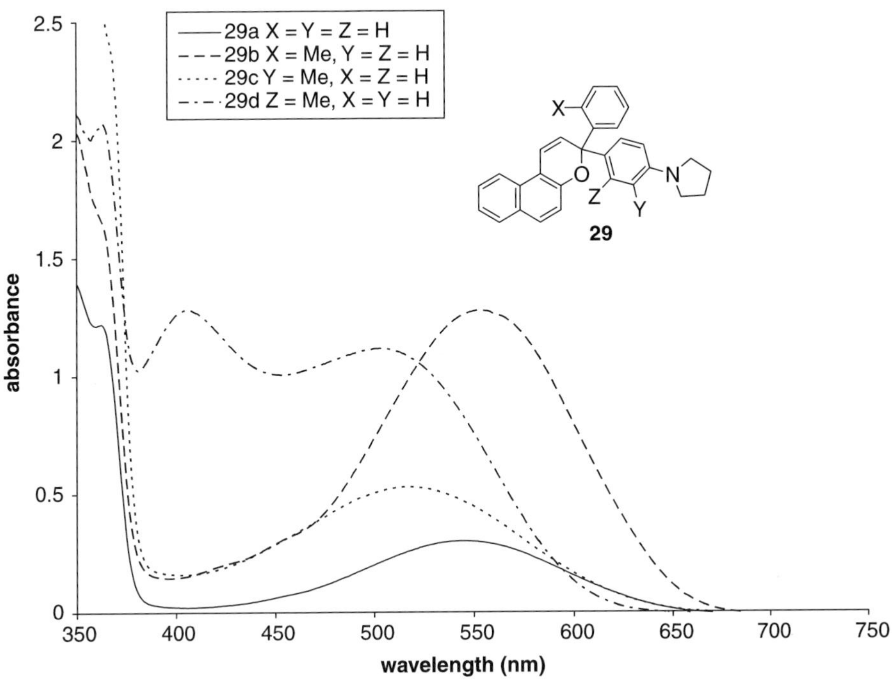

Fig. 7. UV–visible spectra of naphthopyran **29a** and methyl substituted analogues.

Table 5
The influence of peripheral substitution on the spectral parameters of 3, 3-diaryl-3*H*-naphtho[2,1-*b*]pyrans

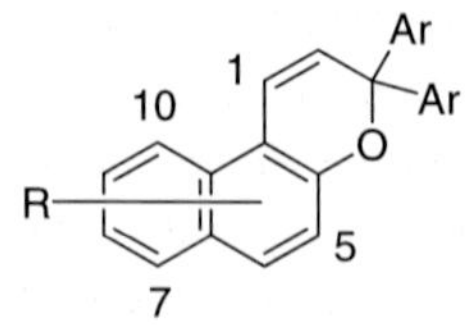

R	Ar	λ_{max} (nm)[a]	IOD
5-MeO	Ph	435	—
6-MeO	Ph	423	—
7-MeO	Ph	435	—
8-MeO	Ph	477	—
9-MeO	Ph	432	—
H	Ph	436	—
H	4-$MeOC_6H_4$	(475)[b]	(0.20)[b]
6-Morpholino	4-$MeOC_6H_4$	[452][c]	[1.95][c]
6-MeO	4-$MeOC_6H_4$	[456][c]	[1.42][c]
8-MeO	4-$MeOC_6H_4$	(502)[b]	(0.55)[b]

[a]Data recorded for aliphatic acrylic polymer [126].
[b]Data recorded in polyurethane.
[c]Data recorded in spectralite [24].

irradiation) follow the introduction of a 6-methoxy [128] or better a 6-amino function (Table 5) [66, 129].

Pd-catalysed cross-coupling has proved useful to effect the conversion of triflates of 3*H*-naphtho[2,1-*b*]pyrans into the *N*-methylpiperazino derivatives. A bathochromic shift of *ca.* 80 nm is observed for the 8-substituted compound, but an amino function in the 9-position has little influence on λ_{max} [130].

A combination of these modes of structural manipulation enables the photochromic properties of naphtho[2,1-*b*]pyrans to be tailored to satisfy a wide range of requirements. Intensely coloured yellow, orange, red and purple photochromes with varying half-lives can be produced by introducing 4-aminophenyl units into the 3-position of 3*H*-naphthopyrans containing an electron-releasing substituent at the 6- or 8- positions [131].

4.2. 2*H*-Naphtho[1,2-*b*]pyrans

In contrast to the [2,1-*b*] isomer, under steady state conditions a higher equilibrium concentration of the open form of 2*H*-naphtho[1,2-*b*]pyran **3** is present, which results in the development of an intense colour that persists for an appreciable period of time. Furthermore, λ_{max} is bathochromically shifted by *ca.* 45 nm relative to 3*H*-naphtho[2,1-*b*]pyran and for the parent compound (**3** Ar = Ph) λ_{max} is at 476 nm and $t_{1/2}$ is greater than 30 min. [in (diethyleneglycol bis(allyl carbonate)] [132]. The differing fade rates of the photoisomers of **2** and **3** have been attributed to the more significant steric interactions between 1-H and 10-H in the photoisomer of **2** compared with those between 4-H and 5-H in the photoisomer of **3** (Scheme 16) [127].

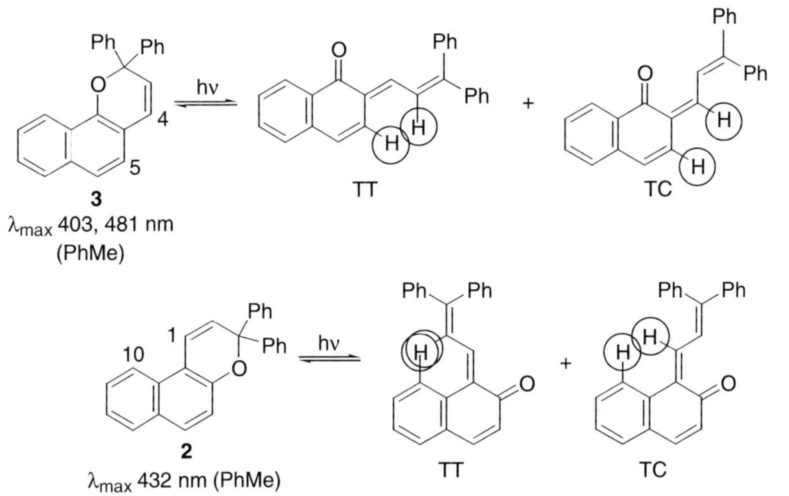

Scheme 16.

In an attempt to mimic the steric interactions present in the photoisomer of **2**, substituents were introduced at the 5- and 6- positions of **3**. These structural changes reported in a *Research Disclosure* [133] promoted a faster ring closure and represented a significant breakthrough in the commercialisation of photochromic naphthopyrans. Thus, the strong colourability of the naphtho[1,2-*b*]pyrans could be linked with the faster fade rate associated with the [2,1-*b*] series as illustrated for compounds of the type **30** in Table 6.

The presence of a 2-aminoaryl substituent in these 5-substituted naphthopyrans leads to intensely coloured red photochromes that are relatively fast fading, e.g. **31** and **32** [135].

Table 6
Photochromic properties of some 5-Substituted 2*H*-naphtho [1, 2-*b*]pyrans

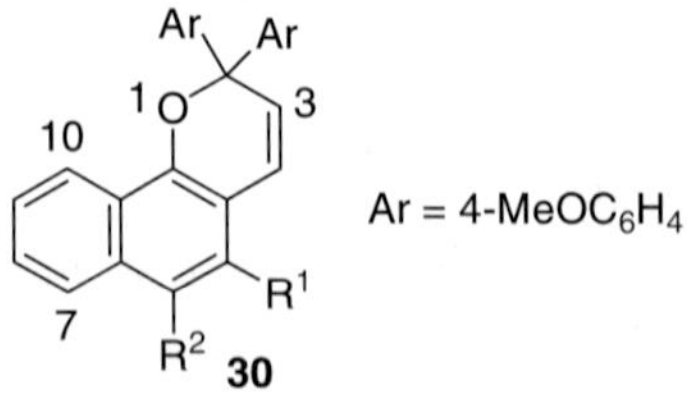

30

R^1	R^2	λ_{max} (nm)[a]	$t_{1/2}$ (s)
H	H	412, 508	>1800
Me	H	416, 496	178
Me	Me	421, 494	66
Ph	Ph	418, 508	20
CO_2Me	H	492	2
CO_2Me	Me	416, 492, [476][b]	4, [217][b]
CO_2Me	Ph	416, 506	7
CO_2Me	OH	[497][b]	[235][b]
CO_2Me	OCH_2CO_2Et	[508][b]	[211][b]

[a]Data recorded for PhMe solutions at 21 °C [24].
[b]Data recorded in diethylene-glycol bis(allyl carbonate) [134].

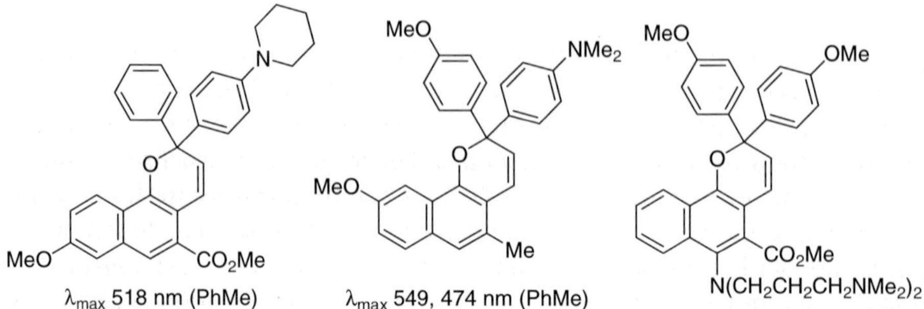

λ_{max} 518 nm (PhMe)
half-life 130 s
31

λ_{max} 549, 474 nm (PhMe)
half-life 63s
32

33

Manipulation of the 6-methoxy function involving reaction with Grignard reagents derived from secondary amines and subsequent conversion to the quaternary ammonium salt confers water solubility on naphtho[1,2-*b*]pyrans, e.g. **33** [136]. In a similar manner, interconversion of a 5-methoxy

Table 7
Absorption data and half-lives for some peripherally-substituted 5-alkoxycarbonyl-2*H*-naphtho[1, 2-*b*]pyrans

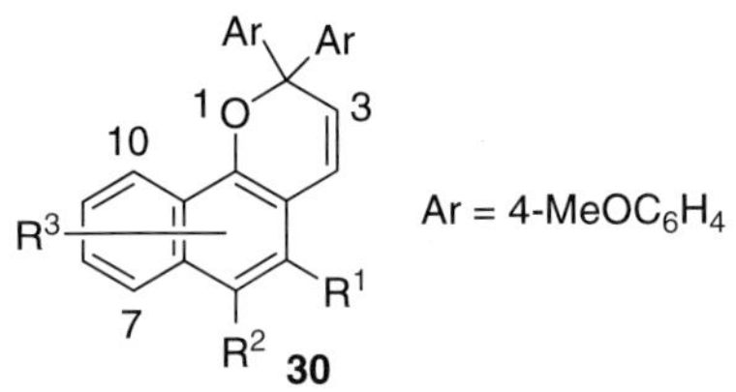

30

R^1	R^2	R^3	λ_{max} (nm)[a]	$t_{½}$ (s)
CO_2Et	H	H	493	3
CO_2Me	MeO	H	502, [510][c]	73, [305][c]
CO_2Et	H	7-MeO	508	3
CO_2Me	H	8-MeO	480	11
CO_2Et	H	9-MeO	505	3
CO_2Me	H	10-MeO	485	21
CO_2Me	Me	H	492[b], [476][c]	4[b], [217][c]
CO_2Me	Me	9-MeO	[522][c]	[376][c]

[a]Data recorded for PhMe solutions [64].
[b]PhMe solutions [24].
[c]Data recorded in diethylene-glycol bis(allyl carbonate) [134].

group and a 6-ester function affords various naphthopyrans which absorb relatively weakly at 406–492 and 487–596 nm [137]. The response to substituents in different positions of the naphthalene unit is shown in Table 7. A methoxy group can bring about either a bathochromic or hypsochromic shift in λ_{max} and slows the fade rate when in the 6-, 8- or 10- positions [64]. These data also illustrate the effect of the matrix on the photochromism. Incorporation into a polymer has a small effect on λ_{max} but slows the fade rate considerably as molecular movement is restricted, hindering the bond rotation necessary for ring closure.

The influence of a halogen atom at the 6-position on λ_{max} is marginal, F causing a small blue shift and Cl and Br somewhat larger red shifts. The major effect is the faster fade rate shown by the less electronegative halogens. All the compounds studied show a slow second fading process indicating that a relatively stable species remains after ring closure of the TC isomer [80].

The presence of an amino substituent at C-9, e.g. **34**, accentuates the two absorption bands with the typically more intense short wavelength band at *ca.* 430–470 nm and the long wavelength band at *ca.* 525–570 nm leading to a range of neutral brown shades [138].

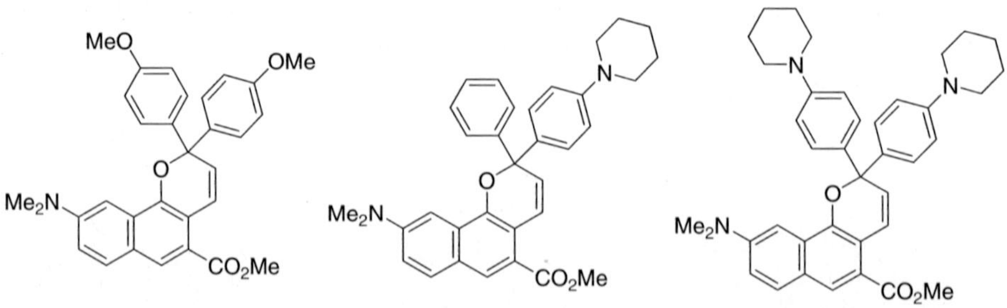

λ_{max} 454, 556 nm (PhMe) **34** — λ_{max} 485, 569 nm (PhMe) **35** — λ_{max} 490, 581 nm (PhMe) **36**

A combination of a 9-amino substituent and 4-aminophenyl groups at C-2 produces photochromes, e.g. **35** and **36**, which absorb in the ranges 460–520 nm and 550–640 nm and as a result they are of various shades of grey, sometimes with a tint of purple, blue or green (Fig. 8) [61].

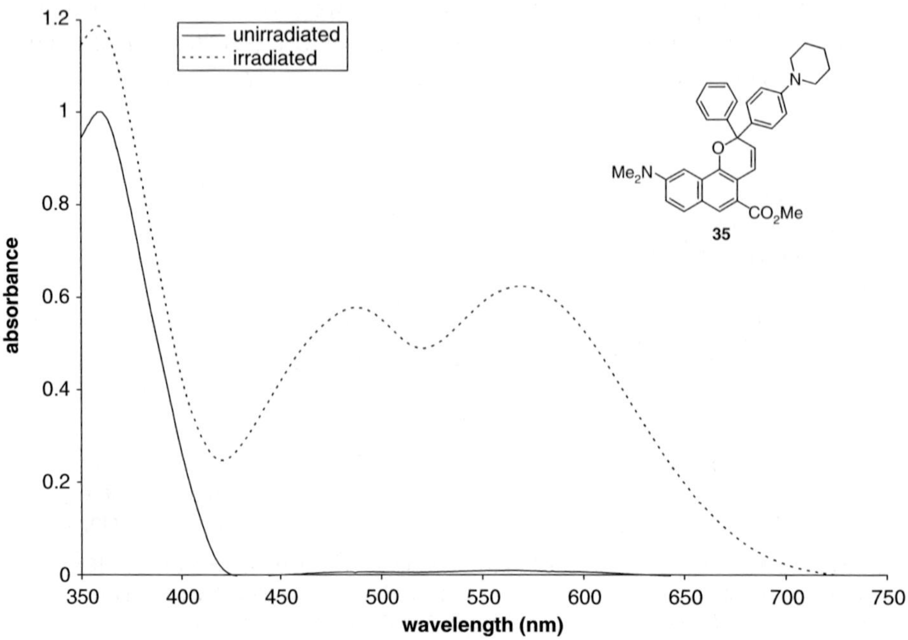

Fig. 8. UV–visible spectra of neutral colouring naphthopyran **35**.

4.3. Carbocyclic-fused naphthopyrans

Simple phenanthropyrans have been obtained from substituted phenanthrols accessed by the application of the Stobbe condensation sequence (Scheme 10) to formyl- or acetyl- naphthalenes. These compounds, e.g. **37** and **38**, display two absorption bands [139]. Several cycloalka-fused derivatives of phenanthropyrans, e.g. **39**, have been obtained by propargylation of phenanthrols derived from naphthylcycloalkanones. The open forms are fast fading irrespective of the mode of fusion and colours range from orange to violet [140].

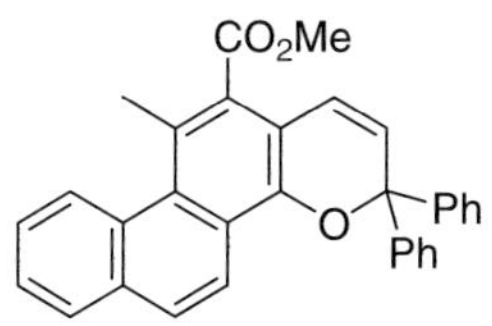

37
λ_{max} 431, 506 nm
(diethyleneglycol dimethyl ether)

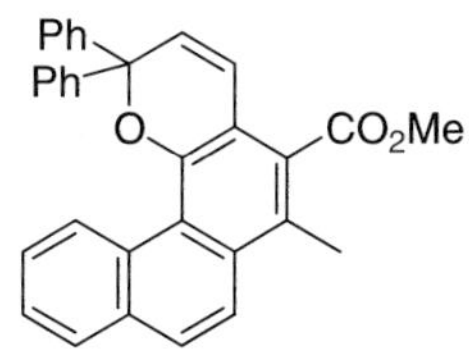

38
λ_{max} 448, 490 nm
(diethyleneglycol dimethyl ether)

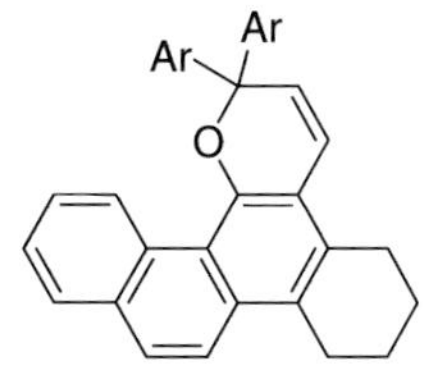

39
Ar = 4-MeOC$_6$H$_4$
λ_{max} 513 nm (THF)

Fusion of alicyclic rings across the 5,6-bond aims to replicate the accelerating effect of individual substituents at these positions. Much work has been reported on 5,6-indeno-fused naphtho[1,2-*b*]pyrans. The synthesis of 4-hydroxy-1-phenylnaphthalene-2-carboxylic acid utilising the Stobbe reaction provides access to the 2-hydroxyfluorenone derivative **40** and thence to the indenonaphthopyran (Scheme 17) [141]. A range of hydroxy-substituted benzo[*c*]fluoren-7-ones has been obtained from dimethoxynaphthaldehydes in which a nucleophilic aromatic substitution is promoted by an adjacent 4,5-dihydrooxazole group (Scheme 18) [142].

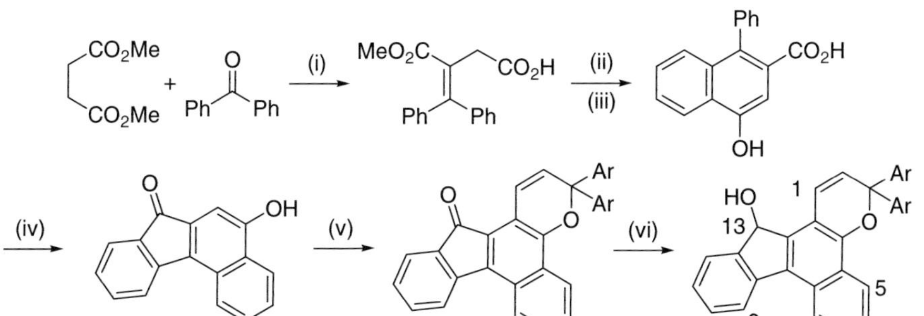

λ_{max} 436, 562 nm, half-life 207 s
(diethyleneglycol dimethyl ether)

Reagents: (i) KO*t*-Bu, PhMe, reflux; (ii) Ac$_2$O, NaOAc, PhMe, reflux; (iii) c. HCl, MeOH, reflux then aq. NaOH, MeOH, reflux; (iv) 85% H$_3$PO$_4$, 200 °C; (v) 1,1-bis(4-methoxyphenyl)-prop-2-yn-1-ol, dodecylbenzene sulfonic acid, PhMe, 50 °C; (vi) LiAlH$_4$, Et$_2$O.

Scheme 17.

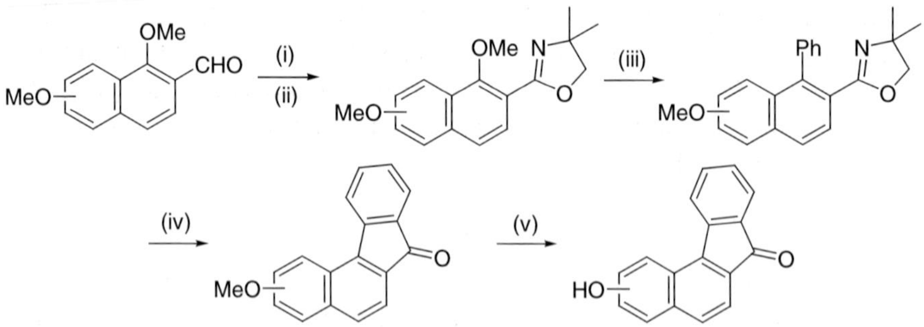

Reagents: (i) $NaOCl_2$, resorcinol, AcOH, dioxane; (ii) (a) $(COCl)_2$, (b) $H_2NC(Me)_2CH_2OH$, (c) $SOCl_2$; (iii) PhMgBr, Et_2O; (iv) (a) MeI, (b) aq. NaOH, (c) H_2SO_4; (v) HBr, AcOH.

Scheme 18.

Compared with the corresponding 5-ethoxycarbonyl-6-phenylnaphtho[1,2-*b*]pyran, the photochromes derived from the hydroxyfluorenones are of improved, though still only of moderate colourability and are slower in fading. The two absorption bands are red shifted to 435–500 nm and 560–634 nm, a feature attributed to the coplanarity of one of the 2-phenyl groups with the naphthopyran fragment enforced by the indene unit [143–148]. Faster fading examples have been obtained by variation of the substituents at C-13 [149, 150]. Several photochromes with a spiro-linked function at C-13 have been described. In some instances there is little difference in λ_{max} of the spiro compound and the corresponding 13-hydroxy derivative [151–153]. Reaction of the 13-ethynyl-13-hydroxyindenonaphthopyran derivative with various naphthols gave photochromes e.g. **41** with a naphthopyran unit spiro-linked at C-13. These show two absorption bands, appearing grey or brown, and have half-lives of the order of 2 minutes [154]. When the 2,2-diphenyl rings are linked by an S bridge, the resulting spiro{indeno[2,1-*f*]naphtho[1,2-*b*]pyran-3,9′-thioxanthene} photochromes, e.g. **42**, absorb at *ca.* 555 nm, some 20 nm to the red of the non-bridged analogues but are weaker coloured and fade significantly faster [57]. The bathochromic shift and increased fading are typical features associated with linking the *gem* diphenyl rings with an S bridge [155].

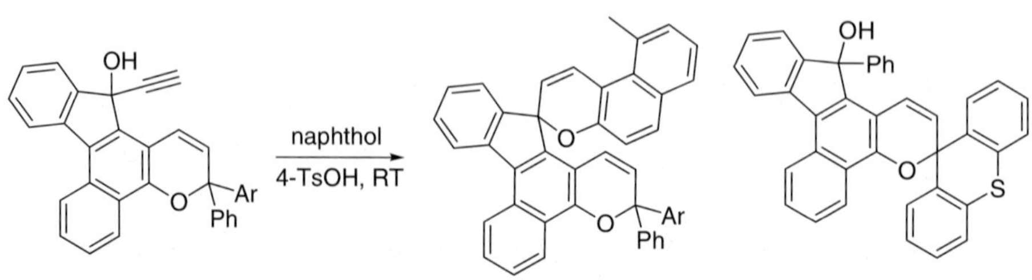

Ar = 4-$Me_2NC_6H_4$

λ_{max} 452, 600 nm, half-life 90 s (polymethacrylate)
41

λ_{max} 558 nm (PhMe)
42

Several examples of photochromes based on the indeno[3′,2′:3,4]naphtho[1,2-*b*]pyran system have been synthesised. They exhibit absorption at 416–480 and 516–560 nm with $t_{½}$ *ca.* 1 min, but are not particularly strong colouring [156]. The open isomer of the fluoranthrapyran **43** absorbs at 532 nm and has a half-life of 44 s [157].

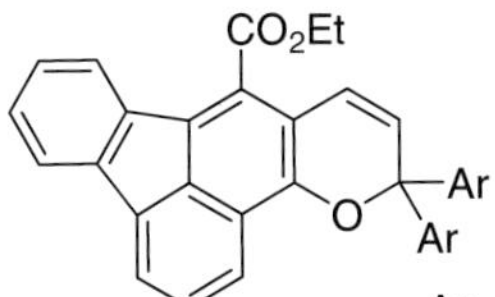

λ_{max} 532 nm
(polymethacrylate)
43

Ar = 4-MeOC$_6$H$_4$

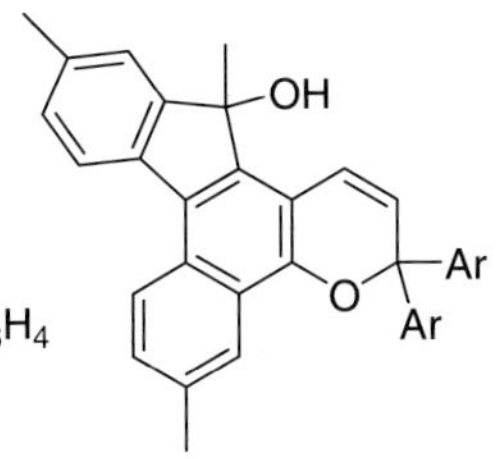

44

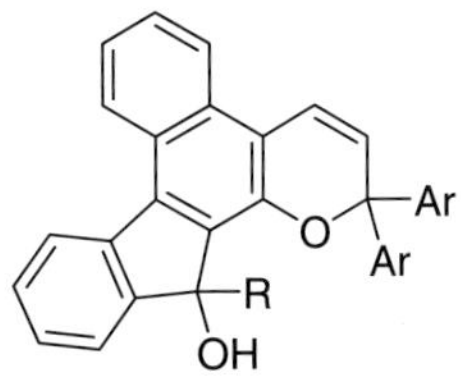

45

The indenonaphthopyran **44** exhibits solvatochromism, the major absorption band shifting to the red as the solvent polarity is increased [105] and, perhaps more importantly, is thermochromic. An open TC species is in thermal equilibrium with the closed form and this is converted into the TT isomer on irradiation with visible light. UV-irradiation of the closed form generates the TT isomer [158].

Derivatives of indeno[1′,2′:4,3]naphtho[2,1-*b*]pyran **45** exhibit a single absorption band at *ca.* 450 nm, have $t_{½}$ of *ca.* 45 s and show moderate initial optical density [159].

1,2,3,4-Tetrahydrophenanthro-9-ol, synthesised from ethyl (2-oxocyclohexyl)ethanoate in three steps, gave the partially reduced phenanthro[9,10-*b*]pyran on reaction with the appropriate propynol (Scheme 19). In an acrylate polymer, the open form absorbs at 504 nm (490 nm in THF), but is a weak colouring and fairly slow to fade [160]. Fusion of a benzenoid ring onto the cyclohexane moiety results in red shifts of λ_{max} and fast fading photochromes in colours ranging from orange to blue can be achieved by varying the substituents [140]. When a methoxy group is present at C-9, e.g. **46**, or a furan ring is attached across the 8,9-bond, the compounds show two absorption bands (429–482 and 548–584 nm) [161, 162]. Other examples incorporate spiro-linked substituents on the cyclohexane ring, e.g. **47** [163].

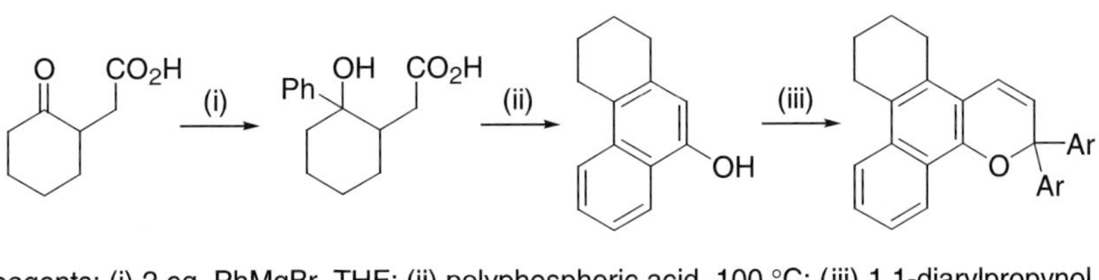

Reagents: (i) 2 eq. PhMgBr, THF; (ii) polyphosphoric acid, 100 °C; (iii) 1,1-diarylpropynol, 4-TsOH, CHCl$_3$

Scheme 19.

Annulation of dihydronaphthalene also produces fast fading photochromes, but these absorb at lower wavelengths (λ_{max} 428–458 nm) [164]. The cyclohepta analogue is shifted 19 nm to the blue (λ_{max} 517 nm; $t_{½}$ 12 s) relative to the dihydronaphthalene [165]. Fusion of a norbornane unit enhances the red shift observed for the cycloalka-fused molecules and in the case of phenanthropyrans, e.g. **48**, absorption is at *ca*. 600 nm [166].

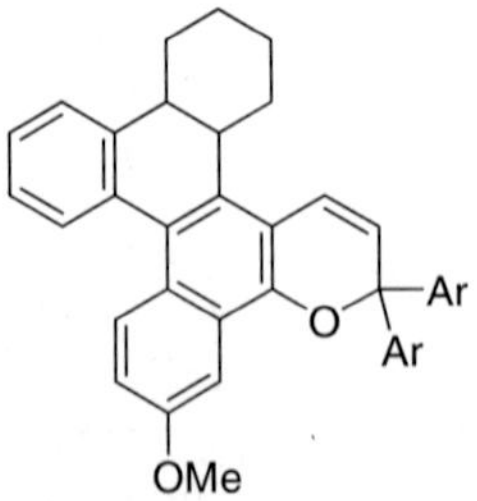

46

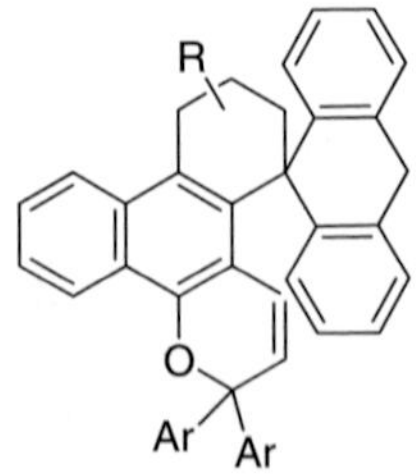

47

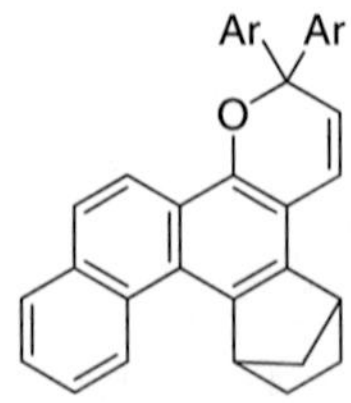

48

4.4. Heterocyclic-fused naphthopyrans

The synthesis of this group of photochromes is generally based on the propargylation of hydroxy-substituted heterocycles and consequently the more interesting chemistry lies in the synthesis of these precursors. The other main approach is by manipulation of preformed naphthopyrans.

Dihydrofuro[2,3-*b*]naphthols are accessible from 3,7-dihydroxy-2-naphthoic acids and are sources of hetero-fused naphtho[2,1-*b*]pyrans through reaction with propynols. The oxacyclic substituent is similar to an alkoxy group and the *j*-fused dihydrofuran derivative **49** absorbs at 481 nm, a red shift of 9 nm relative to the 8-methoxy analogue **50** [167].

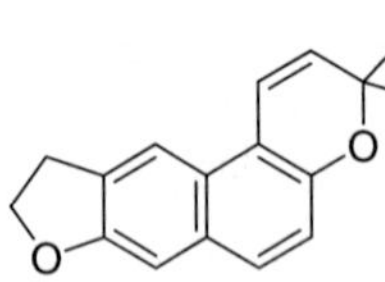

49

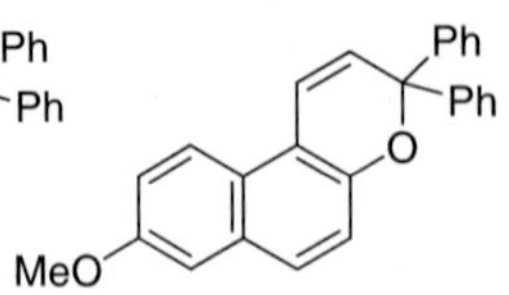

50

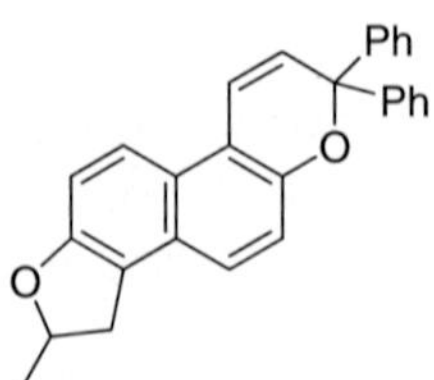

51

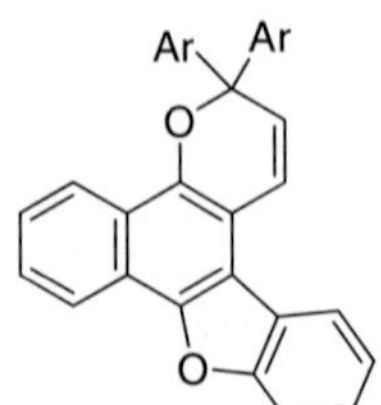

Ar = 4-MeOC$_6$H$_4$

52

Construction of a side-chain onto 6-bromo-2-naphthol allows the formation of naphtho[2,1-*b*]furan-6-ols and hence furo[3,2-*i*]naphtho[2,1-*b*]pyrans **51**. These intense colouring molecules absorb to the red of 8-methoxynaphthopyran in both the closed and open forms and have half-lives of the order of 2 minutes [168].

Red shifts are observed relative to the 5,6-dimethylnaphthopyran **30** R^1 = R^2 = Me (Table 6) together with slightly slower bleaching when a benzofuran ring is fused across the 5,6-bond of naphtho[1,2-*b*]pyrans, e.g. **52**. The compounds are derived from the reaction of naphthoquinone with a methoxyphenol and subsequent propargylation of the resulting naphtho[1,2-*b*][1]benzofuran [169]. In a similar manner, reaction of naphthoquinone with naphthols affords dinaphthofurans from which two differently fused naphthofuronaphtho[1,2-*b*]pyrans have been obtained (Scheme 20), which have λ_{max} at 512 and 583 nm and with half-lives of 34 and 125 s, respectively, but are of only moderate intensity [170].

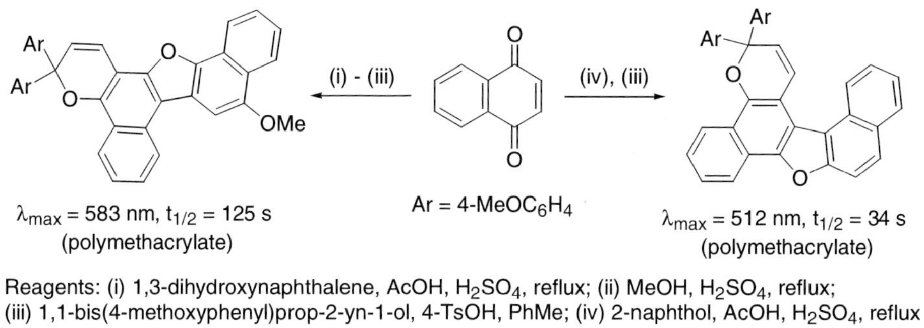

Scheme 20.

Application of the Stobbe reaction to 2-benzoyldibenzofuran leads to two substituted 1-naphthols **53** and **54**, which after cyclisation are substrates for pyran formation by reaction with 1,1-bis(4-methoxyphenyl)prop-2-yn-1-ol (Scheme 21). The resulting heptacyclic photochromes absorb in the range 570–600 nm with half-lives of 20–70 s [171].

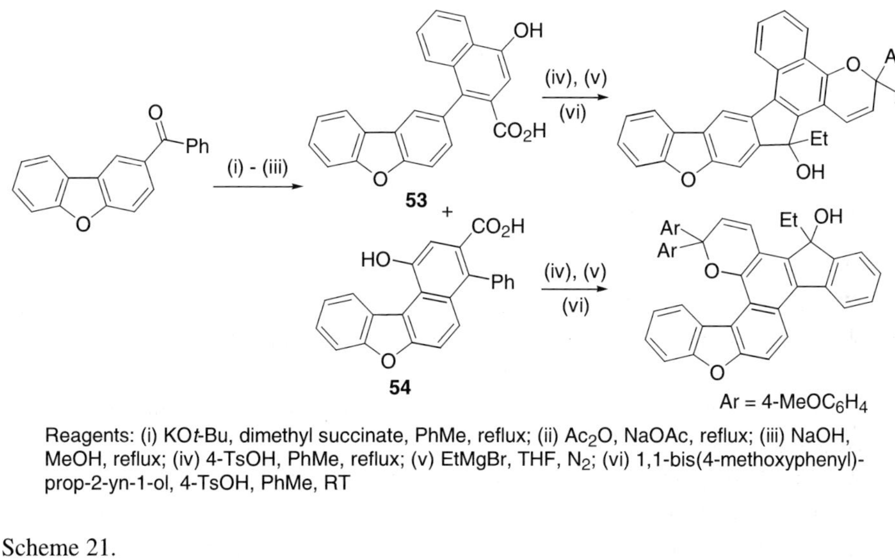

Scheme 21.

Hydrolysis of the ester functions of dimethyl 2,2-bis(4-methoxyphenyl)-2*H*-naphtho[1,2-*b*]pyran-5,6-dicarboxylate and cyclisation of the resulting dicarboxylic acid yields the cyclic anhydride **55** Ar = 4-MeOC_6H_4. Reduction affords a mixture of two isomeric furano-fused naphthopyrans. Treatment of the anhydride with primary amines provides a route to the corresponding pyrrole derivatives. Both types of hetero-fused naphthopyrans show a red shift and a shorter half-life relative to the starting naphthopyran diester [172].

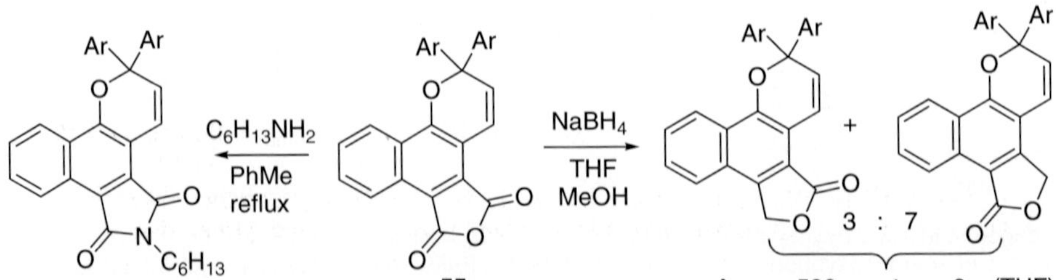

55

λ_{max} = 500 nm, $t_{1/2}$ = 9 s (THF)

A 7-methylene-5-oxofuro[3,4-*f*]naphtho[1,2-*b*]pyran **56** has been obtained *via* reaction of a 6-methoxy-5-methoxycarbonylnaphthopyran with a vinyl Grignard derivative and subsequent Pd-catalysed cyclisation [173].

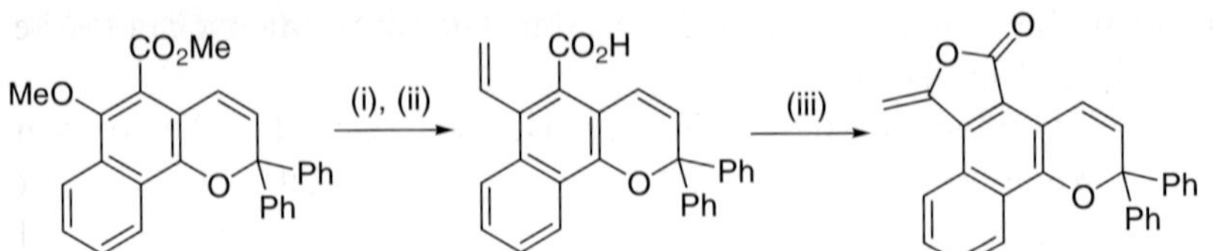

56

λ_{max} = 478 nm, $t_{1/2}$ = 300 s (polymethacrylate)

Reagents: (i) CH_2=CHMgBr, THF, RT; (ii) KOH, EtOH, reflux; (iii) Pd(OAc)$_2$, NaOAc, DMSO, RT

More complex structures can be derived from 1-tetralone through its conversion into (tetrahydro-1-oxo-2-naphthyl)ethanoic acid and subsequent reaction with a heteroaryllithium. Sequential cyclisation to the dibenzofuran or thiophene and propargylation affords fast fading 3,4-dihydronaphtho[2,1-*f*][1]benzofuro[2,3-*h*]naphtho[1,2-*b*]pyrans and thiophene analogues **57** [174].

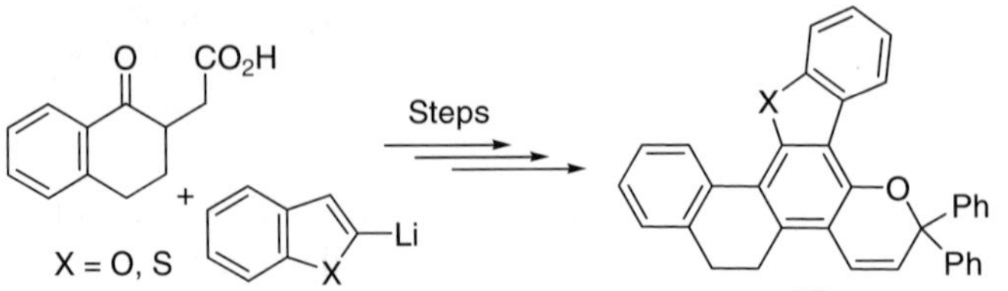

57

λ_{max} = 424, 579 nm
$t_{1/2}$ = 15 s (X = O)
λ_{max} = 445, 584 nm
$t_{1/2}$ = 16 s (X = S)
(polymethacrylate)

The fast fade rate shown by 2-phenyl-2-(4-trifluoromethylphenyl)-5-trifluoromethyl[1]benzofuran[2,3-*i*]naphtho[1,2-*b*]pyran is attributable to the bulky 5-substituent rather than to the fused benzofuran ring [175]. Its synthesis follows from the preparation of 9-hydroxy-7-trifluoromethylbenzo[*b*]naphtho[*d*]furan from 4-chloromethyldibenzofuran (Scheme 22).

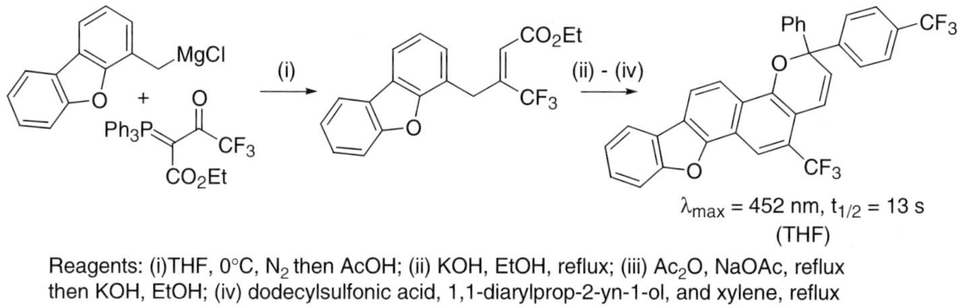

Scheme 22.

A variety of substituted dihydrofuro[2,3-*b*]naphth-1-ols have been derived from 2,3-dihydrobenzofuran and converted into furo[3,2-*j*]naphtho[1,2-*b*]pyrans, e.g. **58**, the open forms of which absorb at 420–440 and 530–540 nm. The former band is the stronger and so these intensely colouring molecules become brown on irradiation [176].

2,3-Dihydrobenzofuran is also the starting point for the synthesis of indeno[3,2-*a*]naphtho[2,3-*b*]furan-12-ols from which dihydrofuro[2,3-*b*]indeno[3,2-*f*]naphtho[1,2-*b*]pyrans **59** have been obtained by propargylation. The consequence of incorporating the O atom in a ring rather than in a methoxy group is a small red shift of both absorption bands [177].

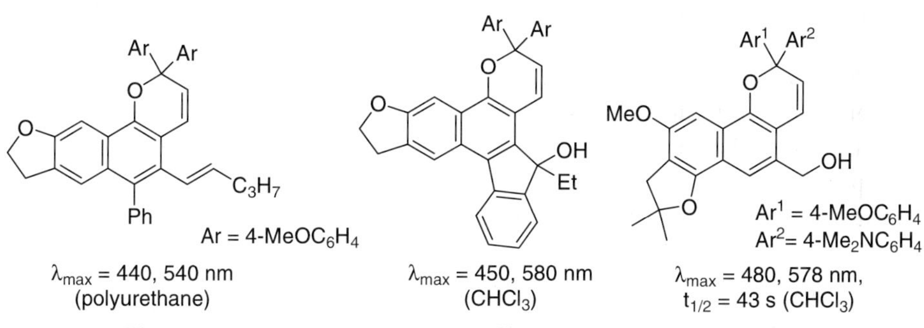

58 **59** **60**

A number of naphtho[1,2-*b*]furan-6-ols have been synthesised using the Stobbe condensation and converted into the furo[2,3-*i*]naphtho[1,2-*b*]pyran system

60. Provided that an amino function is present in one of the 2,2-diaryl rings, irradiation generates blue merocyanines, λ_{max} 480–490 and 575–590 nm, which are strongly coloured and have half-lives of 30–60 s [178]. The 8,9-methylenedioxy- and 8,9-ethylenedioxy- naphtho[1,2-*b*]pyrans, obtained using the same methodology, exhibit similar photochromic properties.

The reaction of 3-aminoprop-2-enoates with 1,4-naphthoquinone affords 5-hydroxybenzo[*g*]indoles and hence pyrano[3,2-*e*]benzo[*g*]indoles that become strongly red coloured on irradiation; the half-lives are similar to the analogous 5,6-dimethylnaphtho[1,2-*b*]pyran **30** $R^1 = R^2 =$ Me (Scheme 23) [179].

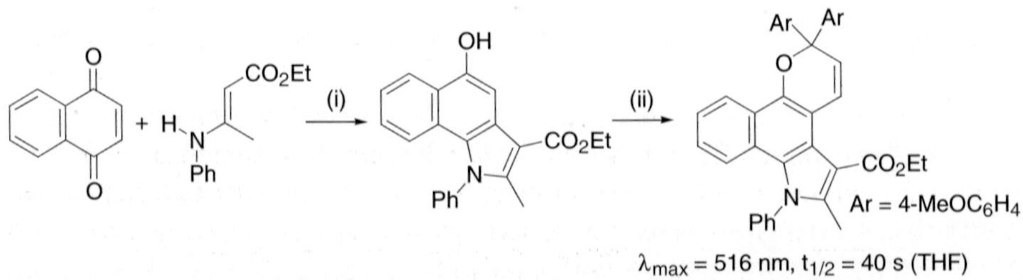

Reagents: (i) $MeNO_2$, 40°C (56%); (ii) 1,1-bis(4-methoxyphenyl)prop-2-yn-1-ol, cat. $BrCH_2CO_2H$, xylene, reflux(70%)

Scheme 23.

The synthesis of a 2-hydroxybenzo[*c*]carbazole, involving a Curtius reaction and carbazole formation by photolytic decomposition of an azide (Scheme 24), provides a route to an *f*-fused indole derivative of naphtho[1,2-*b*]pyran, which absorbs further to the red than both the **30** $R^1 = R^2 =$ Me and the above pyrrole-fused naphthopyrans [180]. Absorption bands are shifted to the red when amino substituents are introduced into the diaryl unit of this and the isomeric *f*-fused indole derivatives; the half-lives are between 2 and 3 minutes [181, 182]. Hydroxybenzo[*a*]carbazoles have been obtained from methoxytetralones and converted into benzopyrano[*a*]carbazoles by reaction with $Ti(OEt)_4$ and an α,β-unsaturated aldehyde. Dual absorption (414 and 517 nm) is observed when the indole unit is fused to the *i*-face of naphtho[1,2-*b*]pyran. Only a single band is shown by the naphtho[2,1-*b*]pyrans with indole fused onto the *i*-face (λ_{max} 485 nm) and *k*-face (λ_{max} 467 nm). The compounds are stronger colouring than the 2*H*-[1]benzopyran analogues and exhibit slower fading [183].

6-Bromo-2-naphthol is a source of 8-hydroxynaphtho[2,1-*b*]pyrans from which pyrano[3,2-*i*]naphtho[2,1-*b*]pyrans are formed on reaction with a 1,1-diarylprop-2-yn-1-ol. These molecules absorb at *ca*. 390 nm in the UV and at *ca*. 490 nm following irradiation [168].

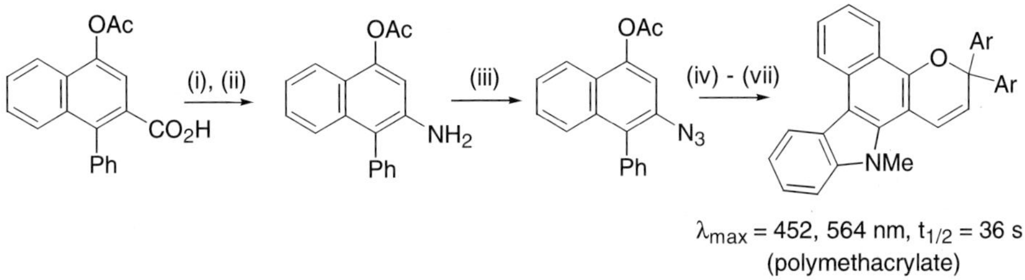

Reagents: (i) $(PhO)_2P(O)N_3$, Et_3N, PhMe, then *t*-BuOH, reflux; (ii) TFA, CH_2Cl_2, PhMe; (iii) $NaNO_2$, HCl, Me_2CO then NaN_3; (iv) irradiation (254 and 365 nm), THF, 4 days; (v) NaOH, THF; (vi) 1,1-bis(4-methoxyphenyl)prop-2-yn-1-ol, $BrCH_2CO_2H$, PhMe, reflux; (vii) NaH, MeI, THF

Scheme 24.

The 4*H*-naphtho[2,1-*c*]pyran-4-one **61** obtained by reaction of methyl 1-hydroxy-3-naphthoate with a propynol readily undergoes a second propargylation to give, after further manipulation, pyranonaphtho[1,2-*b*]pyrans which absorb in the range 521–592 nm with moderate bleaching kinetics [184]. Various benzopyran-fused derivatives and their [2,1-*b*] analogues have been obtained from hydroxy-substituted dibenzo-fused benzopyranones and these generate colours from yellow through to blue on irradiation, with half-lives of a few seconds to several minutes [185].

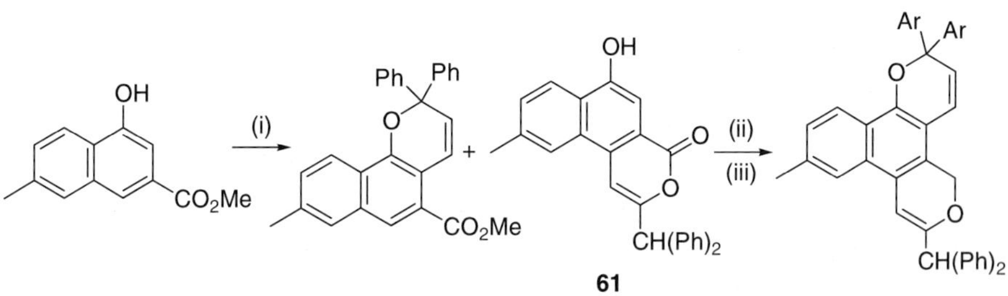

61

Reagents: (i) H_3PO_4, H_3PO_2, PhMe, heptanes, CH_2Cl_2, 1,1-diphenylprop-2-yn-1-ol, reflux; (ii) 1,1-diphenylprop-2-yn-1-ol, dodecybenzenesulfonic acid, PhMe, 50 °C; (iii) $LiAlH_4$, THF

Further structural variation follows from the construction of benzo[*k,l*]xanthen-3-ol, from 9*H*-9-acetylxanthene; propargylation affords the relatively rapid fading, violet-colouring [1]benzopyrano[6,7,8-*k,l*]xanthenes (Scheme 25) [186].

Methyl 10-hydroxybenzo[*b*]naphtho[2,3-*e*][1,4]dioxine-8-carboxylate, synthesised from the reaction of 1,2-dihydroxybenzene with 3,4-difluorobenzaldehyde and subsequent Stobbe condensation, is a source of [1,4]benzodioxinonaphtho[1,2-*b*]pyrans. Similarly, using 2-benzoyl[1,4]benzodioxine in the Stobbe reaction enables indeno analogues to be obtained (Scheme 26). With

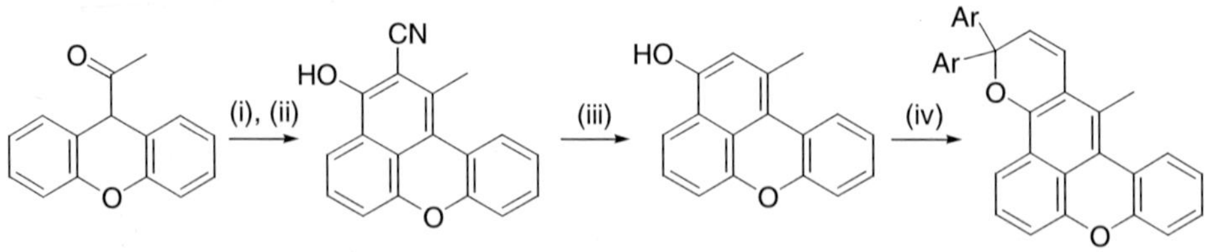

Reagents: (i) ethyl cyanoacetate, NH_4OAc, AcOH, PhMe, reflux; (ii) 200°C; (iii) NaOH, 200°C, 30bar; (iv) 1,1-bis(4-methoxyphenyl)prop-2-yn-1-ol, $BrCH_2CO_2H$, xylene, reflux.

Scheme 25.

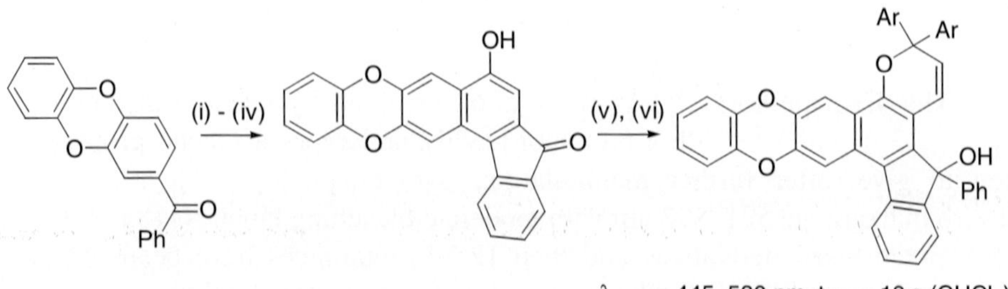

λ_{max} = 445, 580 nm, $t_{1/2}$ = 19 s ($CHCl_3$)

Reagents: (i) KO*t*-Bu, dimethyl succinate, PhMe, reflux; (ii) Ac_2O, KOAc, reflux; (iii) aq. NaOH, MeOH, reflux; (iv) 4-TsOH, PhMe, reflux; (v) 1,1-bis(4-methoxyphenyl)-prop-2-yn-1-ol, 4-TsOH, PhMe, reflux; (vi) PhMgBr, THF

Scheme 26.

appropriate *gem* diaryl substitution in the pyran ring, these molecules show two absorption peaks between 440 and 610 nm and are fast fading [187].

The presence of a 5-ester function combined with a 6-hydroxy or 6-methoxy group in the naphtho[1,2-*b*]pyran system allows construction of *f*-fused heterocyclic derivatives (Scheme 27). Thus reaction of **62** (R^1 = Ph, R^2 = H) with an aldehyde in the presence of a base leads to the dioxinone **63** [188] and with benzamidine a 1,3-oxazin-4-one **64** is produced [189]. The structurally related oxazin-2,4-dione **65** results from the reaction of **62** (R^1 = Me, R^2 = H) with an isocyanate, while **62** (R^1 = R^2 = Me) affords the pyrimidinone **66** with an imino Grignard reagent. The analogous pyranone-fused product **67** is obtained from reaction with a vinylic Grignard reagent and cyclisation of the enoate with Me_3SiCl [188]. For a series of 2,2-diphenyl derivatives, the fused pyrimidine absorbs at the highest wavelength (512 nm) with the other heterocyclic analogues absorbing in the range 460–478 nm.

5-Methoxy-6-methoxycarbonylnaphthopyrans react with a THP-protected Grignard reagent to give benzopyranone-fused naphtho[1,2-*b*]pyrans **68**

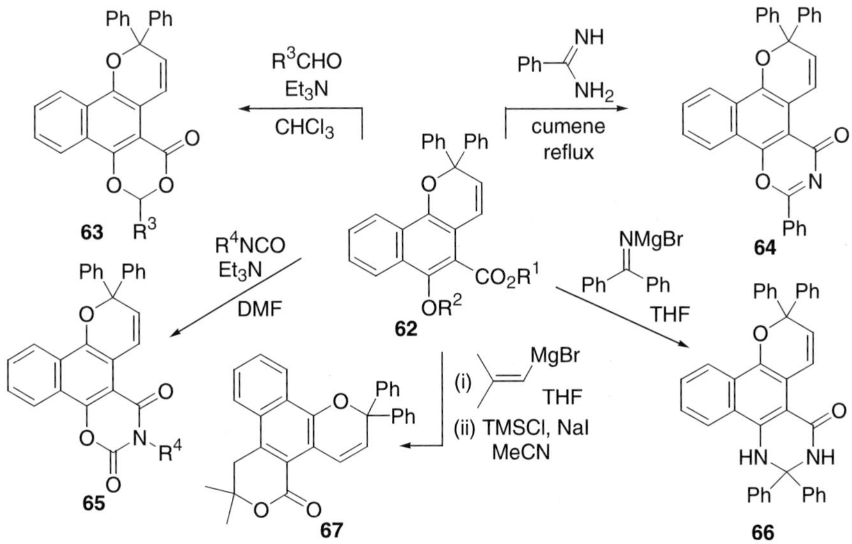

Scheme 27.

(Scheme 28). In a related manner, both 7-methoxy-8-methoxycarbonyl- and 8-methoxy-9-methoxycarbonyl- naphtho[2,1-*b*]pyrans yield benzopyranone-fused naphtho[2,1-*b*]pyrans. The naphthol derived from the reductive cyclisation of 2-(2-methoxycarbonylphenyl)-1,4naphthoquinone is a precursor of other benzopyran-fused naphthopyrans [190].

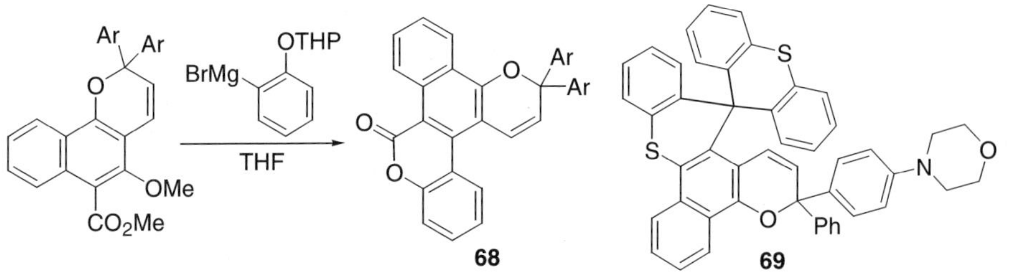

Scheme 28.

Complex spiro hetero-*f*-fused naphtho[1,2-*b*]pyrans, e.g. **69**, show two absorption bands (444–474 and 568–582 nm) and have half-lives of 2–3 minutes [191, 192]. The synthesis of 4-acetoxy-1-phenyl-2-naphthylamine from 4-hydroxy-1-phenylnaphthalene-3-carboxylic acid allows annulation of an isoquinoline unit onto

1-naphthol and subsequent reaction with a propynol yields the fused pyranophenanthridine **70**, λ_{max} 550 nm, $t_{½}$ = 12 s (polymethacrylate) [193].

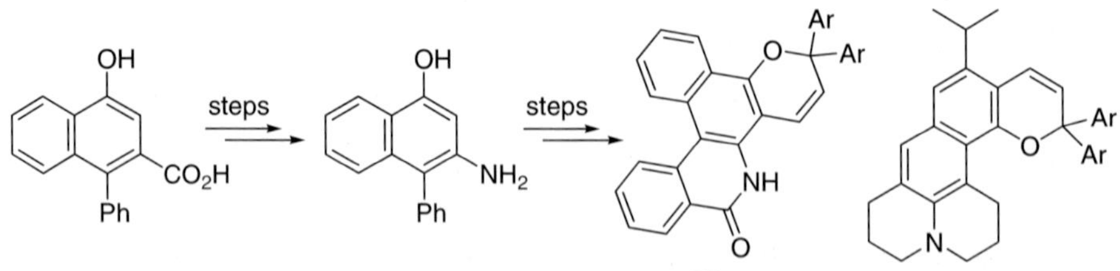

70 **71**

Quinolizine-fused naphthopyrans **71** absorb between 522 and 588 nm depending upon the diaryl substituents and have $t_{½}$ of 2–3 minutes [194]. Other saturated N-containing heterocycles fused across the *i* or *j* faces of the naphtho[2,1-*b*]pyran unit have been described [195].

4.5. Linked photochromic systems

Naphthodipyrans **72–74** have been synthesised by propargylation of 2,3-, 2,6- and 2,7- dihydroxynaphthalenes, respectively [196, 197]. All contain a naphtho[2,1-*b*]pyran unit and are characterised by single absorption bands, low colourability and moderate fade rates. By comparison, the dipyran **75** derived from 1,5-dihydroxynaphthalene, which is a [1,2-*b*] derivative, absorbs at 508 nm and is strongly colouring with a long half-life [196].

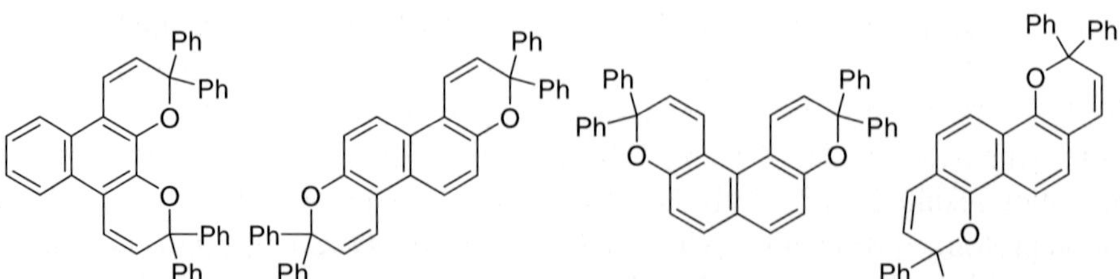

72	**73**	**74**	**75**
λ_{max} = 411 nm	λ_{max} = 481 nm	λ_{max} = 411 nm	λ_{max} = 508 nm
half-life = 608 s	half-life = 283 s	half-life = 155 s	half-life = >1800 s
(ethyl cellulose)	(ethyl cellulose)	(ethyl cellulose)	(ethyl cellulose)

3,3-Diphenyl-8-formyl-3*H*-naphtho[2,1-*b*]pyran, derived from the propargylation of 6-hydroxy-2-naphthaldehyde [79], is readily converted into 8-(1,4-dithiafulven-6-yl) substituted compounds, e.g. **76,** on Wittig–Horner olefination with the phosphonate anion derived from a 1,3-dithiole. Oxidation under a controlled potential resulted in the formation of an electroactive biphotochrome **77**. The monomers show significant red shifts of λ_{max} (*ca.* 100 nm) relative to the simple diphenylnaphthopyran but the dimers **77** are only weakly photochromic [198].

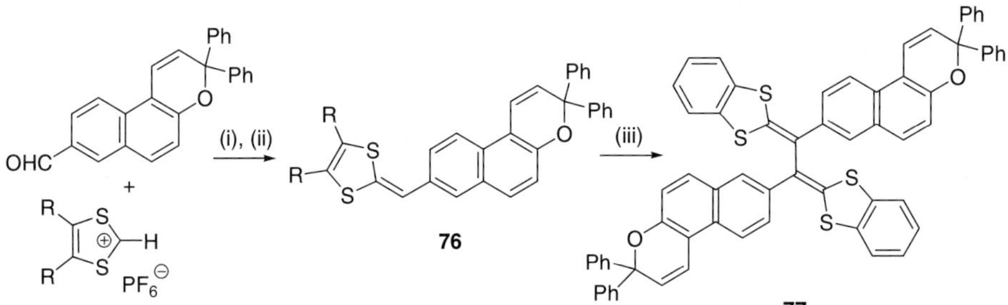

77

Reagents: (i) $P(OEt)_3$, NaI, MeCN, Ar; (ii) *n*-BuLi, THF, 0 °C; (ii) MeCN, $(n\text{-Bu})_4N^+PF_6^-$, controlled potential 0.80 V *vs.* Ag/AgCl

Naphtho[2,1-*b*]pyran moieties have been linked through ethano, hexano and methoxymethano units at the *para* positions of the 3-phenyl groups **78**, but no spectral data are given [199].

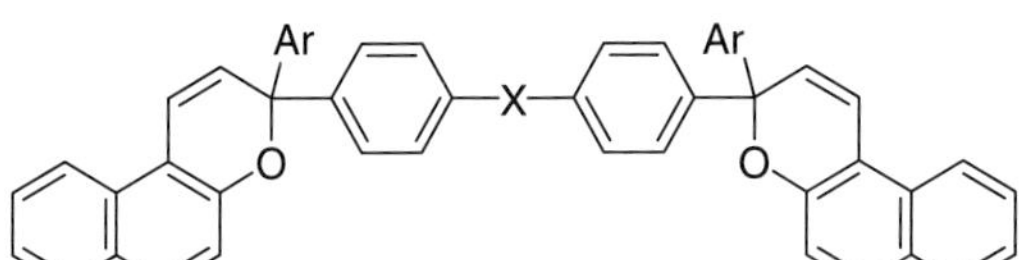

78

$X = (CH_2)_2$
$X = (CH_2)_6$
$X = CH_2OCH_2$

Two 3,3-diphenylnaphtho[2,1-*b*]pyran units have been linked at the 5-position by a *Z*-ethene function and by a flexible ethano group. In the latter case, the two naphthopyran moieties behave independently and show a broad absorption (380–550 nm) on irradiation with a maximum at 431 nm. Both thermal and photochemical fading is observed, indicating the formation of two photo products [200]. The behaviour of the ethene-bridged compound **79**, synthesised from 3-hydroxymethyl-2-naphthol using Wittig methodology [201], is more complex. Irradiation results initially in broad absorption centred around 500 nm, but this is lost on prolonged irradiation. The steady state absorbs at *ca.* 350 and 440 nm and consists of three different species. Of course, this compound is also a diarylethene and there appears to be a competition between the two photochromic systems; the initially formed species is considered to be the ring-closed dihydrophenanthrene. Further irradiation opens both naphthopyran rings and also oxidises the phenanthrene derivative. In the thermal fading process, one of the pyran ring closes and the second is then closed by visible light (Scheme 29). Similar behaviour is shown by a model compound in which 2-methoxynaphthalene is attached to the 5-position of 3,3-diphenylnaphtho[2,1-*b*]pyran through a *Z*-ethene group [202].

There are several examples of naphthopyrans linked in various ways to spirooxazine units, but these are beyond the scope of this chapter [200–203].

Stille coupling of stannylthiophenes with 3-(4-methoxyphenyl)-3-(5-bromothien-2-yl)naphtho[2,1-*b*]pyran has been used to form photochromic ter- and

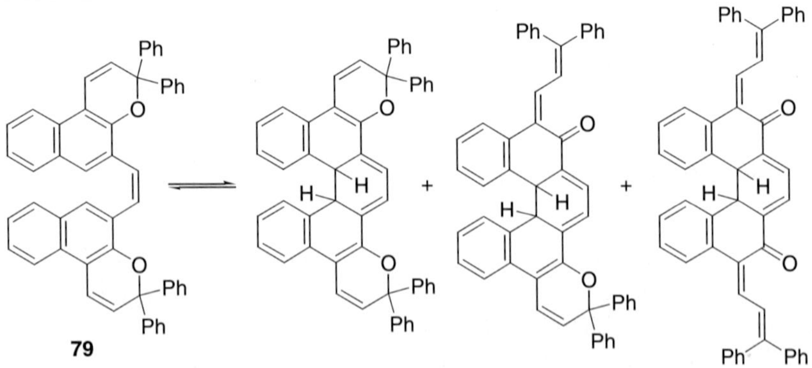

Scheme 29.

quater- thiophenes **80**, the open forms of which show enhanced conductivity [204–206]. However, when two naphthopyran units are connected at the 3-positions through a 5,5′-(2,2′-bithienyl) moiety, the pyran rings are opened sequentially on irradiation at 366 nm. The bis-thiophene unit acts as an electron donor on the unopened system and the initial ring opening generates an absorption band at 517 nm, which is 60 nm to the red of that shown by the monomer, 3-(2-fluorophenyl)-3-thienylnaphtho[2,1-*b*]pyran. On continued irradiation, this absorption dies away and is replaced by a new stronger band at 580 nm as a fully conjugated system is achieved (Scheme 30). Fading is the reverse process in which the first ring closure is faster than the second and a band at 520 nm appears before complete decoloration is observed [207].

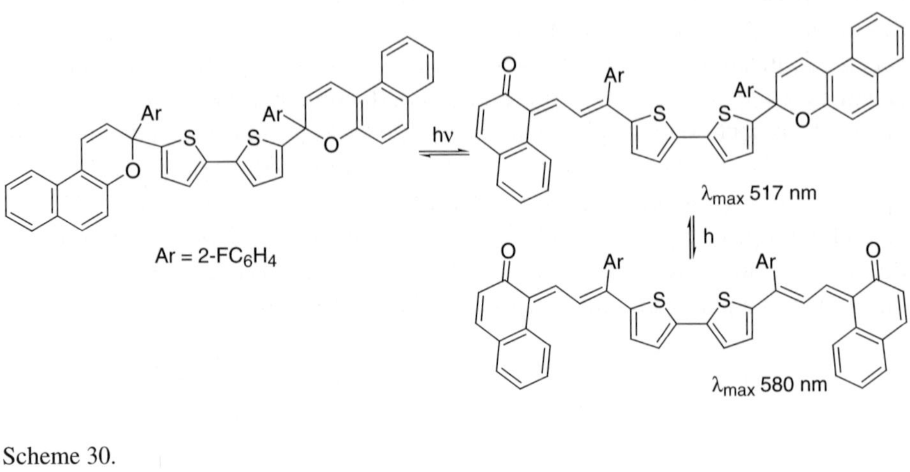

Scheme 30.

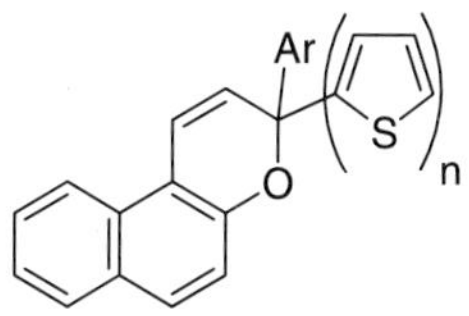

Ar = 4-MeOC$_6$H$_4$ n = 1 - 4

80

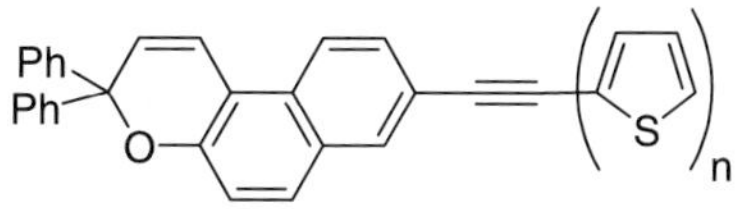

n = 1,2

81

In a different vein, application of Suzuki cross-coupling methodology to thiophene boronates and bromo- or triflate- functionalised naphthols or naphthopyrans enables (2-thienyl)$_n$ substituents to be introduced around the periphery of 3,3-diphenyl-3*H*-naphtho[2,1-*b*]pyrans (Scheme 31) [206, 208].

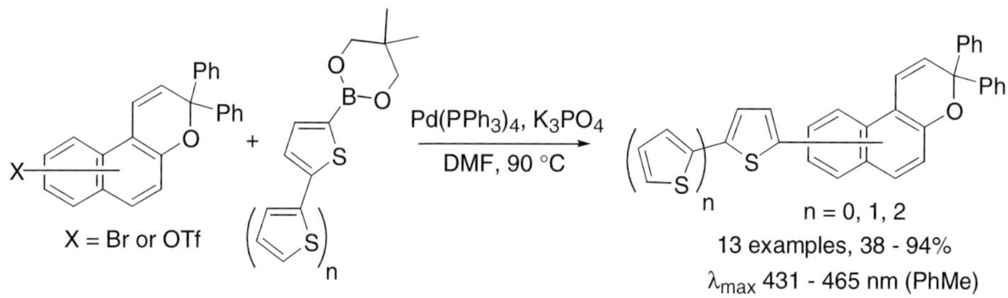

Scheme 31.

The synthesis of naphtho[2,1-*b*]pyrans linked to thiophene units through an alkyne function **81** has been achieved using Sonogashira coupling. These compounds have similar photochromic properties to the simple 3,3-diphenyl derivative apart from a red shift of 22 nm shown by the 8-ethynyl compound [200], which is further shifted by 17 nm and 31 nm in the corresponding (2-thienyl)ethynyl and the (2,2′-bithien)-5-ylethynyl derivatives, respectively. The terphenyl analogue shows very weak photochromism [209]. When two naphtho[2,1-*b*]pyran units are linked by an ethyne–(thiophene)$_n$–ethyne bridge, e.g. **82**, fluorescence and intersystem crossing are in competition with photochromism. The length of the oligomeric thiophene chain controls the photochromism. When n = 1, irradiation rapidly results in the opening of one pyran ring (λ_{max} 480 nm) and after prolonged irradiation at 228 K the second ring opens (λ_{max} 550 nm) [87]. For the bis-thienyl, both photochromism and fluorescence are shown, but when n = 3 the compound acts as a light converter rather than as a photochromic compound and is highly fluorescent [210].

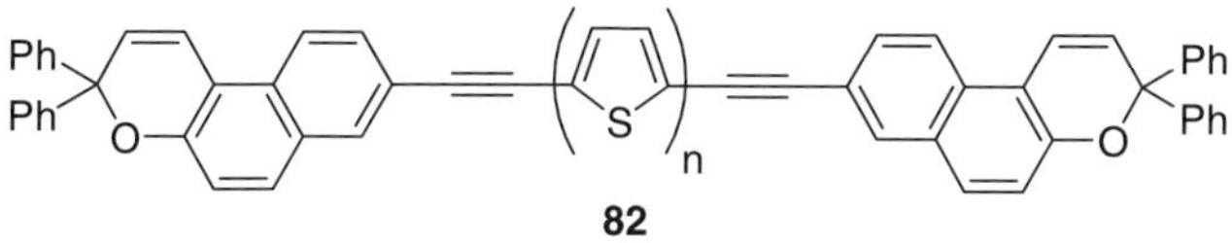

82

4.6. Supramolecular assemblies and transition metal complexes

Naphtho[2,1-*b*]pyrans have been combined with various ion-responsive molecules, using the reactive 5-bromomethyl group to join the two units. In the case of monoazacrown ethers, e.g. **83**, the addition of alkali metal ions does not generate the merocyanine to any significant extent unlike the behaviour shown by crowned spirobenzopyrans. However, such thermal isomerisation becomes appreciable with alkaline earth metal ions, particularly in the 15-crown-5 and 18-crown-6 systems. The effect of the metal ions on the photoisomerisation is variable, with Li^+ and Na^+ facilitating the process and bringing about red shifts. The Group II metal ions enhance these changes, which are associated with stabilisation through a strong interaction of the ion with the ring-opened isomer [211]. Both the closed and open isomers of the monoaza 15-crown-5 ether **84** form 1:1 complexes with Ca^{2+} in which the metal ion is bound to the macrocyclic portion. Complexation stabilises the open form but leads to a blue shift of *ca.* 100 nm confirming that the Ca^{2+} ion interacts with the crown fragment. Similar complexation with the simple 3-(4-dimethylaminophenyl)-3-phenylnaphthopyran induces a red shift of over 100 nm indicating interaction of the metal ion with the oxygen atom of the electron-accepting naphthalenone moiety [212].

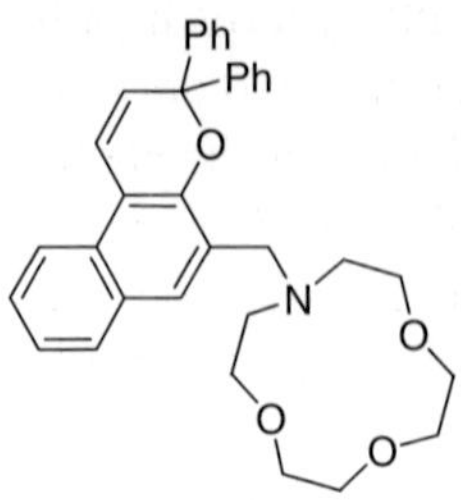

83

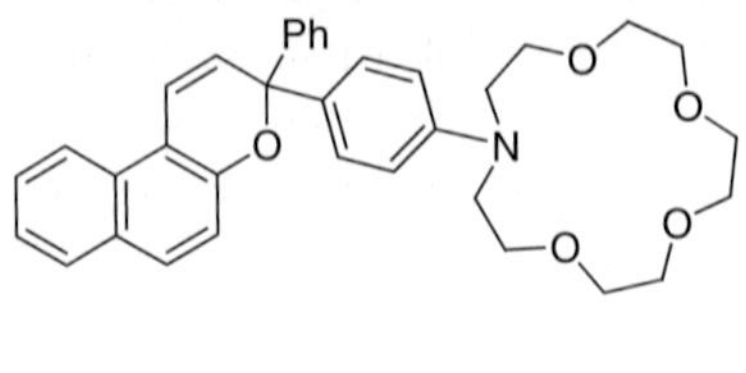

84

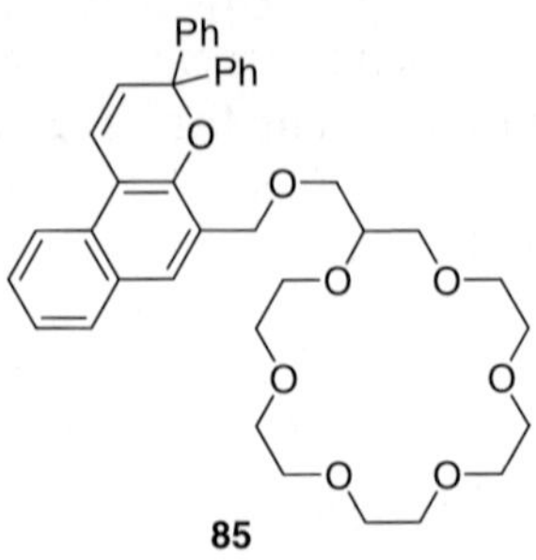

85

The oxymethylcrowned naphthopyrans, e.g. **85**, show similar behaviour towards metal ions as the azacrowned derivatives, with Li^+, Na^+ and K^+ showing the maximum influence with 12-crown-4, 15-crown-5 and 18-crown-6, respectively. The alkaline earth metal ions do not discriminate between the different crown ring sizes, although notably Mg^{2+} influences the photoisomerisation of only the 18-crown-6 derivative. This different behaviour indicates that the Group I ions interact predominantly with the crown ether moiety, whereas, ions from Group II interact with the carbonyl group of the open form of the naphthopyran. Of particular interest is the observation that the stoichiometry of the Na^+ complex with 12-crown-4 changed from 1:2 (Na^+:ligand) for the closed naphthopyran to 1:1 in the ring-opened isomer following irradiation with UV light [213].

3,3-Diphenylnaphtho[2,1-*b*]pyran units have also been attached to several *t*-butylcalix[4]arenes through a 5-oxymethyl group. The response of the closed isomers to metal ions is again negligible, but spectral changes were observed following irradiation. As the number of naphthopyran units bonded to the calixarene increased to a maximum of $n = 4$ in **86**, so the ion that most facilitated the photoisomerisation and stabilised the open form changed from Li^+ to K^+ and from Mg^{2+} to Ba^{2+}, a feature attributed to the lariat effect of the naphthopyran moieties. It is noteworthy that photoisomerisation was also facilitated by this increase even in the absence of metal ions [214].

The complex between Pb^{2+} and 15-crown-5 naphtho[2,1-*b*]pyran **87** (λ_{max} 476 nm) is dissociated on irradiation. The loss of cation binding ability is attributed to the withdrawal of electron density from the crown ether unit on opening the pyran ring. Subsequent ring closure by visible light allows recomplexation [215].

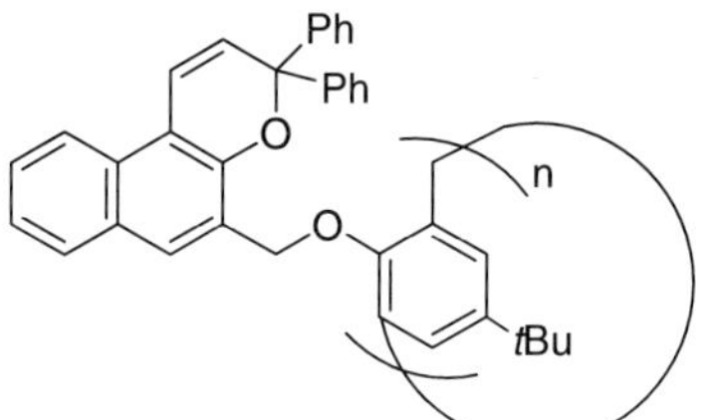

86

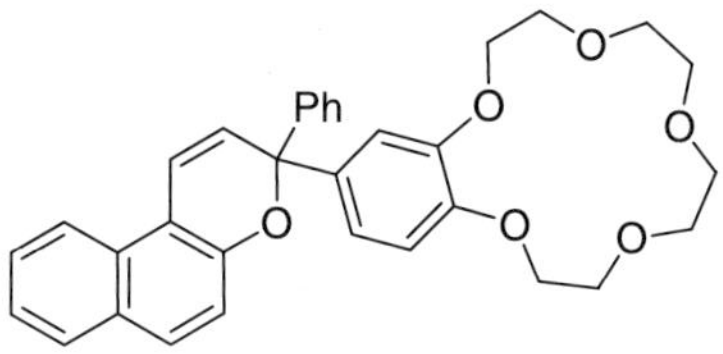

87

Replacement of one of the phenyl rings in diphenylnaphthopyrans by a ferrocenyl group has a marked effect on the photochromic properties. Notably, both the [1,2-*b*] and [2,1-*b*] compounds exhibit two absorption bands on irradiation, with the higher wavelength band occurring above 600 nm. The two bands are of similar intensity and consequently extended coverage of the visible region is seen. The compounds exhibit solvatochromism. The fade rate is increased in toluene and acetonitrile solution but ethanol appears to stabilise the merocyanines derived from the naphtho[1,2-*b*]pyran. Their synthesis utilises 1-ferrocenylpropynols (Scheme 32) [216]. The behaviour of some ferrocenyl-methyl naphthopyrans in lower alcohols has been interpreted as stabilisation of a *cis-*, *cis-*, *cis-* open isomer by chelation of the Fe to the quinonoid oxygen with further stabilising interactions involving the alcohol [217].

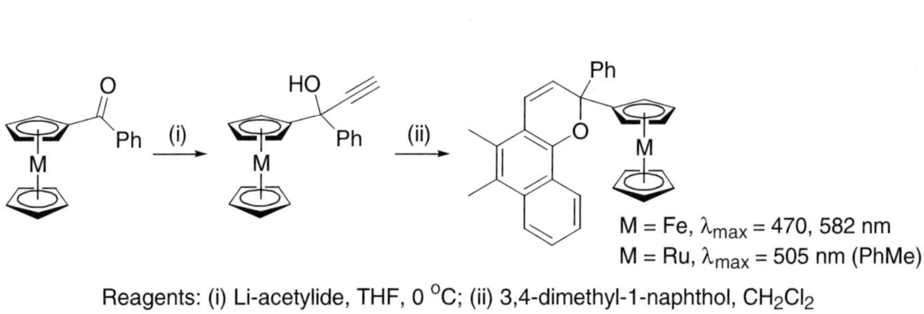

Scheme 32.

Using a similar synthetic approach, naphthopyrans with ruthenocyl and osmocenyl moieties adjacent to the hetero O atom have been prepared. Unlike the ferrocene derivative, these metallocenyl compounds show only a single absorption band at *ca.* 500 nm on irradiation. All exhibit a red shift and show enhanced fade rates [218]. Attempts to prepare η^6-Cr tricarbonyl complexes from 3,3-diphenyl-3*H*-naphtho[2,1-*b*]pyran were unsuccessful [219].

An *N*-acyl-11-aminoundecanoic acid group has been introduced at the 8-position of 3,3-diphenylnaphtho[2,1-*b*]pyran *via* the 8-hydroxymethyl derivative through reaction with isocyanates and using peptide coupling methods. Hot solutions of the Na salts of **88** in DMF or DMSO form stable gels on cooling and solutions of the related carboxylic acids gel on treatment with NaOH involving self aggregation of the naphthopyrans. Irradiation of the gels not only generates the expected colour of the ring-opened isomer but also breaks down the gel. Gelation can be reproduced by a heating–cooling cycle after cessation of irradiation [220].

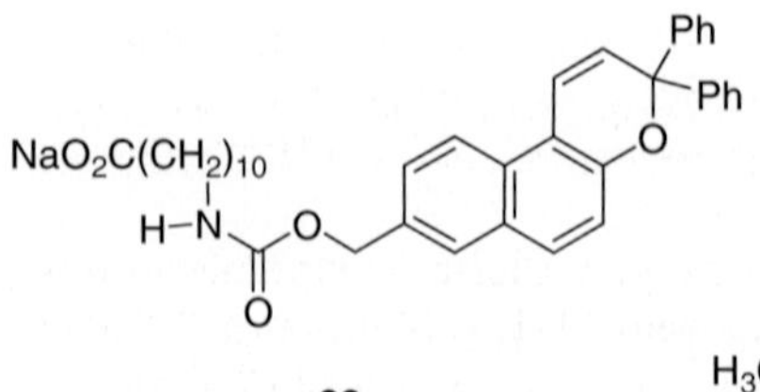

88

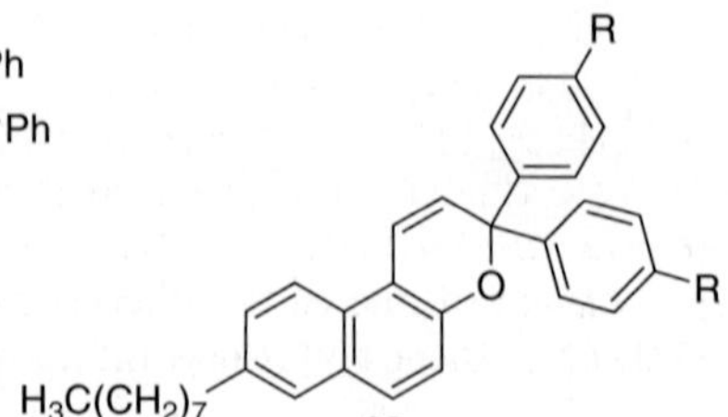

89

89, R =
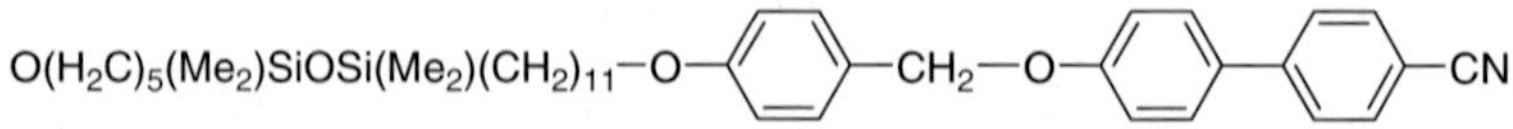

Alkyl and siloxane spacers separate the naphtho[2,1-*b*]pyran and cyanobiphenyl units of the first photochromic liquid crystals of this type, e.g. **89**. Although greater absorption is shown in the UV by these molecules than the precursor naphthopyrans, there is little difference in the absorption of the red ring-opened forms or in the biexponential fading kinetics. In the mesophase of the liquid crystal with two mesogens, λ_{max} of the open isomer is red shifted by 14 nm but the fade rates remain the same in the solid state [221].

5. CONCLUSION

Significant advances in our understanding of the photochromic properties of naphthopyrans have been made in the past decade. It is now possible to achieve almost any shade of any colour, such is the structural diversity of naphthopyrans now available. Furthermore, control of the fade rate, with half-lives of less than a second to many hours is possible, supplementing the fast and strong coloration and increasing the potential applications of these long-lived materials.

Their facile synthesis from readily available precursors has allowed the commercial development of naphthopyrans to proceed at a good pace through the efforts of a small number of research groups linked to international manufacturers.

The major end user is the ophthalmic industry. Brown or grey sun lenses can now be based on a single photochromic naphthopyran, whereas less than 20 years ago, multi-component lenses were the norm. Naphthopyrans are used in security inks and as identity markers in a variety of products. Future uses in cosmetic applications and as devices are under development.

REFERENCES

1. Cuong, N.K., Fournier, F. & Basselier, J.-J., *Bull. Soc. Chim. Fr.*, (1974) 2117.
2. Merlini, L., *Adv. Heterocyc. Chem.*, **18** (1975) 159.
3. Prokof'ev, E.P. & Krasnaya, Zh.A., *Izv. Akad. Nauk SSSR, Ser. Khim.*, (1980) 1011.
4. Kuthan, J., *Adv. Heterocyc. Chem.*, **34** (1983) 249.
5. Kuthan, J., Šebek, P. & Böhm, S., *Adv. Heterocyc. Chem.*, **62** (1995) 77.
6. Ishar, M.P.S., Singh, R., Kumar, K., Singh, G., Velmurugan, D., Pandi, A.S., Raj, S.S.S. & Fun, H.K., *J. Org. Chem.*, **67** (2002) 2234.
7. Ellis, G.P., Ed., *Chromenes, Chromanones and Chromones*, Wiley, New York (1977); Ellis, G.P. & Lockhart, I.M., Eds., *Chromans and Tocopherols*, Wiley, New York (1981).
8. Bowers, W.S., Ohta, T., Cleere, J.S. & Marsella, P.A., *Science*, **193** (1976) 542; Bowers, W.S., *Pontif. Acad. Sci. Scr. Varia.*, **41** (1977) 129.
9. Venkateswarlu, Y., Faulkner, D.J., Steiner, J.L.R., Corcoran, E. & Clardy, J., *J. Org. Chem.*, **56** (1991) 6271; Blunt, J.W., Copp, B.R., Munro, M.H.G., Northcote, P.T. & Prinsep, M.R., *Nat. Prod. Rep.*, **21** (2004) 1; Olsen, B.S. & Trauner, D., *Synlett*, (2005) 700.
10. Wipf, P. & Weiner, W.S., *J. Org. Chem.*, **64** (1999) 5321.
11. Venkateswarlu, Y., Faulkner, D.J., Steiner, J.L.R., Corcoran, E. & Clardy, J., *J. Org. Chem.*, **56** (1991) 6271; Hamann, M.T., Scheuer, P.J. & Kelly-Borges, M., *J. Org. Chem.*, **58** (1993) 6565.
12. Balaban, A.T., Dinculescu, A., Dorofeenko, G.N., Fischer, G.W., Koblik, A.V., Mezheritskii, V.V. & Schroth, W., *Adv. Heterocyc. Chem.*, **S2** (1982) 1.
13. Ellis, G.P., *Comp. Heterocyc. Chem., 1st edn.*, **3** (1984) 662.
14. Brown, G.H., Ed., *Photochromism, Techniques of Chemistry*, Wiley Interscience, New York, Vol. 3 (1971).
15. McArdle, C.B., Ed., *Applied Photochromic Polymer Systems*, Blackie, London (1992).
16. Crano, J.C. & Guglielmetti, R.J., Eds., *Organic Photochromic and Thermochromic Compounds, Volume 1: Main Photochromic Families*, Plenum Press, New York (1998).
17. *Organic Photochromic and Thermochromic Compounds, Volume 2: Physicochemical Studies, Biological Applications and Thermochromism*, Crano, J.C. & Guglielmetti, R.J., Eds., Plenum Press, New York (1998).
18. Irie, M., *Chem. Rev.*, **100** (2000) 1685.
19. Bouas-Laurent, H. & Dürr, H., *Pure Appl. Chem.*, **73** (2001) 639.
20. Kobatake, S. & Irie, M., *Ann. Rep. Prog. Chem. Sect. C*, **99** (2003) 277.
21. *Photochromism Molecules and Systems*, Durr, H. & Bouas-Laurent, H., Eds., Elsevier, Amsterdam (2003).
22. Becker, R.S. & Michl, J., *J. Am. Chem. Soc.*, **88** (1966), 5931; Becker, R.S., *US Patent*, US 3,567,605 (1971).
23. Van Gemert, B., *Organic Photochromic and Thermochromic Compounds, Volume 1: Main Photochromic Families*, Crano, J.C. & Guglielmetti, R.J., Eds., Plenum Press, New York (1998) p. 111.

24. Hepworth, J.D., Gabbutt, C.D. & Heron, B.M., *Colour Science '98; Volume 1: Dye and Pigment Chem.*, Griffiths, J., Ed., University of Leeds, UK (1999) p. 161.
25. Schweizer, E.E. & Meeder-Nycz, D., *Chem. Heterocyc. Compd.*, **3** (1977) 11.
26. Hepworth, J.D., *Comp. Heterocyc. Chem., 1st edn.*, **3** (1984) 757.
27. Hepworth, J.D., Gabbutt, C.D. & Heron, B.M., *Comp. Heterocyc. Chem., 2nd edn.*, **5** (1996) 351.
28. Livingstone, R., Miller, D. & Morris, S., *J. Chem. Soc.*, (1960) 5148.
29. Cottam, J. & Livingstone, R., *J. Chem. Soc.*, (1964) 5228.
30. Cotterill, W.D., Livingstone, R. & Walshaw, M.V., *J. Chem. Soc. C*, (1970) 1758.
31. Brogden, P.J. & Hepworth, J.D., *J. Chem. Soc., Perkin Trans. 1*, (1983) 827; Bethune, R.J., Gabbutt, C.D., Guinot, S.G.R., Hepworth, J.D. & Heron, B.M., *J. Chem. Soc., Perkin Trans. 1*, (1994) 1925; Tanaka, T., Imura, S., Tanaka, K. & Kida, Y., *US Patent*, US 5,106,998 (1992).
32. Kabbe, H.J., *Synthesis* 1978, 886; Kabbe, H.J. & Widdig, A., *Angew. Chem. Int. Ed. Engl.*, **21** (1982) 247.
33. Wawzonek, S., Nagler, R.C. & Carlson, L.J., *J. Am. Chem. Soc.*, **76** (1954) 1080.
34. Talley, J.J., *Synthesis*, (1983) 845.
35. Heller, H.G., Levell, J.R., Hibbs, D.E., Hughes, D.S. & Hursthouse, M.B., *Mol. Cryst. Liq. Cryst.*, **297** (1997) 123.
36. Uchida, M. & Irie, M., *Chem. Lett.*, (1991) 2159; Uchida, M. & Irie, M., *Chem. Lett.*, (1992) 2257; Uchida, M. & Irie, M., *J. Am. Chem. Soc.*, **115** (1993) 6442; Uchida, M., Kume, M. & Irie, M., *Bull. Chem. Soc. Jpn.*, **69** (1996) 1023.
37. Sartori, G., Casiraghi, G., Bolzoni, L. & Casnati, G., *J. Org. Chem.*, **44** (1979) 803.
38. Pozzo, J.-L., Lokshin, V. A. & Guglielmetti, R., *J. Chem. Soc., Perkin Trans. 1*, (1994) 2591.
39. Pozzo, J.-L., Samat, A., Guglielmetti, R., Dubest, R. & Aubard, J., *Helv. Chim. Acta*, **80** (1997) 725.
40. Koch-Pomeranz, U., Hansen, H.J. & Schmid, H., *Helv. Chim. Acta*, **56** (1973) 2981; Moghaddam, F.M., Sharifi, A. & Saidi, M.R., *J. Chem. Res., (S)*, (1996) 338.
41. Iwai, I. & Ide, J., *Chem. Pharm. Bull.* **10** (1962) 926; Iwai, I. & Ide, J., *Chem. Pharm. Bull.*, **11** (1963) 1042.
42. Godfrey, J.D., Jr., Mueller, R.H., Sedergran, T.C., Soundararajan, N. & Colandrea, V.J., *Tetrahedron Lett.*, **35** (1994) 6405.
43. Bigi, F., Carloni, S., Maggi, R., Muchetti, C. & Sartori, G., *J. Org. Chem.*, **62** (1997) 7024.
44. Van Gemert, B. & Bergomi, M.P., *US Patent*, US 5,066,818 (1991).
45. Gabbutt, C.D., Hepworth, J.D., Heron, B.M. & Partington, S.M., *Dyes and Pigments*, **47** (2000) 73.
46. Queiroz, M.-J.R.P., Dubest, R., Aubard, J., Faure, R. & Guglielmetti, R., *Dyes and Pigments*, **47** (2000) 219.
47. Queiroz, M.-J.R.P., Plasencia, P.M.S., Dubest, R., Aubard, J., Guglielmetti, R., *Tetrahedron*, **59** (2003) 2567.
48. Tanaka, K., Aoki, H., Hosomi, H. & Ohba, S., *Org. Lett.*, **2** (2000) 2133.
49. Gabbutt, C.D., Heron, B.M., Instone, A.C., Thomas, D.A., Partington, S.M., Hursthouse, M.B. & Gelbrich, T., *Eur. J. Org. Chem.*, (2003) 1220.
50. Carvalho, L.M., Silva, A.M.S., Martins, C.I., Coelho, P.J. & Oliveira-Compos, A.M.F., *Tetrahedron Lett.*, **44** (2003) 1903.
51. Rupe, H., *Helv. Chim. Acta*, **9** (1920) 672.
52. Swaminathan, S. & Narayanan, K.V., *Chem. Rev.*, **71** (1971) 429.
53. Gabbutt, C.D., Heron, B.M., Instone, A.C., Horton, P.N. & Hursthouse, M.B., *Tetrahedron*, **61** (2005) 463.
54. Mitsunobu, O., *Synthesis*, (1981) 1.

55. Zhao, W. & Carreira, E.M., *Org. Lett.*, **5** (2003) 4153.
56. Gabbutt, C.D., Hepworth, J.D., Heron, B.M., Partington, S.M. & Thomas, D.A., *Dyes and Pigments*, **49** (2001) 65.
57. Coelho, P.J., Salvador, M.A., Oliveira, M.M. & Carvalho, L.M., *Tetrahedron,* **60** (2004) 2593.
58. Rauss-Godineau, J., Chodkiewicz, W. & Cadiot, P., *Bull. Soc. Chim. Fr.*, (1966) 2885.
59. Johnson, W.S. & Daub, G.H., *Org. React.*, **6** (1951) 1; Sargent, M.V. & Stransky, P.O., *J. Chem. Soc., Perkin Trans., 1* (1982) 1605.
60. Kumar, A., Knowles, D.B. & Van Gemert, B., *US Patent*, US 5,651,923 (1997).
61. Clarke, D.A., Heron, B.M., Gabbutt, C.D., Hepworth, J.D., Partington, S.M. & Corns, S.N., *PCT Int. Appl. WO* 18755 (2000).
62. Nelson, C.M., Chopra, A., Knowles, D.B., Van Gemert, B. & Kumar, A., *PCT Int. Appl. WO* 19812 (2001).
63. Lin, J., *PCT Int. Appl. WO* 51483, (2001).
64. Gabbutt, C.D., Hepworth, J.D., Heron, B.M. Thomas, D.A., Kilner, C. & Partington, S.M., *Heterocycles*, **63** (2004) 567.
65. Knowles, D.B., *US Patent*, US 5,514,817 (1996).
66. Rickwood, M., Smith, K.E., Gabbutt, C.D. & Hepworth, J.D., *PCT Int. Appl. WO* 22850 (1994).
67. Van Gemert, B. & Knowles, D.B., *US Patent*, US 5,552,090 (1996).
68. Benson, S.C., Lam, J.Y.L., Upadhya, K.G., Radel, P.A., Zhen, W. & Menchen, S.M., *US Patent*, US 6,051,719 (2000).
69. Wizinger, R. & Wenning, H., *Helv. Chim. Acta*, **23** (1940) 247.
70. Abson, D., Bartle, K.D., Bryant, J., Livingstone, R. & Watson, R.B., *J. Chem. Soc.*, (1965) 2978.
71. Bradley, E., Cotterill, W.D., Livingstone, R. & Walshaw, M., *J. Chem. Soc. C*, (1971) 3028.
72. Zwanenburg, D.J. & Maas, Th.A.M.M., *Recl. Trav. Chim. Pays-Bas.*, **94** (1975) 8.
73. Aiken, S., Gabbutt, C.D., Heron, B.M., Instone, A.C., Horton, P.N. & Hursthouse, M.B., *Adv. Col. Sci. Technol.*, **7** (2004) 55.
74. Hlubucek, J., Ritchie, E. & Taylor, W.C., *Aust. J. Chem.*, **24** (1971) 2347; Arnone, A., Cardillo, G., Merlini, L. & Mondelli, R., *Tetrahedron Lett.*, **8** (1967) 4201.
75. Batterham, T.J., *NMR Spectra of Simple Heterocycles*, Taylor, E.C.& Weissberger, A., Eds., Wiley, New York (1973) p. 385.
76. Thomas, D.A., Ph.D. thesis, University of Leeds, UK (2003) p. 97.
77. Barberis, C., Campredon, M., Lokshin, V., Giusti, G. & Faure, R., *Magn. Reson. Chem.*, **33** (1995) 977.
78. Delbaere, S., Teral, Y., Bochu, C., Campredon, M. & Vermeersch, G., *Magn. Reson. Chem.*, **37** (1999) 159.
79. Chamontin, K., Lokshin, V., Rossollin, V., Samat, A. & Guglielmetti, R., *Tetrahedron*, **55** (1999) 5821.
80. Arnall-Culliford, J.C., Teral, Y., Sgarabotto, P., Campredon, M. & Giusti, G., *J. Photochem. Photobiol. A*, **159** (2003) 7.
81. Oliveira, M.M., Carvalho, L.M., Moustrou, C., Samat, A., Guglielmetti, R. & Oliveira-Campos, A.M.F., *Magn. Reson. Chem.*, **39** (2001) 637.
82. Gabbutt, C.D., Gelbrich, T., Hepworth, J.D., Heron, B.M., Hursthouse, M.B. & Partington, S.M., *Dyes and Pigments*, **54** (2002) 79.
83. Gabbutt, C.D., Heron, B.M. & Instone, A.C., *Heterocycles*, **60** (2003) 843.
84. Hepworth, J.D., Gabbutt C.D. & Heron, B.M., *Comp. Heterocyc. Chem., 2nd edn.*, **5** (1996) 317.
85. Delbaere, S., Luccioni-Houze, B., Bochu, C., Teral, Y., Campredon, M. & Vermeersch, G., *J. Chem. Soc., Perkin Trans., 2* (1998) 1153.
86. Gabbutt, C.D., Heron, B.M., Thomas, D.A., Light, M.E. & Hursthouse, M.B., *Tetrahedron Lett.*, **45** (2004) 6151.

87. Ortica, F., Moustrou, C., Berthet, J., Favaro, G., Samat, A., Guglielmetti, R., Vermeersch, G. & Mazzucato, U., *Photochem. Photobiol.*, **78** (2003) 558.
88. Berthet, J., Delbaere, S., Lokshin, V., Samat, A., Micheau, J.C. & Vermeersch, G., *Int. J. Photoenergy*, **6** (2004) 215.
89. Sallenave, X., Delbaere, S., Vermeersch, G., Saleh, A. & Pozzo, J.-L., *Tetrahedron Lett.*, **46** (2005) 3257.
90. Delbaere, S. & Vermeersch, G., *Tetrahedron Lett.*, **44** (2003) 259.
91. Delbaere, S. & Vermeersch, G., *J. Photochem. Photobiol. A*, **159** (2003) 227.
92. Gabbutt, C.D., Hepworth, J.D., Heron, B.M. & Rahman, M.M., *J. Chem. Soc. Perkin Trans. 1*, (1994) 1733.
93. Aldoshin, S., *Organic Photochromic and Thermochromic Compounds, Volume 2: Physicochemical Studies, Biological Applications and Thermochromism*, Crano, J.C. & Guglielmetti, R. J., Eds., Plenum Press, New York (1998) p. 297.
94. Aldoshin, S., Chuev, I. & Utenyshev, A., *Acta Crystallogr.*, **C51** (1995) 141.
95. Aldoshin, S., Chuev, I., Utenyshev, A., Filipenko, O., Pozzo, J.-L., Lokshin, V. & Guglielmetti, R., *Acta Crystallogr.*, **C52** (1996) 1834.
96. Birukov, B.P. & Unkovskij, B.V., *Kristallokhimia*, **11** (1976) 132.
97. Huang, H.-M., Ding, Y., Chen, P.-L. & Meng, J.-B., *Acta Crystallogr.*, **C60** (2004) 709.
98. Aiken, S., Gabbutt, C.D. & Heron, B.M., Unpublished Results.
99. Hannesschlager, P., Brun, P. & Pepe, G., *Acta Crystallogr.*, **C55** (1999) 200.
100. Gabbutt, C.D., Heron, B.M., Thomas, D.A., Partington, S.M., Light, M.E. & Hursthouse, M.B., *Mol. Cryst. Liq. Cryst.*, **431** (2005) 57.
101. Hepworth, J.D., Gabbutt, C.D. & Heron, B.M., *Comp. Heterocyc. Chem., 2nd edn.*, **5** (1996) 338.
102. Hepworth, J.D., Gabbutt, C.D. & Heron, B.M., *Comp. Heterocyc. Chem., 2nd edn.*, **5** (1996) 325.
103. Thomas, D.A., Ph.D. thesis, University of Leeds, UK (2003) p. 99; Instone, A.C., Ph.D. thesis, University of Leeds, UK (2004) p. 71.
104. Harié, G., Samat, A., Guglielmetti, R., De Keukeleire, D., Sayens, W. & Van Parys, I., *Tetrahedron Lett.*, **38** (1997) 3075.
105. Ortica, F., Romani, A., Blackburn, F. & Favaro, G., *Photochem. Photobiol. Sci.*, **1** (2002) 803.
106. Clarke, D.A., Heron, B.M., Gabbutt, C.D., Hepworth, J.D., Partington, S.M. & Corns, S.N., *PCT Int. Appl. WO* 31081 (1999).
107. Goto, Y., Sugita, K., Takahashi, T., Ohga, Y. & Asano, T., *Chem. Lett.*, **32** (2003) 618.
108. Sugita, K., Goto, Y., Ono, M., Yamashita, K., Hayase, K., Takahashi, T., Ohga, Y. & Asano, T., *Bull. Chem. Soc. Jpn.*, **77** (2004) 1803.
109. Zayat, M. & Levy, D., *J. Mat. Chem.*, **13** (2003) 727.
110. Pardo, R., Zayat, M. & Levy, D., *J. Mat. Chem.*, **15** (2005) 703.
111. Ottavi, G., Favaro, G. & Malatesta, V., *J. Photochem. Photobiol. A*, **115** (1998) 123.
112. Görner, H. & Chibisov, A.K., *J. Photochem. Photobiol. A*, **149** (2002) 83.
113. Jockusch, S., Turro, N.J. & Blackburn, F.R., *J. Phys. Chem. A*, **106** (2002) 9236.
114. Hobley, J., Malatesta, V., Millini, R., Giroldini, W., Wis, L., Goto, M., Kishimoto, M. & Fukumura, H., *Chem. Commun.*, (2000) 1339.
115. Christie, R.M., Hepworth, J.D., Gabbutt, C.D. & Rae, S., *Dyes and Pigments,* **35** (1997) 339.
116. Hobley, J., Malatesta, V., Hatanaka, K., Kajimoto, S., Williams, S.L. & Fukumura, H., *Phys. Chem. Chem. Phys.*, **4** (2002) 180.
117. Kaneko, T., Akutsu, H., Yamada, J. & Nakatsuji, S., *Org. Lett.*, **5** (2003) 2127.
118. Harié, G., Samat, A., Guglielmetti, R., Van Parys, I., Saeyens, W., De Keukeleire, D., Lorenz, K. & Mannschreck, A., *Helv. Chim. Acta,* **80** (1997) 1122.

119. Beach, S.F., Hepworth, J.D., Jones, P., Mason, D., Sawyer, J., Hallas, G. & Mitchell, M.M., *J. Chem. Soc. Perkin Trans.*, *2* (1989) 1087.
120. Guinot, S.G.R., Hepworth, J.D. & Wainwright, M., *J. Chem. Soc. Perkin Trans.*, *2* (1998) 297.
121. Azzaro, M., Gal, J.F., Geribaldi, S. & Videau, B., *J. Chem. Soc. Perkin Trans.*, *2* (1983) 57.
122. Nash, C.P. & Maciel, G.E., *J. Phys. Chem.*, **68** (1964) 832.
123. Effenberger, F., Fischer, P., Schoeller, W.W. & Stohrer, W.-D., *Tetrahedron*, **34** (1978) 2409.
124. Gabbutt, C.D., Hepworth, J.D. & Heron, B.M., Unpublished Results.
125. Van Gemert, B., Bergomi, M. & Knowles, D., *Mol. Cryst. Liq. Cryst.*, **246** (1994) 67.
126. Crano, J.C., Flood, T., Kumar, A. & Van Gemert, B., *Pure Appl. Chem.*, **68** (1996) 1395.
127. Van Gemert, B., Kumar, A. & Knowles, D.B., *Mol. Cryst. Liq. Cryst.*, **297** (1997) 131.
128. Rickwood, M., Marsden, S.D., Hepworth, J.D. & Gabbutt, C.D., *US Patent*, US 5,520,853 (1996).
129. Momoda, J., Matsuoka, S. & Nagou, H., *US Patent*, US 6525194 (2003).
130. Demadrille, R., Moustrou, C., Samat, A. & Guglielmetti, R., *Heterocyc. Commun.*, **5** (1999) 123.
131. Clarke, D.A., Heron, B.M., Gabbutt, C.D., Hepworth, J.D., Partington, S.M. & Corns, S.M., *PCT Int. Appl. WO* 45281 (1998).
132. Kumar, A., Van Gemert, B. & Knowles, D.B., *Mol. Cryst. Liq. Cryst.*, **344** (2000) 217.
133. Anonymous, *Research Disclosure* No. **361** (1994) 266.
134. Kumar, A., Van Gemert, B. & Knowles, D.B., *US Patent*, US 5,458,814 (1995).
135. Clarke, D.A., Heron, B.M., Gabbutt, C.D., Hepworth, J.D., Partington, S.M. & Corns, S.N., *PCT Int. Appl. WO* 42695 (1998).
136. Kumar, A., *US Patent*, US 5,879,592 (1999).
137. Kumar, A., *US Patent*, US 6,353,102 (2002).
138. Clarke, D.A., Heron, B.M., Gabbutt, C.D., Hepworth, J.D., Partington, S.M. & Corns, S.N., *PCT Int. Appl. WO* 42693 (1998).
139. Knowles, D.B., *US Patent*, US 5,514,817 (1996).
140. Chan, Y.-P., Jean, P. & Breyne, O., *PCT Int. Appl. WO* 15628 (2000).
141. Van Gemert, B., *PCT Int. Appl. WO* 14596 (1996).
142. Coelho, P.J., Salvador, M.A., Oliveira, M.M. & Carvalho, L.M., *Synlett*, (2004) 1015.
143. Van Gemert, B., *US Patent*, US 5,645,767 (1997).
144. Melzig, M., Mann, C. & Weigand, U., *PCT Int. Appl. WO* 15518 (1999).
145. Momota, J. & Komuro, Y., *Japanese Patent*, JP11067 (2001).
146. Nelson, C.M., Chopra, A., Petrovskaia, O.G., Knowles, D.B., Van Gemert, B. & Kumar, A., *PCT Int. Appl. WO* 19813 (2001).
147. Mann, C., Melzig, M. & Weigand, U., *PCT Int. Appl. WO* 22594 (2002).
148. Martins, C.I., Coelho, P.J., Carvalho, L.M., Oliveira-Campos, A.M.F., Samat, A. & Guglielmetti, R., *Helv. Chim. Acta,* **86** (2003) 570.
149. Van Gemert, B., Chopra, A. & Kumar, A., *US Patent*, US 6,113,814 (2000).
150. Izumi, S., Kawabata, Y., Takeda, Y., Momoda, J. & Nagoh, H., *European Patent*, EP1184379 (2002).
151. Mann, C., Melzig, M. & Weigand, U., *German Patent*, DE 19902771 (1999).
152. Momoda, J. & Kawabata, Y., *European Patent*, EP1054010 (2000).
153. Mann, C., Weigand, U. & Melzig, M., *PCT Int. Appl. WO* 94336 (2001).
154. Momoda, J. & Komuro, Y., *US Patent*, US 6,469,076 (2002).
155. Coelho, P.J., Carvalho, L.M., Abrantes, S., Oliveira, M.M., Oliveira-Campos, A.M.F., Samat, A. & Guglielmetti, R., *Tetrahedron,* **58** (2002) 9505.
156. Petrovskaia, O.G. & Kumar, A., *PCT Int. Appl. WO* 053553 (2002).
157. Kumar, A., *US Patent*, US 5,891,368 (1999).

158. Favaro, G., Ortica, F. & Romani, A., *New J. Chem.*, **27** (2003) 639.
159. Lin, J. & Van Gemert, B.,*US Patent*, US 5,869,658 (1999).
160. Lin, J., *US Patent*, US 5,783,116 (1998).
161. Izumi, S. & Nagoh, H., *PCT Int. Appl. WO* 090342 (2002).
162. Bremand, R., Breyne, O. & Chan, Y.-P., *PCT Int. Appl. WO* 099172 (2004).
163. Melzig, M., Mann, C. & Weigand, U., *PCT Int. Appl. WO* 67234 (1999).
164. Breyen, O., Chan, Y.-P. & Jean, P., *PCT Int. Appl. WO* 36406 (2001).
165. Breyne, O., Chan, Y.-P. & Jean, P., *PCT Int. Appl. WO* 30916 (2002).
166. Chan, Y.-P. & Jean, P., *US Patent*, US 6,210,608 (2001).
167. Melzig, M. & Weigand, U., *PCT Int. Appl. WO* 24438 (1999).
168. Knowles, D.B. & Van Gemert, B., *US Patent*, US 5,674,432 (1995).
169. Chan, Y.-P., *PCT Int. Appl. WO* 28289 (1998).
170. Kumar, A., Knowles, D.B. & Van Gemert, B., *PCT Int. Appl. WO* 21698 (1997).
171. Kumar, A., *PCT Int. Appl. WO* 32037 (1998).
172. Chan, Y.-P., Breyne, O. & Jean, P., *PCT Int. Appl. WO* 32661 (2001).
173. Lin, J., *US Patent*, US 5,811,034 (1998).
174. Breyne, O., Chan, Y.-P. & Jean, P., *PCT Int. Appl. WO* 77007 (2000).
175. Chan, Y.-P., Breyne, O. & Jean, P., *PCT Int. Appl. WO* 36424 (2001).
176. Qin, X., *US Patent*, US 0215024 (2004).
177. Qin, X., *PCT Int. Appl. WO* 044022 (2003).
178. Qin, X., *PCT Int. Appl. WO* 025638 (2003).
179. Breyne, O., *PCT Int. Appl. WO* 31080 (2000).
180. Bourchteine, K. & Breyne, O., *PCT Int. Appl. WO* 23071 (1999).
181. Momota, J. & Nago, H., *Japanese Patent*, JP229974 (2000).
182. Momota, J. & Nago, H., *Japanese Patent*, JP229975 (2000).
183. Oliveira, M.M., Salvador, M.A., Coelho, P.J. & Carvalho, L.M., *Tetrahedron*, **61** (2005) 1681.
184. Van Gemert, B. & Selvig, C.D., *PCT Int. Appl. WO* 28323 (1999).
185. Kumar, A., *PCT Int. Appl. WO* 02884 (2000).
186. Chan, Y.-P., *PCT Int. Appl. WO* 15631 (2000).
187. Qin, X. & Ippoliti, J.T., *US Patent*, US 0168645 (2003).
188. Kumar, A., *PCT Int. Appl. WO* 02883 (2000).
189. Kumar, A., *US Patent*, US 6,153,126 (2000).
190. Kumar, A., *US Patent*, US 6,149,841 (2000).
191. Nago, H. & Momota, J., *Japanese Patent*, JP344761 (2000).
192. Kawabata, Y. & Momota, J., *Japanese Patent*, JP344762 (2000).
193. Breyne, O., *US Patent*, US 6,379,591 (2002).
194. Momota, J. & Omuro, Y., *Japanese Patent*, JP229976 (2000).
195. Mann, C., Melzig, M. & Weigand, U., *PCT Int. Appl. WO* 080595 (2003).
196. Knowles, D.B. & Van Gemert, B., *US Patent*, US 5,464,567 (1995).
197. Hughes, F.J., *US Patent*, US 5,840,926 (1998).
198. Navarro, A.-E., Moggia, F., Moustrou, C., Heynderickx, A., Fages, F., Leriche, P. & Brisset, H., *Tetrahedron*, **61** (2005) 423.
199. Melzig, M. & Zinner, H., *PCT Int. Appl. WO* 01884 (1996).
200. Favaro, G., Levi, D., Ortica, F., Samat, A., Guglielmetti, R. & Mazzucato, U., *J. Photochem. Photobiol. A*, **149** (2002) 91.
201. Samat, A., Lokshin, V., Chamontin, K., Levi, D., Pepe, G. & Guglielmetti, R., *Tetrahedron*, **57** (2001) 7349.
202. Ortica, F., Levi, D., Brun, P., Guglielmetti, R., Mazzucato, U. & Favaro, G., *J. Photochem. Photobiol. A*, **139** (2001) 133.

203. Berthet, J., Delbaere, S., Brun, P., Guglielmetti, R. & Vermeersch, G., *J. Chem. Soc. Perkin Trans.*, *2* (2002) 2118.
204. Yassar, A., Rebière-Galy, N., Frigoli, M., Moustrou, C., Samat, A., Guglielmetti, R. & Jaafari, A., *Synth. Met.*, **124** (2001) 23.
205. Yassar, A., Garnier, F., Jaafari, H., Rebière-Galy, N., Frigoli, M., Moustrou, C., Samat, A. & Guglielmetti, R., *Appl. Phys. Lett.*, **80** (2002) 4297.
206. Rebiere, N., Moustrou, C., Meyer, M., Samat, A., Guglielmetti, R., Micheau, J.-C. & Aubard, J., *J. Phys. Org. Che.*, **13** (2000) 523.
207. Zhao, W. & Carreira, E.M., *J. Am. Chem. Soc.*, **124** (2002) 1582.
208. Frigoli, M., Moustrou, C., Samat, A. & Guglielmetti, R., *Eur. J. Org. Chem.*, (2003) 2799.
209. Frigoli, M., Moustrou, C., Samat, A. & Guglielmetti, R., *Helv. Chim. Acta*, **83** (2000) 3043.
210. Coen, S., Moustrou, C., Frigoli, M., Julliard, M., Samat, A. & Guglielmetti, R., *J. Photochem. Photobiol. A*, **139** (2001) 1.
211. Ahmed, S.A., Tanaka, M., Ando, H., Iwamoto, H. & Kimura, K., *Eur. J. Org. Chem.*, (2003) 2437.
212. Ushakov, E.N., Nazarov, V.B., Fedorova, O.A., Gromov, S.P., Chebun'kova, A.V., Alfimov, M.V. & Barigelletti, F., *J. Phys. Org. Chem.*, **16** (2003) 306.
213. Ahmed, S.A., Tanaka, M., Ando, H., Iwamoto, H. & Kimura, K., *Tetrahedron*, **60** (2004) 3211.
214. Ahmed, S.A., Tanaka, M., Ando, H., Iwamoto, H. & Kimura, K., *Tetrahedron*, **59** (2003) 4135.
215. Stauffer, M.T., Knowles, D.B., Brennan, C., Funderburk, L., Lin, F.-T. & Weber, S.G., *Chem. Commun.*, (1997) 287.
216. Anguille, S., Brun, P., Guglielmetti, R., Strokach, Y.P., Ignatin, A.A., Barachevsky, V.A. & Alfimov, M.V., *J. Chem. Soc. Perkin Trans.*, *2* (2001) 639.
217. Brun, P., Guglielmetti, R., Pepe, G. & Anguille, S., *J. Photochem. Photobiol. A*, **156** (2003) 77.
218. Brun, P., Guglielmetti, R. & Anguille, S., *Appl. Organometal. Chem.*, **16** (2002) 271.
219. Hannesschlager, P. & Brun, P., *Appl. Organometal. Chem.*, **13** (1999) 447.
220. Ahmed, S.A., Sallenave, X., Fages, F., Mieden-Gundert, G., Muller, W.M., Muller, U., Vogtle, F. & Pozzo, J.-L., *Langmuir*, **18** (2002) 7096.
221. Frigoli, M. & Mehl, G.H., *Chem. Commun.*, (2004) 2040.

Functional Dyes
Sung-Hoon Kim (Editor)

Chapter 4

Cyanine dyes as fluorescent non-covalent labels for nucleic acid research

Todor Deligeorgiev and Aleksey Vasilev

Faculty of Chemistry, University of Sofia, Sofia, Bulgaria

1. INTRODUCTION

Nucleic acid molecules are the structural supports of genetic materials and therefore the key factors in many vital cellular processes. Deoxyribonucleic acid (DNA) and ribonucleic acid (RNA) encode the biological information through their linear sequence of nucleotides to specify the composition of proteins, and through their shapes to control their assembly with other cellular macromolecules [1].

DNA is a biopolymer and an essential constituent of the cell nucleus. It is the carrier of hereditary information. It is made up of nucleotides, which are the individual units of nucleic acids. The nucleotides contain three essential components: a phosphate group, a sugar, and a base. The sugar is deoxyribose, a pentose with one hydroxyl group less than ribose. The base is any one of four nucleobases – the heterocyclic compounds adenine (A), guanine (G), cytosine (C), and thymine (T). Adenine and guanine are purines; thymine and cytosine are pyrimidines. Nucleotides are joined by the phosphate group and the deoxyribose to form a continuous chain. This chain pairs up with a second chain through the lateral bases to form a double helix. The two chains are not identical, but possess a complementary relationship. The base pairs are always adenine–thymine (AT) and cytosine–guanine (CG), linked through hydrogen bonds.

Molecular biology and genetic engineering are based on DNA. Thus, proteins are produced when the double helix is split to reveal the genetic code. The process of transferring information is called transcription, the outcome of which is the corresponding RNA, a chain in which the ribose replaces deoxyribose and the base uracyl (U) replaces thymine. Thymine and uracyl differ by a methyl group. The RNA parts from the DNA, exits from the nucleus and finds its way to the ribosomes in the protoplasm, where protein synthesis occurs. Since RNA acts as a messenger between the source of information (DNA) and the site of synthesis, it is called messenger RNA or mRNA. Transfer RNAs guide amino acids into place during protein synthesis and are denoted tRNA. DNA is found in chromosomes and RNA is found in the nucleus and cytoplasm.

It is very important for researchers to explore the secret of life, to develop new functional medicines for curing different diseases. In the recent decades, the investigations of the reaction mechanism between small molecules and nucleic acids and the development of rapid and convenient assays for nucleic acids is an active area in bio-analytical chemical research [2].

The direct use of the intrinsic fluorescence [3,4] and ultraviolet absorption [5,6] of nucleic acids for their determination and structural study has been severely limited by the low sensitivity and serious interferences of biological samples. Therefore, many techniques, based on the interaction between nucleic acids and extrinsic reagents (probes or labels) have been established for the research related to nucleic acids, such as spectrophotometric, fluorimetric, and radioactive labeling methods [7]. However, radioactive probes have short shelf life, which is dangerous for humans, and have high disposal costs. In addition, they may be unstable. So, extensive efforts to develop alternative labeling techniques have resulted in spectrophotometric, fluorescent, resonant light scattering, and chemiluminescent assay formats.

Generally, the interaction of small molecules with nucleic acids involves four modes, namely intercalative binding, groove binding [8], electrostatic binding [9], and long-range assembly on the molecular surface of nucleic acids that is not involved in intercalative or groove binding [10].

Fluorescence methods have higher sensitivity than absorption, so they are often used for studying nucleic acids. It was mentioned that the natural fluorescence intensity of nucleic acids is so weak [3] that their fluorescence properties cannot be used directly in structural and quantitative analysis [11], and an extrinsic fluorescent probe should be introduced to study nucleic acids [12]. If a small molecular probe intercalates into the base pairs of a nucleic acid, the fluorescence spectrum of the probe will redshift and the fluorescence polarization will increase. According to the different types of luminescent molecules, the fluorescent probes of nucleic acids are of five types: organic dyes, rare-earth ions, metal ion complexes, photochemical fluorescent substances, and molecular beacon probes.

Studies on binding of organic dyes with DNA are very important for designing of novel and more efficient drugs targeted to DNA [13, 14], and for exploring the biological function of nucleic acids and the interaction mechanism of some drugs.

Intercalating dyes are in general aromatic cations with planar structure that insert between stacked base pairs on the DNA duplex, which provides an environmentally dependent fluorescence enhancement for dye molecules, and creates a large increase of the fluorescence signal relative to the free dye in solution. The signal enhancement provides a proportional response, allowing direct quantitative DNA measurements. The most common intercalating agents are ethidium bromide [15], acridine orange [16], proflavine [17], and the cationic cyanine dye thiazole orange (TO) and its analogs [18], such as TOMEHE.

Acridine orange (Scheme 1) interacts with RNA and DNA by intercalation or electrostatic attraction. This typically cationic dye fluoresces with an emission maximum at 525 nm when bound to DNA, but upon association with RNA it shows a red fluorescence around 650 nm [19].

The bisbenzimidazole dyes (Scheme 2) – Hoechst 33258, Hoechst 33342, and Hoechst 34580 are cell membrane permeable minor groove binders, fluorescing

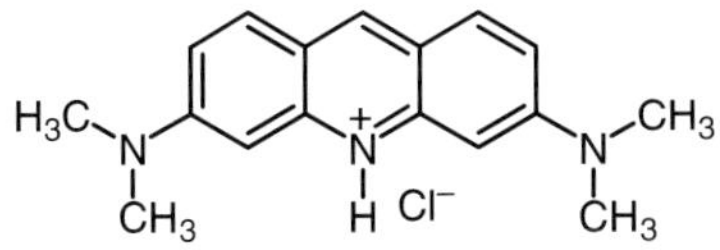

Scheme 1: Acridine orange.

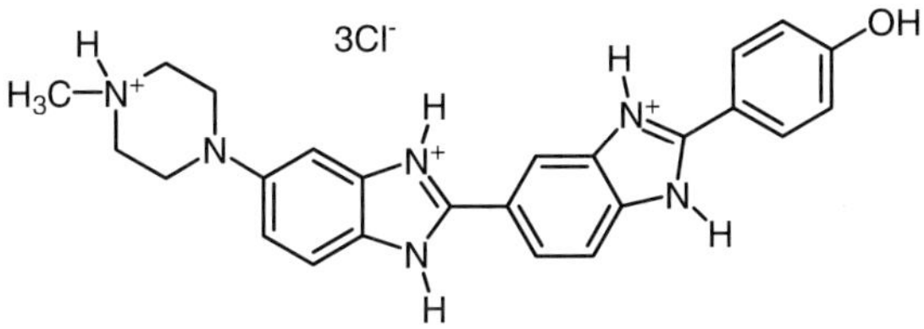

Hoechst 33258

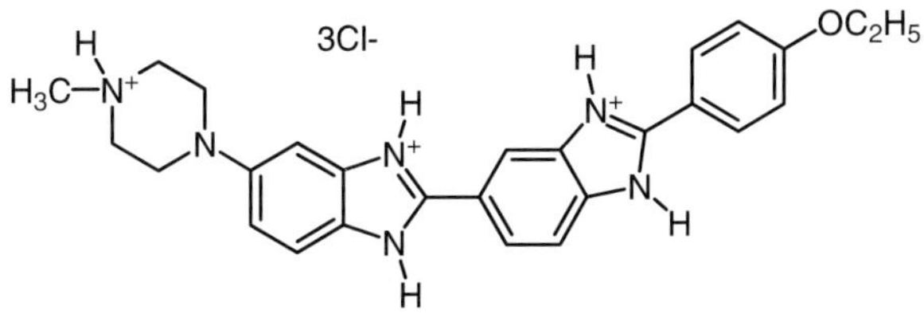

Hoechst 33342

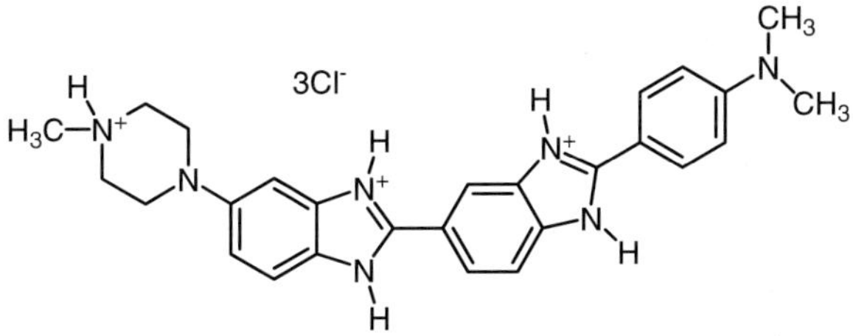

Hoechst 34580

Scheme 2: Structures of dyes Hoechst 33258, Hoechst 33342, and Hoechst 34580.

bright blue upon binding to DNA [20]. These dyes can be excited with UV light sources and exhibit relatively large Stokes shifts – excitation at 350 nm and emission at 460 nm. They bind to DNA with AT preference.

The phenantridinium intercalotors' ethidium bromide (EtBr) and propidium iodide (PI) (Scheme 3) are generally cell membranes impermeable with little or no sequence preference.

The fluorescence is enhanced 20- to 30-fold and exhibits large Stokes shifts upon binding to nucleic acids, but these dyes have relatively low molar absorptivities. They can be excited in the UV region and with argon laser at 488 nm. Ethidium bromide is still the most general nucleic acid gel stain.

Another phenantridinium dye, hexidium iodide (Scheme 3), is moderately lipophilic and permeable to molecular cells. This dye selectively stains almost all Gram-positive bacteria in the presence of Gram-negative bacteria [19].

4,6′-Diamidino-2-phenylindole (DAPI) (Scheme 4) is a minor groove binder fluorescing blue. It can be excited with UV light sources and shows AT preference with 20-fold fluorescence enhancement. The DAPI–RNA complex exhibits longer wavelength fluorescence emission than the DAPI–dsDNA complex (at 460 nm) [19].

Below, we discuss mainly the preparation of cyanine dyes and their applications as fluorescent non-covalent labels for nucleic acid research.

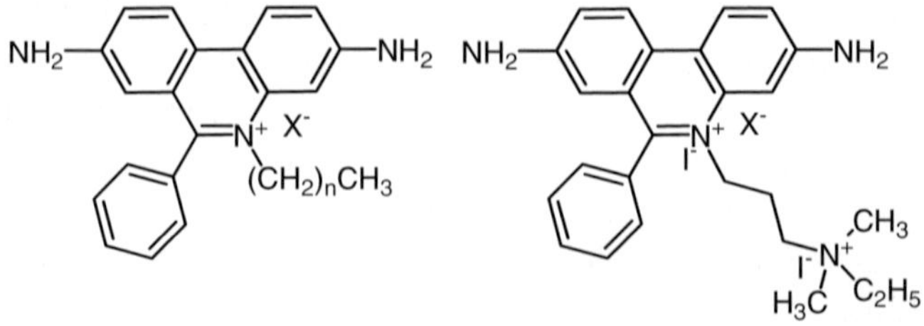

Ethidium bromide (X = Br, n = 1)
Hexidium iodide (X = I, n = 5)

Propidium iodide

Scheme 3: Structures of EtBr, hexidium iodide, and PI.

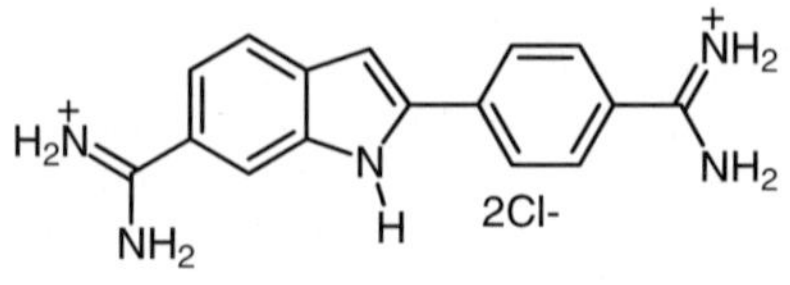

Scheme 4: 4,6′-Diamino-2-phenylindol (DAPI).

2. SYNTHESIS OF CYANINE DYES AS NON-COVALENT NUCLEIC ACID PROBES

2.1. Synthesis of monomeric monomethine and trimethine cyanine dyes as nucleic acid probes

Since the discovery of Lee and co-workers [21, 22] that an old photographic dye (they called it thiazole orange) is an excellent non-covalent nucleic acid probe, many new dyes of this class have been synthesized and investigated [19]. Thiazole orange (Scheme 5) has a 14-fold higher molar absorptivity at about 509 nm than ethidium bromide with strong enhancement (over 1,000-fold) of the fluorescence on binding to DNA and has a characteristic of a monointercalator [21]. It is cell permeable and can be excited with one of the lines of the argon laser, and is 50-fold more sensitive than ethidium bromide [23].

These valuable properties have stimulated the researchers and novel representatives based mainly on TO and oxazole yellow (YO) chromophores (Scheme 5) (YO – an oxazole analog of TO) are designed, synthesized, and commercialized [19].

Generally, all monomethine cyanine dyes are synthesized by the reaction of 2-methylthiobenzothiazolium or 2-methylthiobenzoxazolium salts with quaternized quinolinium salts with active methyl group (Scheme 5).

By this method, the researchers in Molecular Probes [19] have synthesized and commercialized some of the dyes with two positive charges in the chromophore – PO-PRO-1, BO-PRO-1, YO-PRO-1, and TO-PRO-1 [19,24]. They are typically cell membrane impermeable dyes (Scheme 6).

The same method was used for the production of the cell membrane permeable SYTO dyes (Scheme 7), the cell membrane impermeable dyes SYTOX, and the ultrasensitive SYBR Gold, SYBR Green I (SG), and SYBR Green II [25, 26], for which up to 0.8–0.9 fluorescence quantum yields are reported upon binding to DNA [26].

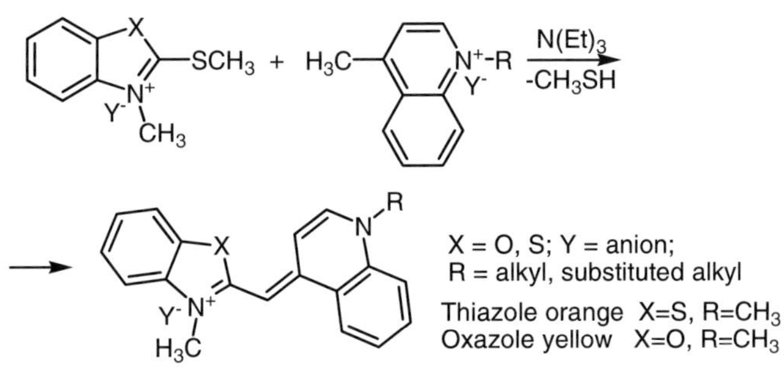

Scheme 5.

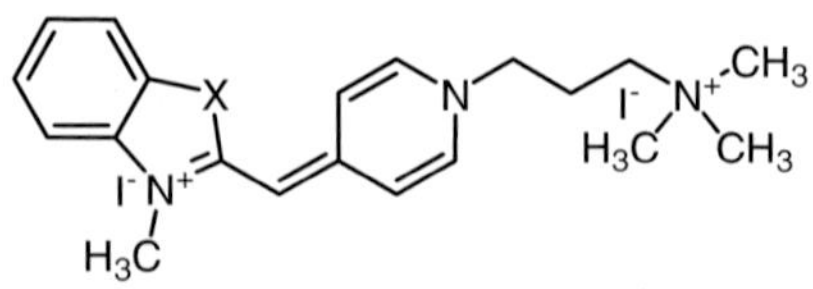

X = O, PO-PRO-1, $\lambda_{ex/em}$ = 435/455 nm; X = S, BO-PRO-1, $\lambda_{ex/em}$ = 462/481 nm

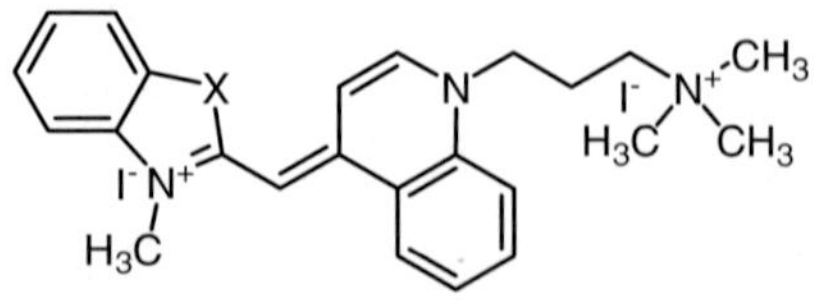

X = O, YO-PRO-1, $\lambda_{ex/em}$ = 491/509 nm; X = S, TO-PRO-1, $\lambda_{ex/em}$ = 515/531 nm

Scheme 6.

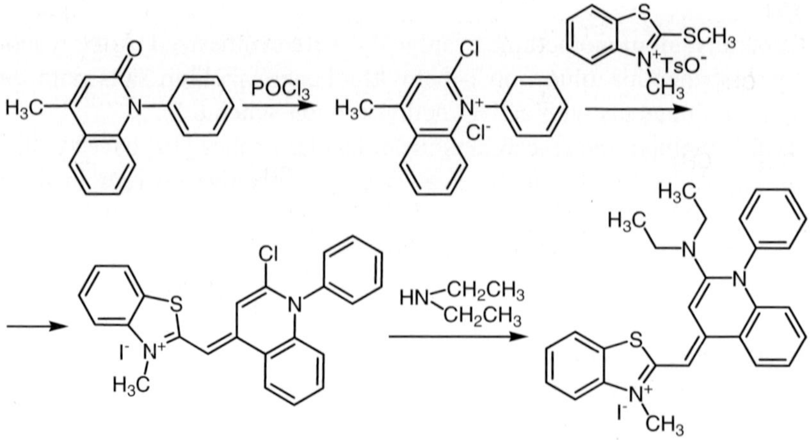

Scheme 7.

Nothing was known about the structures of these dyes, but recently German researchers [27] have found and proved the structures of SG and Pico Green (PG) (Scheme 8).

Researchers [28] have designed novel fluorescent dyes, having a cationic chain, with strong binding affinity and a high fluorescence enhancement upon binding to nucleic acids.

The dye illustrated in Scheme 9 is useful in the detection of dsDNA in gel electrophoresis and in solution at a substantially higher sensitivity, compared to some other known dyes. Because this general method for the preparation of monomethine cyanines has the disadvantage of evolution of methyl thiol – a

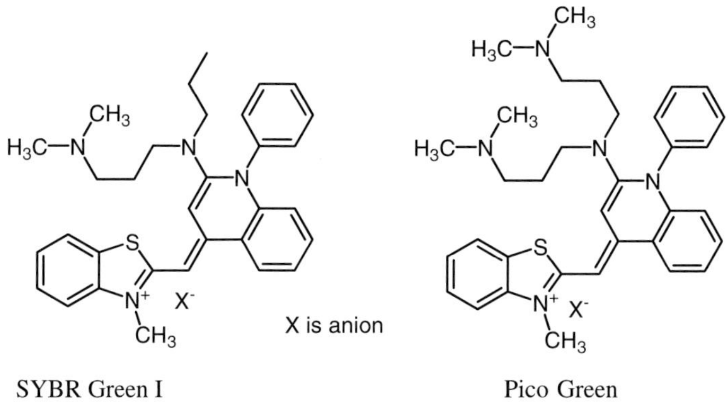

Scheme 8.

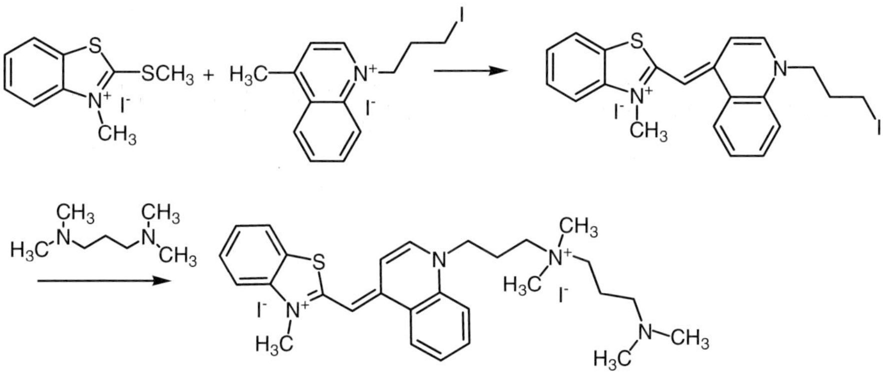

Scheme 9.

strong pollutant and a toxic agent with unpleasant odor – some alternative methods have also been used.

Condensation of 2-imino-3-methylbenzothiazolines with 1-alkyl-4-methylquinolinium salts [29] is such an alternative route (Scheme 10). This method avoids the main drawback of the preceding synthesis – the evolution of methyl thiol and permits, by simple melting of the components, dyes with substituents at the 6th-position in the benzothiazole moiety to be synthesized.

A novel method for the preparation of symmetric and asymmetric monomethine cyanine dyes based on the solvent-free condensation of various active methyl group-containing quaternized heterocycles with zwitterionic quinolinium compounds is reported [30] (Scheme 11). The condensation is carried out in the absence of a basic agent. Heterocycles sensitive to hydrolysis, such as benzoxazolium or benzizoxazolium salts, give higher yields.

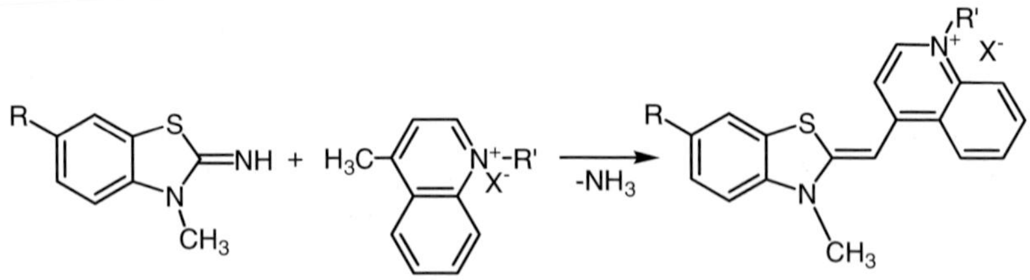

R = H, Cl, NO_2, CH_3O, CH_3CONH, HOC_2H_4O, R′ = alkyl or substituted alkyl

Scheme 10.

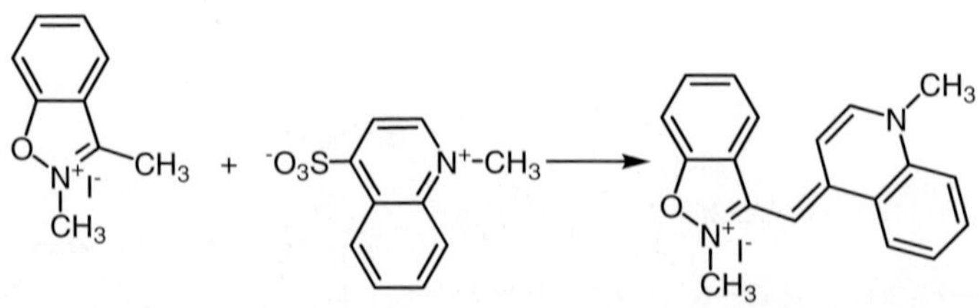

Scheme 11.

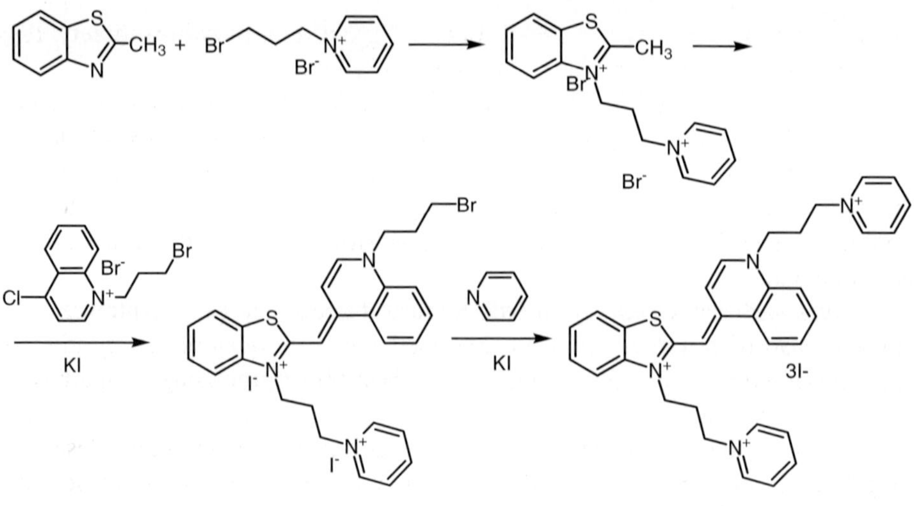

Scheme 12.

Dyes with two or three positive charges in the chromophore can be synthesized by condensation of 4-chloroquinolinium salts with benzothiazolium salts and by additional quaternization with ω-bromoalkyl substituents [31, 32] (Scheme 12).

Different approaches have been demonstrated by Yarmoluk and Kryvorotenko [33]. They made a primary aliphatic amine to react with a

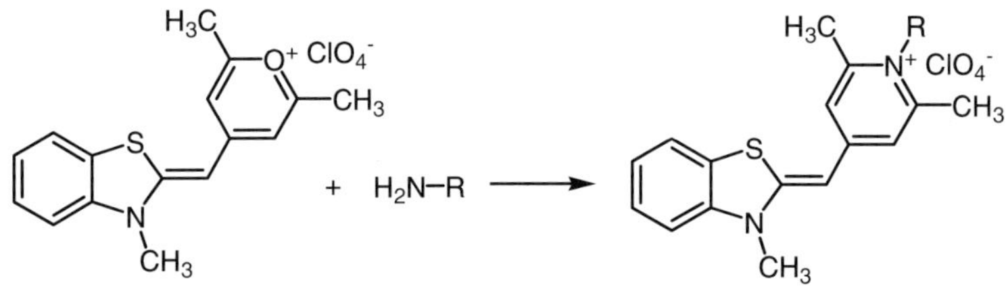

Scheme 13.

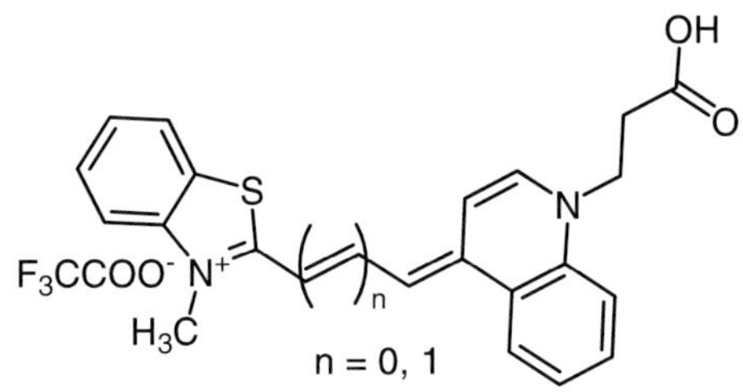

Scheme 14.

monomethine cyanine dye in which one end-group is a pyrylium moiety, thus obtaining a dye with a pyridinium end-group (Scheme 13).

A small-scale solid-phase synthesis of asymmetric monomethine and trimethine dyes is described in [34]. The authors claim if sufficient condensation time is given the purification step becomes unnecessary. The carboxylic linker of the intermediate is coupled with the polystyrene resin and, after condensation, the dye (Scheme 14) is cleaved by 95% trifluoracetic acid and the solvent is evaporated.

Dyes with acetylthio and mercapto substituents have been synthesized by condensation of novel heterocyclic intermediates with quaternized pyridinium and quinolinium salts [35] (Scheme 15); they show promising properties as nucleic acid probes.

Preparation of aza-benzazolium asymmetric cyanine dyes is also described [36]. The key intermediates are 4-methyl-2-methylthiooxazolo[4,5-b]pyridinium tosylate [37] and 4-methyl-2-methylthiothiazolo[4,5-b]pyridinium tosylate [38] (Scheme 16).

The monomethine dyes are prepared by a well-known method (Scheme 17).

Recently, an improved method for the preparation of this class of compounds having two or three positive charges was published [39] (Scheme 18). The dyes show a molar absorptivity higher than 100,000 $l\ mol^{-1}\ cm^{-1}$.

These dyes have been investigated as nucleic acid probes and some of them are showing a higher sensitivity than SG when excited in the visible

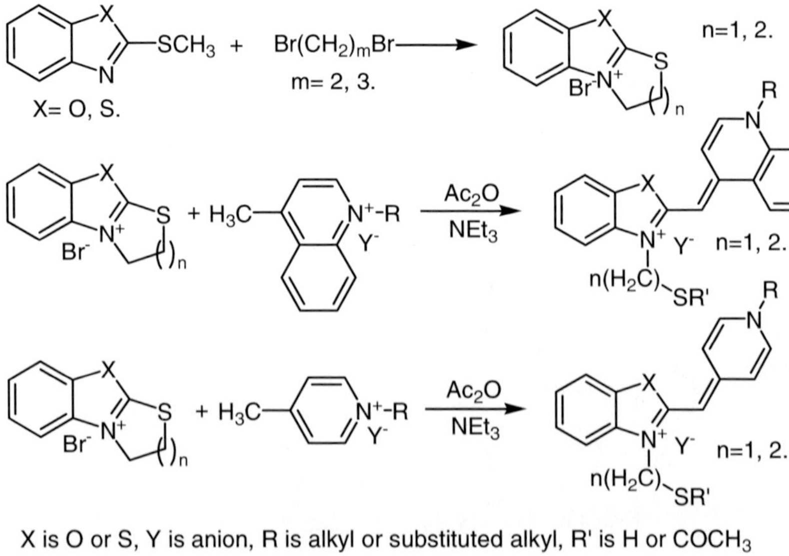

Scheme 15.

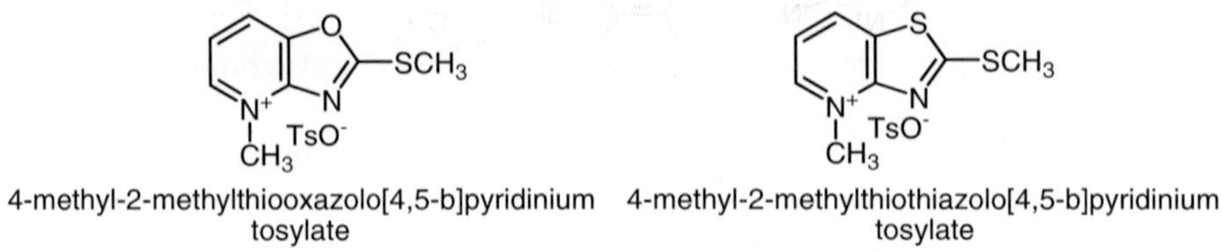

Scheme 16.

region [40]. Some high-affinity dyes of different types, including TO and YO chromophores, are patented by Abbot Laboratories [41]. The dyes based on TO chromophore are synthesized [41] by the classical cyanine method [42] (Scheme 19).

The same procedure is claimed for the synthesis of YO chromophore, which is difficult because of the instability of the quaternized benzoxazolium salts in alkaline media.

SYBR Green I is probably the most widely used dye in real-time polymerase chain reaction (PCR) [43] for clinical analysis and melting-curve analysis [44]. It was found that the dye binds predominantly with the minor groove and has a lower sensitivity for ssDNA [27]. Because of the importance of such minor groove binders, some novel dyes have been investigated. Swedish researchers have studied some monomethine cyanine dyes having a crescent-shape molecule

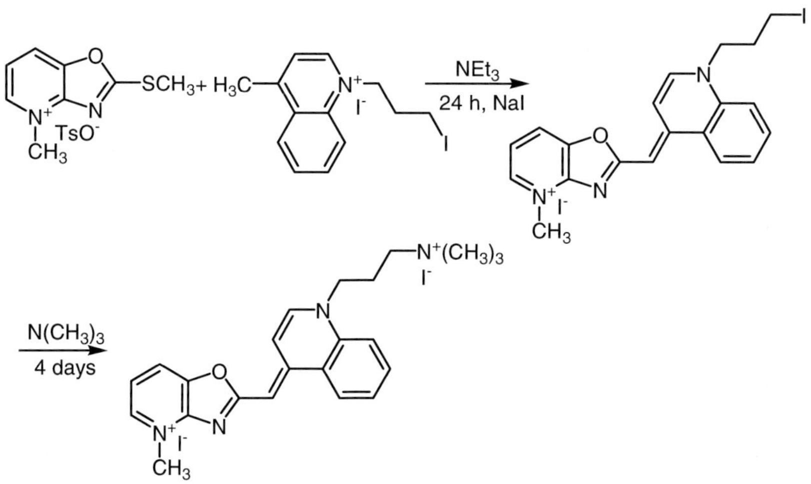

Scheme 17.

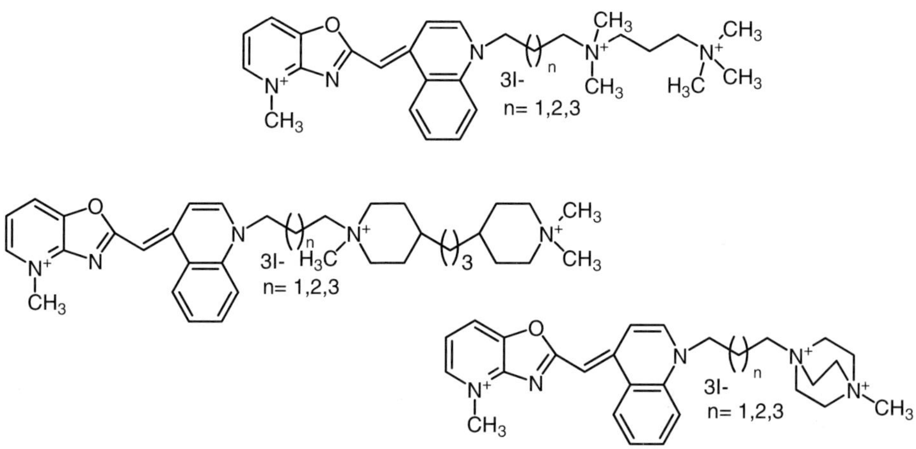

Scheme 18.

similar to the well-known minor groove binder Hoechst 32250. They synthesized [45, 46] a monomethine dye extended with a benzothiazole substituent (BEBO, Scheme 20), generating a signal with sufficient fluorescence without inhibiting the PCR. SYBR Green I and BEBO behave in a similar manner in all important aspects [47].

It was found that the structure modification has induced a shift in the binding mode from intercalation toward minor groove binding. The intercalation and

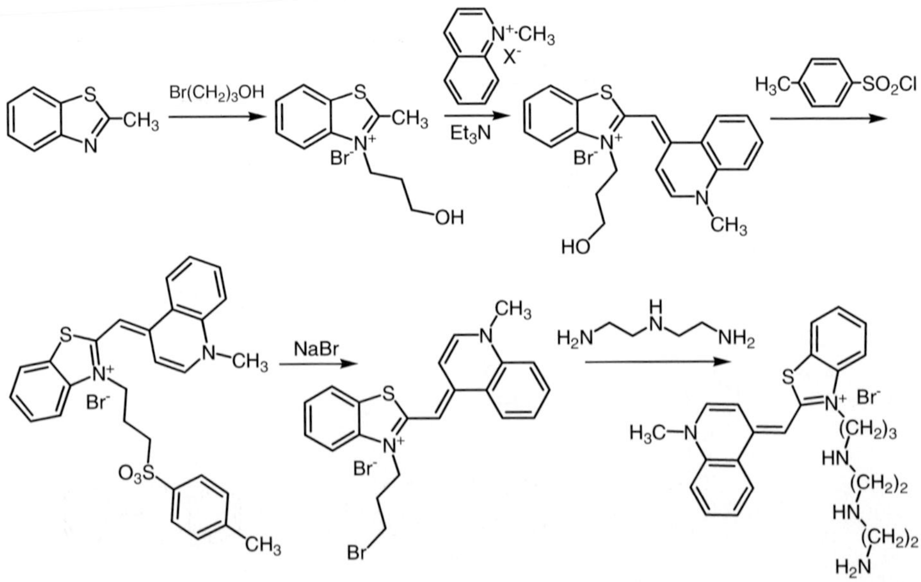

Scheme 19.

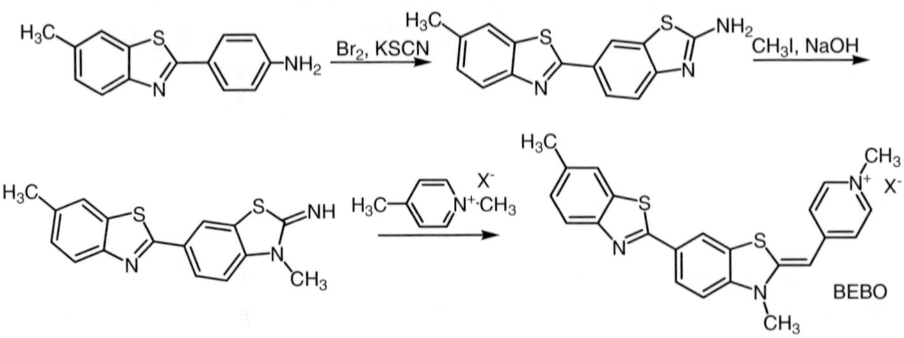

Scheme 20.

groove binding depend on the dsDNA sequence [46, 48]. The same researchers synthesized other crescent-shape BETO and BOXTO dyes [49] (Scheme 21). It was found that BOXTO has a higher preference for the minor groove than BETO. The increase in quantum yield upon binding to DNA is relatively high: 14-fold for BETO and 50-fold for BOXTO (at a dye:base ratio of 1:100). Owing to the blueshift of the emission maxima upon binding to DNA, the increase in fluorescence intensity when measured at 530 nm is greater than the quantum yield enhancement: 260-fold for BOXTO and 130-fold for BETO.

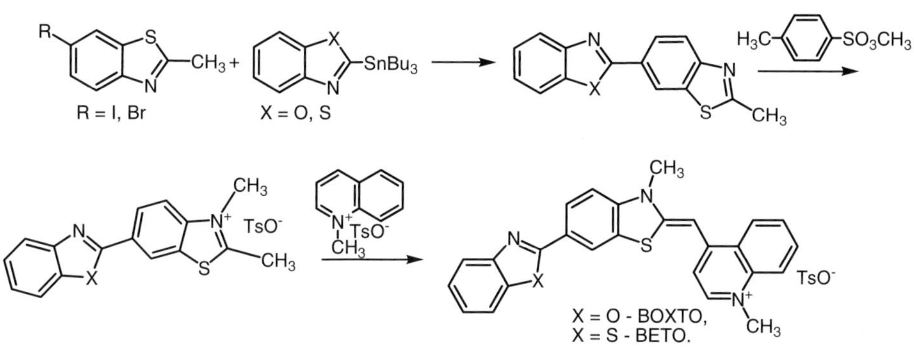

Scheme 21.

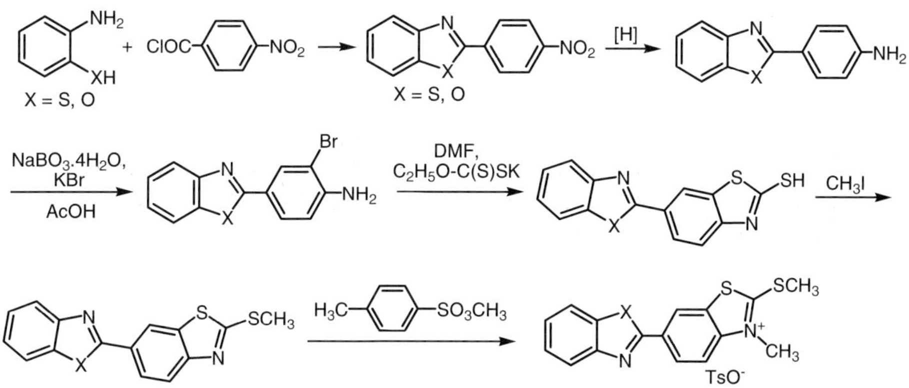

Scheme 22.

A new convenient synthetic route to this type of cyanine dyes is developed [50]. These dyes are analogs to the minor groove binding asymmetric cyanine dye BEBO reported by the same authors [45, 46].

The remarkable steps in Scheme 22 are the aromatic bromination without protection of the amine group [51] and the preparation of 2-mercaptobenzothiazole by reaction of 2-bromoanilines with potassium O-ethyl dithiocarbonate (ethylxantogenate) in dimethyl formamide (DMF) [52, 53].

Among the dyes of this series (Scheme 23), BOXTO exhibits the most pronounced tendency to monomeric minor groove binding. The 2-pyridinium dyes (BEBO, H-BEBO, and BOXTO) show a stronger tendency to intercalation going from higher to lower hydrophobicity [50]. This increase in the degree of intercalation can be a purely structural effect and is also dependent on the size of the dyes. In BEBO [45, 46], the methyl group on the extending benzothiazol moiety might impose a more favorable conformation for minor groove binding. Relative

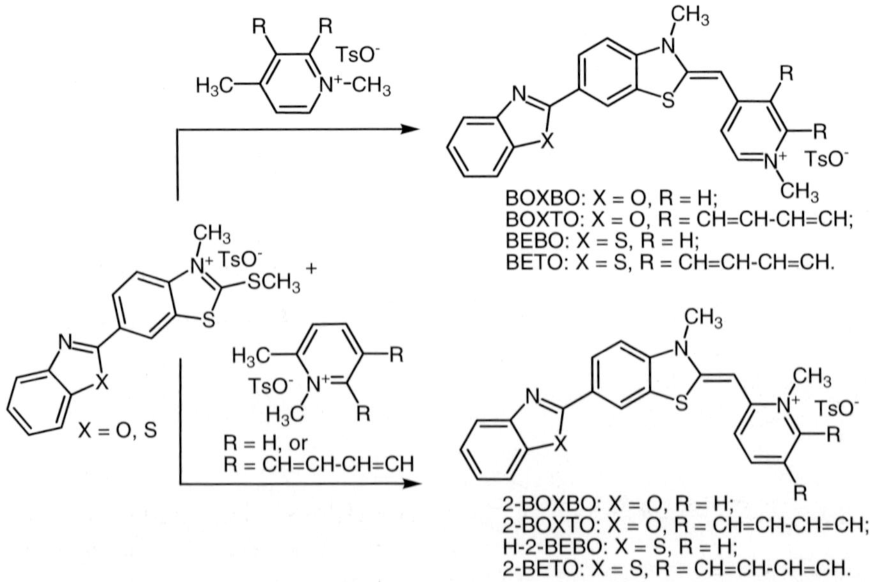

Scheme 23.

to 4-pyridinium dyes, the 2-pyridinium dyes (2-BEBO, H-2-BEBO, and 2-BOXBO) have a stronger tendency to minor groove binding. In particular, two of the benzoxazole substituent-containing dyes (BOXTO and 2-BOXTO) show a high degree of minor groove binding and fluorescence quantum yields of 0.52 and 0.32, respectively, when bound to DNA. Upon binding to ctDNA, the dyes exhibit a strong enhancement in fluorescence quantum yield of varying values [50].

In the exploration of gene organization and function, there is a growing need for rapid and cost-efficient methods to detect specific nucleic acid sequences. A new approach is demonstrated with probes for homogeneous assays – the so-called light-up probes [54]. They are composed of TO chromophore conjugated to peptide nucleic acid (PNA) and combine the excellent hybridization properties of PNA [55] with the extraordinary fluorescence enhancement of asymmetric cyanine dyes upon binding to nucleic acids [19]. The large enhancement in fluorescence upon hybridization makes the light-up probes particularly suitable for the detection of specific nucleic acids in diagnostic assays. With a fluorescence quantum yield of about 0.1 in hybridized state, picomolar concentrations of target nucleic acid can be detected using light-up probes in a regular spectrofluorometer [54] (Scheme 24).

The TO chromophore is covalently bound to the N-terminus of the PNA oligomer by a linker ($-(CH_2)_5-$ or $-(CH_2)_{10}-$) attached either to the benzothiazole or to the quinoline nitrogen of the TO dye. The light-up probes can be synthesized [54]

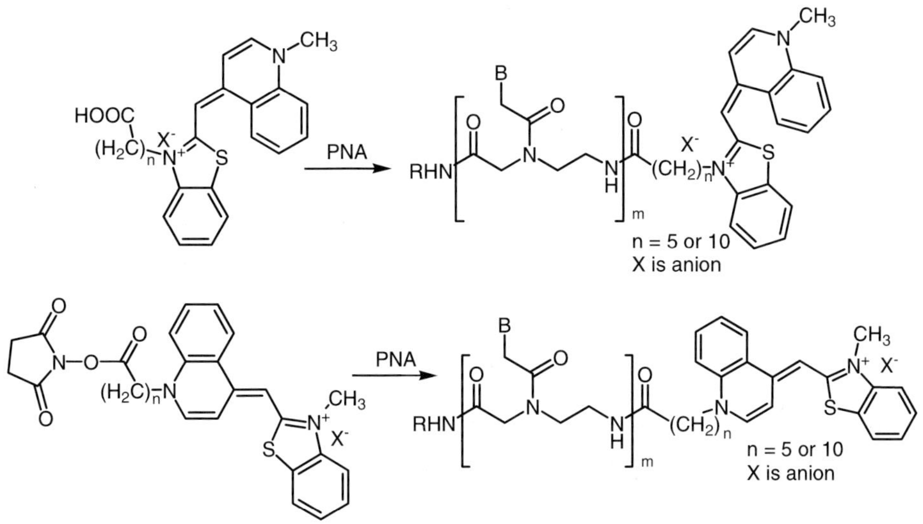

Scheme 24.

either in aqueous solution by attaching a TO succinimidyl ester derivative to PNA modified with an amino linker or by coupling a carboxylic group of the linker directly to the terminal PNA residue during the solid-phase synthesis. The linker in the light-up probe is flexible, allowing the dye to interact with the target nucleic acid upon hybridization. The dye can also fold back, interacting with the bases in the free probe, which might give rise to a residual free probe fluorescence. One reason for light-up probes to be based on PNA instead of normal oligodeoxyribonucleotides is that the electrostatic attraction of the cationic TO dye is eliminated, minimizing "back-binding" [54]. It was found [56] that the fluorescence enhancement of light-up probes upon hybridization to a target nucleic acid depends on the probe sequence, mainly due to large variations in free probe fluorescence. The fluorescence in the free state varies due to the TO intramolecular back-binding with the PNA bases. It was demonstrated [57] that the light-up probes are also suitable for real-time PCR applications.

Other reactive dyes based on TO chromophore have also been patented [58] (Scheme 25).

A wide variety of applications is mentioned. The dyes can be immobilized on a surface or substrate, such as polymeric micro particles or a polymeric membrane. They can be used as high-affinity nucleic acid capture reagents for nucleic acid purification or detection. Bound to the surface, the dye acts as a quantitative or qualitative indicator (test strip or dip stick) of nucleic acids in solution. The dye covalently bound to a nucleic acid can be used to analyze nucleic acid–protein or nucleic acid–drug interactions. Many other applications are also mentioned [58].

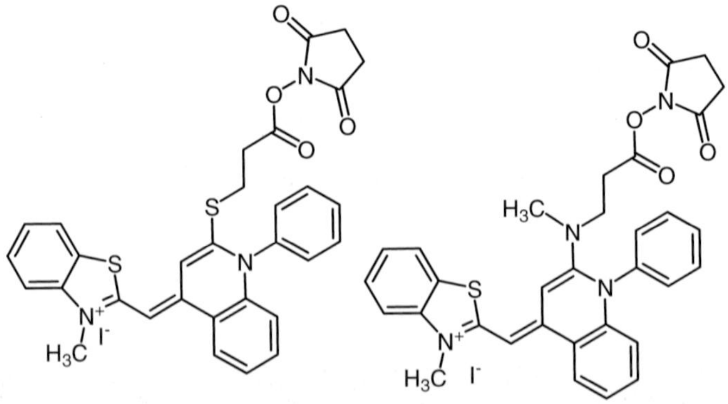

Scheme 25.

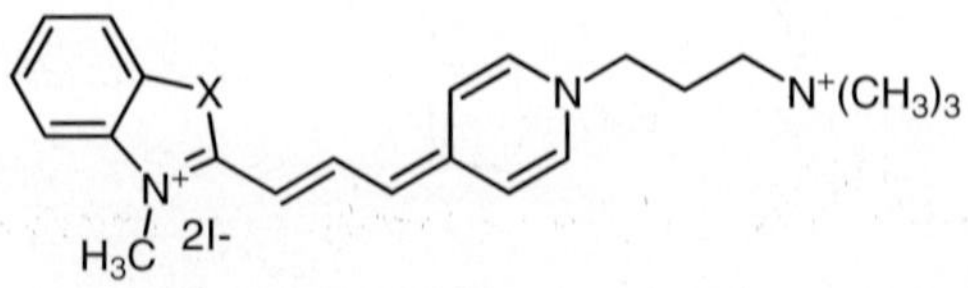

X = O, PO-PRO-3, λex/em in nm – 539/567; X = S, BO-PRO-3, λex/em in nm – 575/599

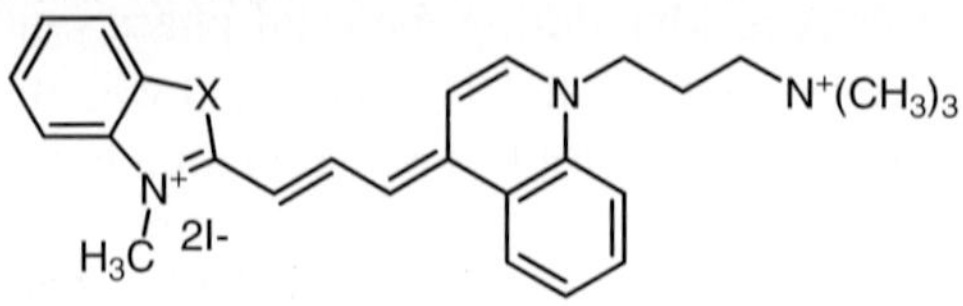

X = O, YO-PRO-3, λex/em in nm – 612/631; X = S, TO-PRO-3, λex/em in nm – 642/661

Scheme 26.

The trimethine cyanine dyes useful as nucleic acid probes are not so numerous as the monomethine cyanines. When the dyes are bound to nucleic acids, the intensity of the fluorescence is lowered with increasing length of the polymethine chain. The commercial products of this type are the cell membrane impermeable dyes of Molecular Probes Inc. [19, 24] PO-PRO-3, BO-PRO-3, YO-PRO-3, and TO-PRO-3 (Scheme 26).

A Commercially available dye TO-PRO-5 absorb in the NIR region, but the DNA-induced fluorescence enhancement is not as strong as that observed with other cyanine dyes (Scheme 27).

The synthetic route to the preparation of asymmetric trimethine cyanine dyes is well known [42] (Scheme 28).

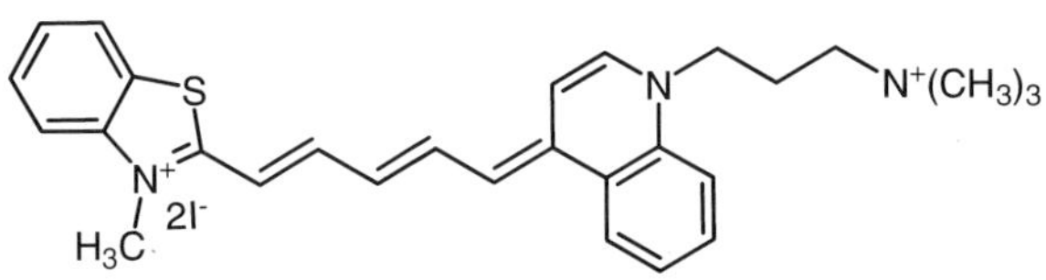

TO-PRO-5, λex/em in nm – 745/720

Scheme 27.

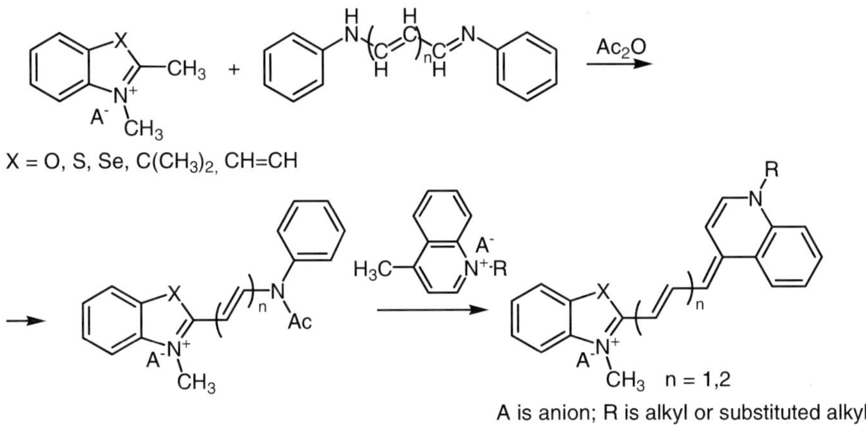

Scheme 28.

A modification of this method is the solid-phase synthesis to asymmetric trimethine cyanine dyes published in [59] (Scheme 29). The chemistry is designed to minimize the requirements for purification steps.

A key step is the preparation of a sulfonamide intermediate by reacting the hemicyanine dye with polymer-bound sulfonyl chloride, which immobilizes the hemicyanine dye. Reaction of the immobilized hemicyanine dye with a quinolinium intermediate leads to a pure cyanine dye in solution [59]. A similar method for the preparation of monomethine cyanine dyes and trimethine cyanine dyes was already mentioned [32].

Analysis of red blood cells and enumeration of reticulocytes is useful in the diagnostics of hemorrhage and anemia, for monitoring bone marrow transplantation, and for patients undergoing chemotherapy. Because reticulocytes contain RNA, these cells if stained with RNA binding excitable dyes, fluoresce when illuminated by a light source of appropriate wavelength. RNA binding dyes have been used to distinguish reticulocytes from mature red blood cells that lack RNA. Usually, suitable for flow cytometry analysis of red blood cells are the trimethine cyanine dyes excitable in the red region with relatively inexpensive diode or

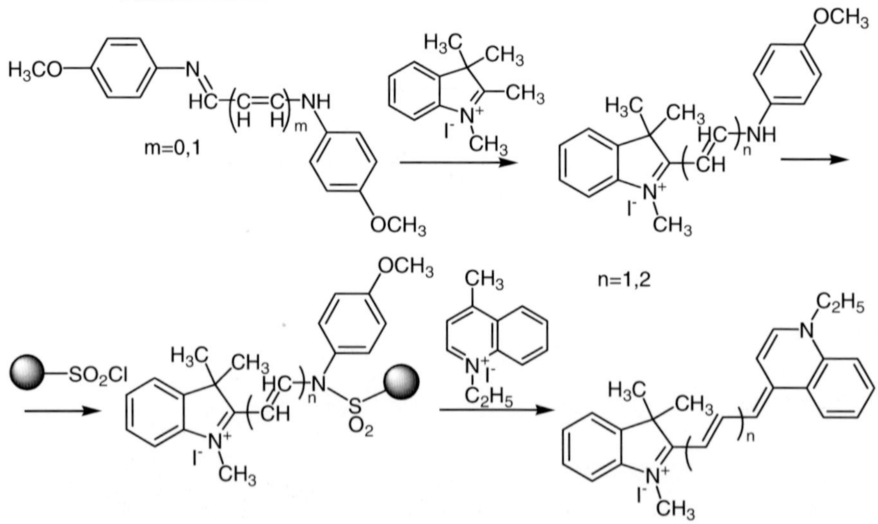

Scheme 29.

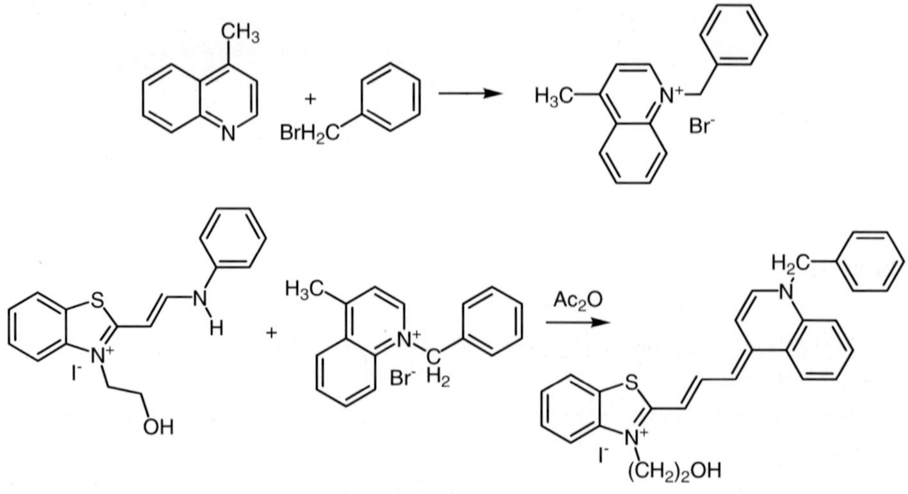

Scheme 30.

HeNe laser. Blue trimethine cyanine dyes with short incubation periods are patented [60] (Scheme 30).

Similar dyes for reticulocyte analysis by flow cytometry are also published in [61] (Scheme 31).

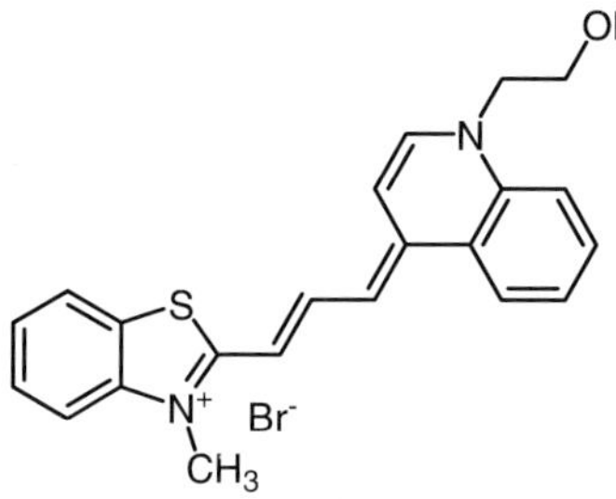

Scheme 31.

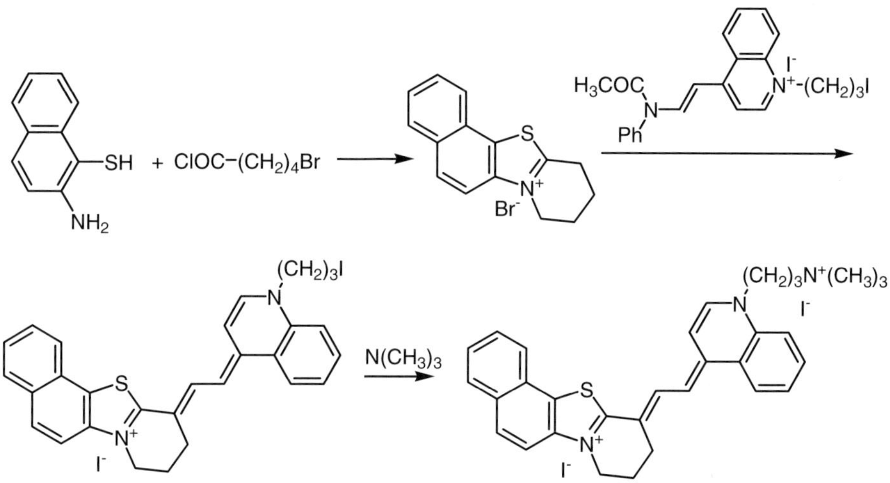

Scheme 32.

Novel intercalating trimethine cyanine dyes with modified structure are patented [62] (Scheme 32).

Reactive asymmetric trimethine cyanine dyes have also been proposed [58] (Scheme 33).

Trimethine cyanine dyes with more than one positive charge in the molecule, having high affinity to nucleic acids and used for high-sensitivity detection of dsDNA, are prepared [28] (Scheme 34).

2.2. Synthesis of homodimeric cyanine dyes as nucleic acid probes

In 1978, Le Pecq et al. [63] stuck together two ethidium bromide (EB) molecules with a short linker and made ethidium homodimer (EH) (Scheme 35).

The result was an improvement of some of its properties over the monomeric ethidium bromide dye. Ethidium homodimer binds to dsDNA, ssDNA, RNA, and

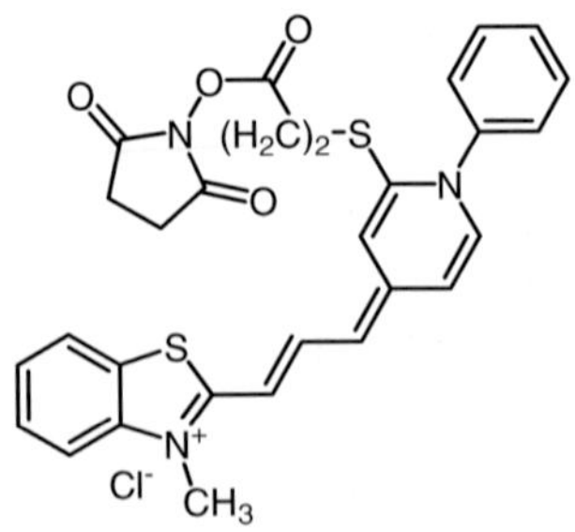

Scheme 33.

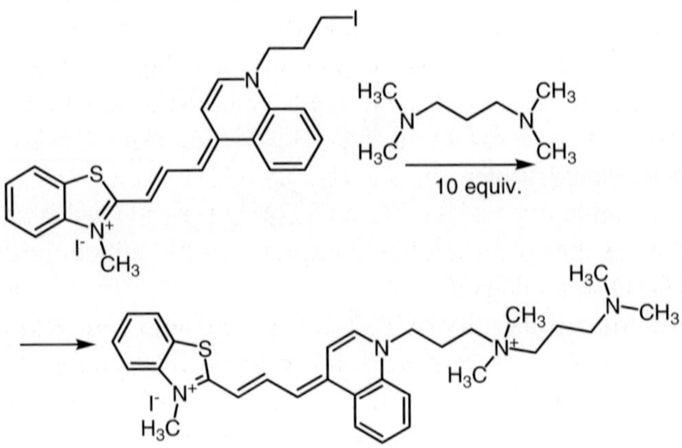

Scheme 34.

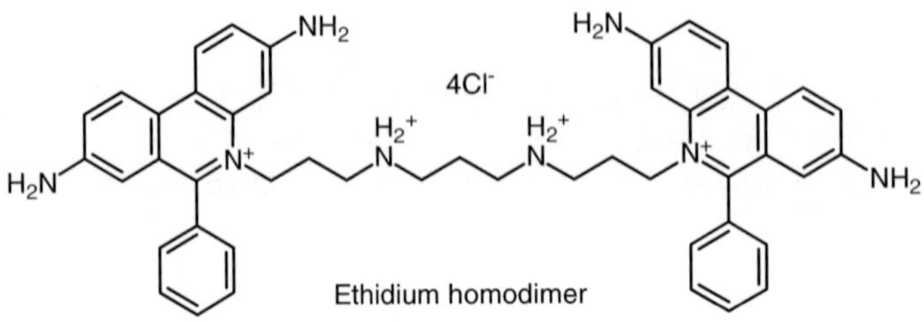

Scheme 35.

oligonucleotides up to 1,000-fold stronger than EB. A 40-fold increase in the fluorescence quantum yield of the dimer is observed on binding to dsDNA. EH is a typical *bis*-intercalator. Such a *bis*-intercalation becomes possible when the spacer connecting the two chromophores is flexible and equal to or longer than 10 Å.

The same strategy was adopted later [64, 65], but it was started with the chromophores of better dyes, YO and TO, to make the homodimers YOYO-1 and TOTO-1 (Scheme 36).

By the same reaction scheme, homodimeric monomethine and trimethine cyanine dyes having benzothiazole, benzoxazole, and quinolinium or pyridinium end-groups are synthesized [66] (Scheme 37).

Monomethine homodimeric dyes with pyrido[4,5-b]oxazolium, pyrido[4,5-b]thiazolium, and quinolinium end-groups are also synthesized [36] and commercialized [19] (Scheme 38). Similar pyrido[4,5-b]oxazolium homodimeric dyes with extended methylene bridges are also recently prepared [39].

Chemical modifications produce dramatic shifts in the chromophore absorption spectra and, with the longer polymethine chain, the fluorescence quantum yield of the nucleic acid-bound dyes is reduced, as compared to the monomethine dyes. Little or no change in their high affinity for DNA is observed. The spectra of these dyes at dye:base ratios of less than 1:1 are essentially the same for the corresponding dye-ssDNA and dye-RNA complexes. At higher dye:base ratios, however, ssDNA and RNA complexes of all of the monomethine cyanine dyes of the TO and TOTO series [19, 67] have redshifted emissions, whereas the corresponding complexes of the trimethine cyanines (TO-PRO-3 and TOTO-3) analogs do not.

The extraordinary stability of TOTO-1 nucleic acid complexes [64] ensures that the dye-DNA association remains stable even during electrophoresis and samples can be prestained with nanomolar dye concentrations prior to electrophoresis [68]. In contrast, the binding of TO (the parent compound of TOTO-1) is rapidly reversible, limiting the sensitivity of mono- and dicationic monomeric dyes and rendering its nucleic acid complex unstable to electrophoresis [69]. In addition to their superior binding properties, the homodimeric dyes are essentially

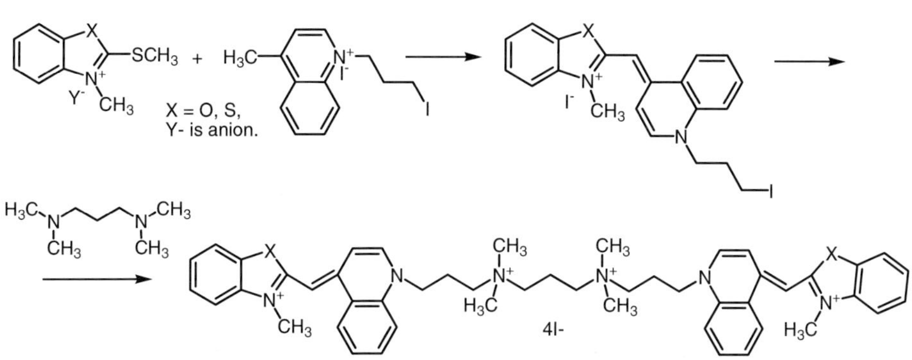

X = O, YOYO-1 λ_{ex} 491 nm, λ_{em} 509 nm; Y = S, TOTO-1 λ_{ex} 514 nm, λ_{em} 533 nm

Scheme 36.

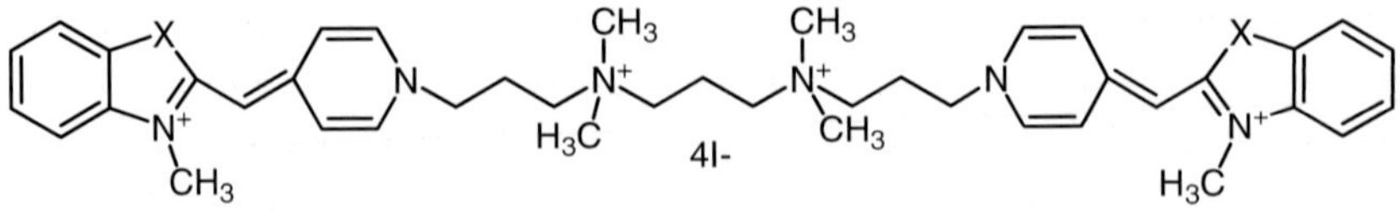

	λ_{ex} nm	λ_{em} nm
X = O POPO-1	434	456
X = S BOBO-1	462	481

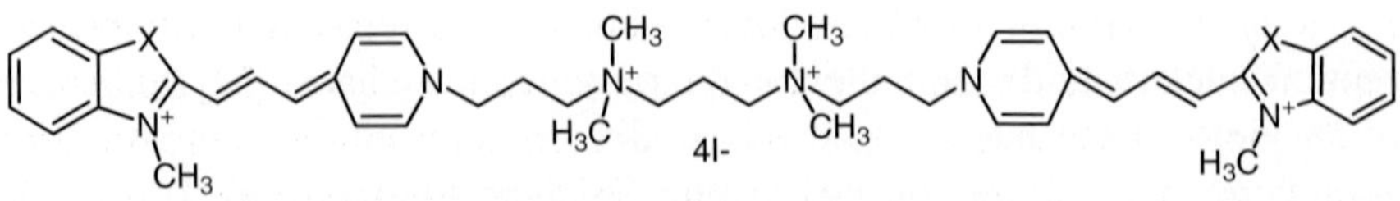

	λ_{ex} nm	λ_{em} nm
X = O POPO-3	534	570
X = S BOBO-3	570	602

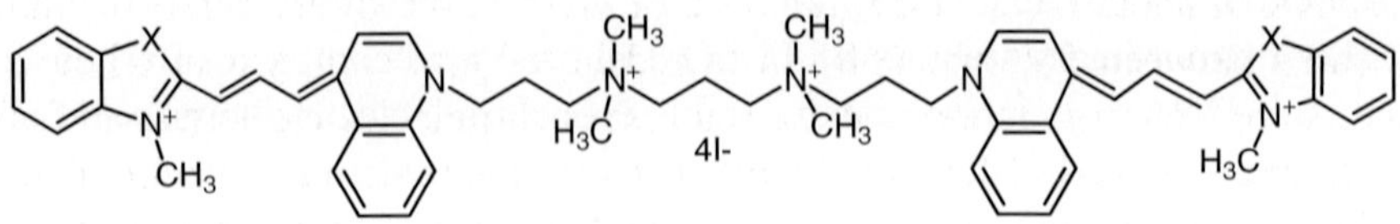

	λ_{ex} nm	λ_{em} nm
X = O YOYO-3	612	631
X = S TOTO-3	642	660

Scheme 37.

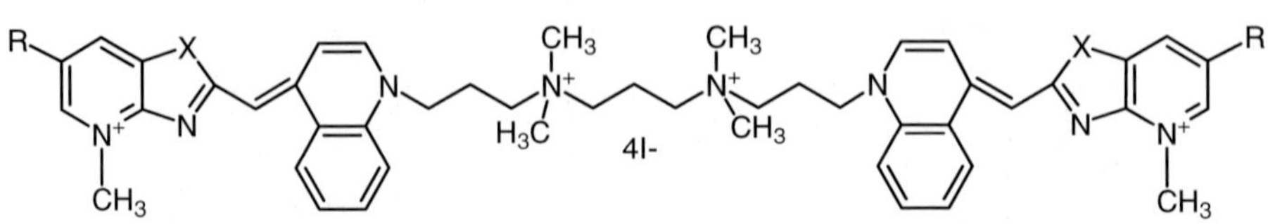

	λ_{ex} nm	λ_{em} nm
R = H, X = O JOJO-1	529	545
R = Br, X = S LOLO-1	565	579

Scheme 38.

non-fluorescent in the absence of nucleic acids and exhibit significant fluorescence enhancements (100- to 1,000-fold) upon DNA binding [64, 70]. Furthermore, the fluorescence quantum yields of the monomethine cyanine homodimers are high (generally between 0.2 and 0.6) and their molar absorptivities are usually > 100,000 l mol^{-1} cm^{-1} [19].

It was established that the staining of nucleic acids by BOBO-1 and POPO-1 dyes is much faster (occurring within minutes) than staining by the YOYO-1 or TOTO-1 dyes (which can take several hours to reach equilibrium under the same experimental conditions) [71].

A series of novel homodimeric monomethine dyes similar to TOTO-1 are prepared by an improved synthetic method and their absorption and fluorescence spectral characteristics in the presence of nucleic acids are listed [72] (Scheme 39).

The novel dyes do not exhibit any fluorescence in solution at room temperature and show a fluorescence quantum yield of less than 0.001. They have absorption at 505 nm with molar absorptivity 130,000–180,000 l mol^{-1} cm^{-1}. Their fluorescence maxima in the presence of dsDNA are located at 530–534 nm with quantum yields in the range 0.5–0.9 (with the exception of the dye with a viologen moiety in the linker – TOTOBIPY) considerably higher than the commercial product TOTO-1 which has a fluorescence quantum yield in the presence of dsDNA of about 0.35. In the presence of ssDNA, the fluorescence maximum is located between 560 and 650 nm (a redshifted emission typical of the TO-monomethine chromophore) and the fluorescence quantum yields were 0.3–0.8.

In another study [73], novel homodimeric monomethine cyanine dyes of the YOYO-1 type are synthesized (Scheme 40) and their fluorescence spectral properties in the presence of nucleic acids are investigated.

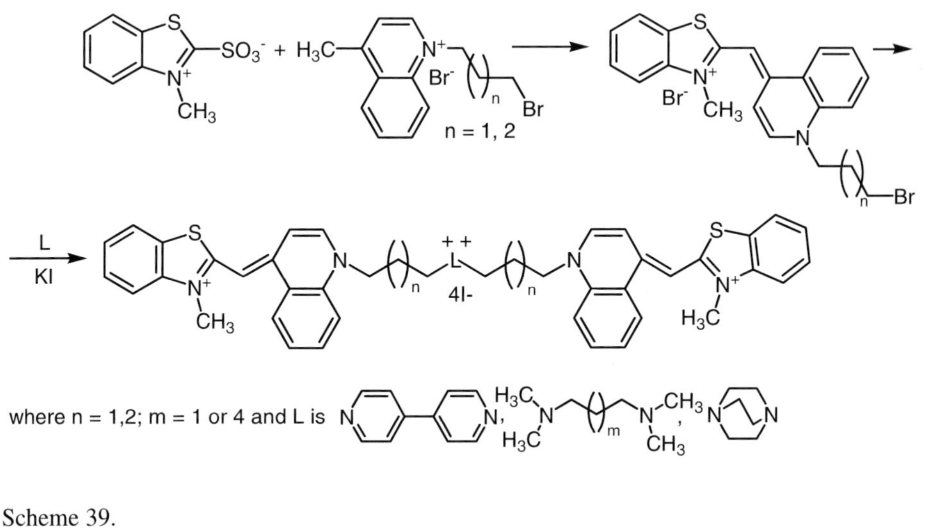

Scheme 39.

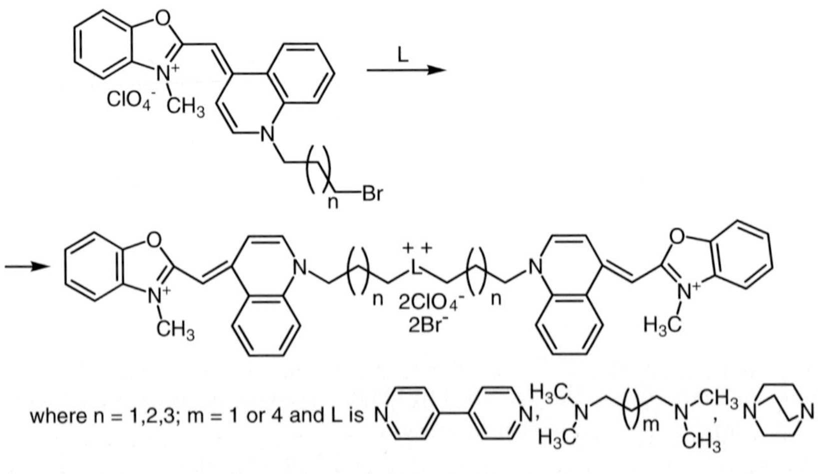

Scheme 40.

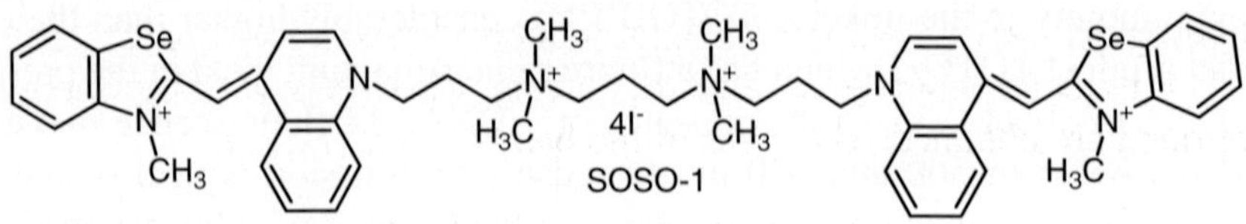

Scheme 41.

The dyes absorb in the region 478–485 nm and the corresponding molar absorptivities are between 140,000 and 180,000 l mol^{-1} cm^{-1}. The dyes have no fluorescence of their own, but become strongly fluorescent after binding to dsDNA. The fluorescence maximum of the dye–dsDNA complexes are around 505 nm and the fluorescence quantum yields are between 0.3 and 0.7, except for the dye having a viologen moiety in the linker.

A homodimeric monomethine cyanine dye with benzoselenazolium and quinoline end-groups (SOSO-1) is synthesized [74] (Scheme 41).

In the same investigation, an analog of TOTO-1 with six positive charges (TOTO-1-6C) is also synthesized [74] (Scheme 42).

The absorption maximum of SOSO-1 is at 495 nm with a shoulder at 495 nm and the corresponding molar absorptivities are 100,000 and 74,000 l mol^{-1} cm^{-1}, respectively. For TOTO-1-6C, the maximum is at 505 nm and the molar absorptivity is 147,000 l mol^{-1} cm^{-1}. The fluorescence maximum of the complex SOSO-1/dsDNA is at 540 nm, while that of TOTO-1-6C/dsDNA is at 534 nm. The fluorescence quantum yields are 0.25 and 0.35, respectively. The stability of the dye-dsDNA complexes has been studied. Since their formation is

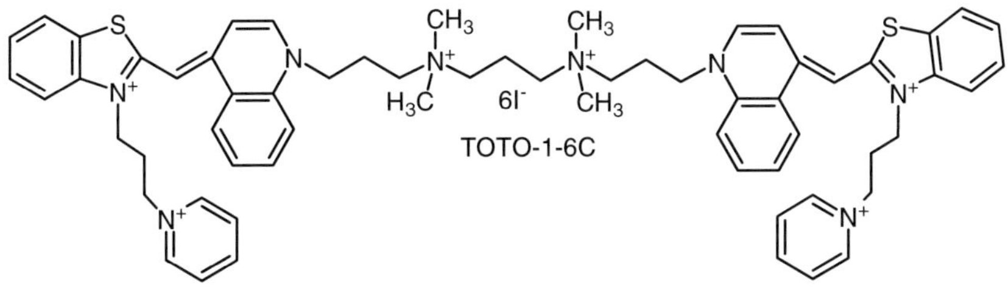

Scheme 42.

too fast, the measurements are performed immediately after mixing the dye with the dsDNA solution. The experimental results show that the complexes are stable for at least 48 h when kept in the sunlight – no photobleaching was observed.

Melting curve analysis is performed by heating dsDNA in the presence of SOSO-1 and TOTO-1-6C. The fluorescence intensity diminishes and the position of the fluorescence maximum shifts to the right by approximately 20 nm. After cooling for 1 min, the fluorescent characteristics became identical with those of the starting dye-dsDNA complexes in the solution. These complexes are also stable when used for agarose gel electrophoresis. Clear gels with no fluorescent background are obtained. The color of the bands is bright green.

2.3. Heterodimeric nucleic acid cyanine dyes suitable as non-covalent probes

Benson et al. [75–77] designed new heterodimeric dyes (Scheme 43), in which two different chromophores (with different spectral characteristics) are connected with a spacer longer than 10 Å, allowing *bis* intercalation of both dye fragments. Both dye fragments – of TO and thiazole blue (a trimethine cyanine dye) – are chosen to serve as a donor–acceptor pair to generate a heterodimeric dye molecule, which could be efficiently excited at a single-wavelength, but will fluoresce with a well-separated emission maximum. The choice of the chromophores, paired in the heterodymeric dyes, is based on the following considerations. The spectroscopic properties of the chromophores are such that the emission spectrum of the dye, chosen as donor, overlaps extensively the absorption spectra of the acceptor and both chromophores are in close proximity (10 Å). In the heterocyclic TOTAB dye, efficient fluorescence resonance energy transfer (FRET) is found, as it is shown by the quenching exceeding 90% of the donor emission and by a large enhancement of the acceptor chromophore fluorescence emission on binding to dsDNA.

Another TO-ethidium heterodimer (TOED) is obtained, as it is outlined in the following reaction Scheme 44.

Some other heterodimeric dyes (TOTIN) with improved properties have been synthesized by the same researchers [78, 79] (Scheme 45). These cyanine heterodimeric dyes for FRET are designed with TO as the common donor and

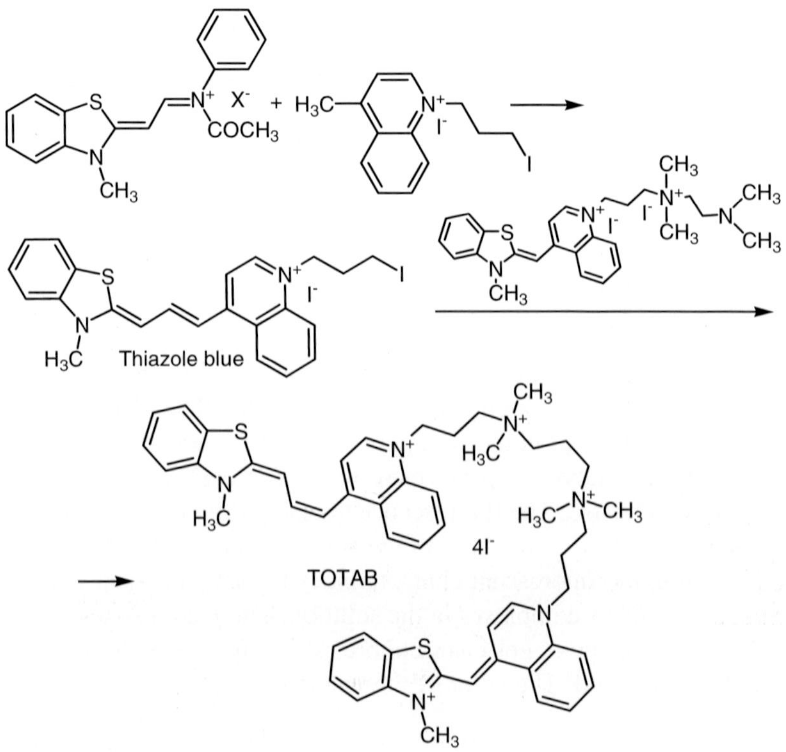

Scheme 43.

thiazole indolenine (a pentamethine dye) as acceptor. The acceptor has a large molar absorptivity and a fluorescence emission maximum at about 650 nm. The affinity to dsDNA and the quenching of the donor fluorescence are optimized by varying the length of the linker between the donor and acceptor. The stability to electrophoresis of the dsDNA complexes of the optimized heterodimeric dyes and their fluorescent emission properties are considerably superior to those of similar previously described complexes. The TO emission is quenched by 83% in the TOTIN dyes and this is a measure of the efficiency of the energy transfer from the donor to the acceptor molecule. The long-wavelength emission of the heterodimer is almost 25-fold higher than that of the monomeric pentamethine fragment. The linker length strongly affects the intensity of the emission of DNA-bound dyes. The best dye is butyl-TOTIN, $n = 2$, presumably because of a more favorable geometry for energy transfer from the donor to the acceptor chromophore.

Japanese scientists [80] have synthesized heterodimeric dyes with ethidium chromophore as donor and thiazole blue as acceptor (Scheme 46).

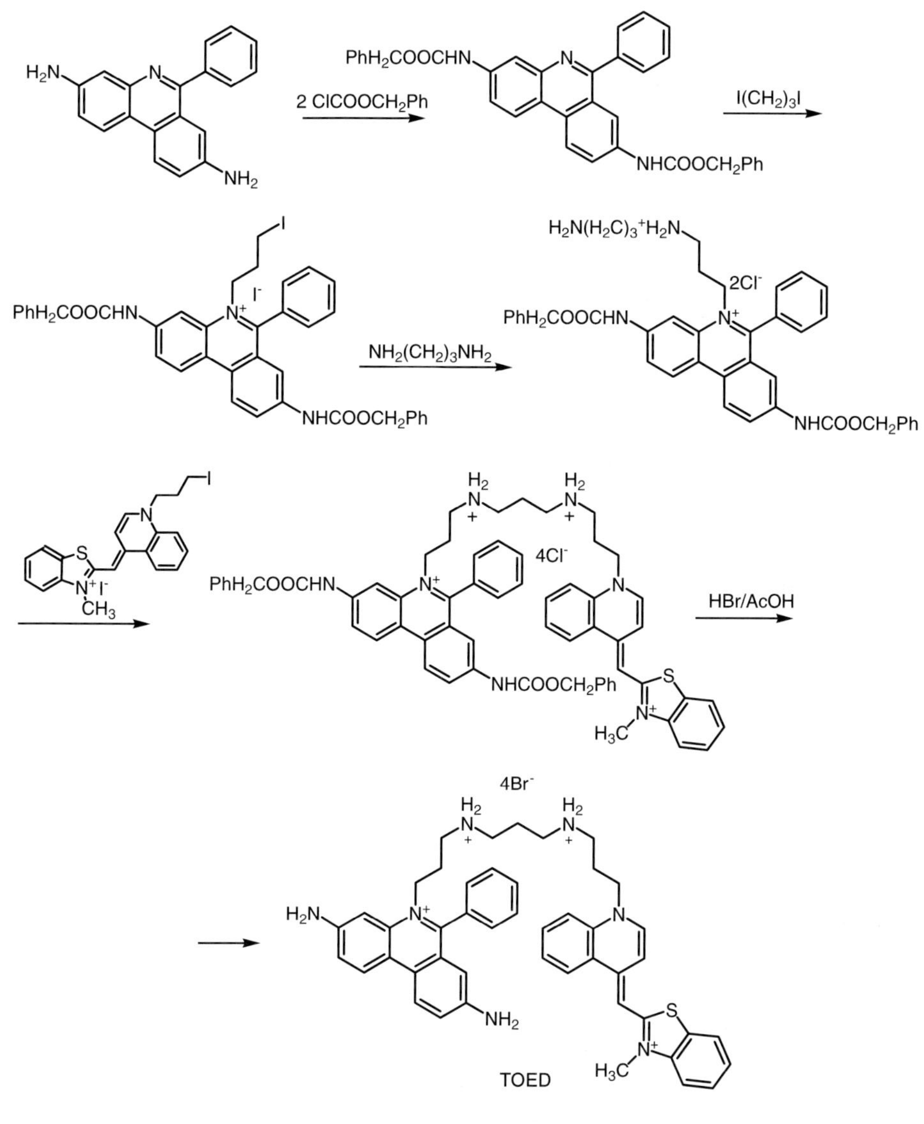

Scheme 44.

An interesting development [80] is the synthesis of reactive heterodimeric dyes capable of reacting covalently with a DNA oligomer (Scheme 47).

Up to 58-fold enhancement is observed of the fluorescence intensity of the acceptor dye upon binding or conjunction to DNA or oligomeric DNA.

An important question related to the use of such dyes as fluorescent labels is the origin of the drastic difference in quantum yield in the free and the bound form. Carlson et al. [81] have investigated the photophysical properties of the

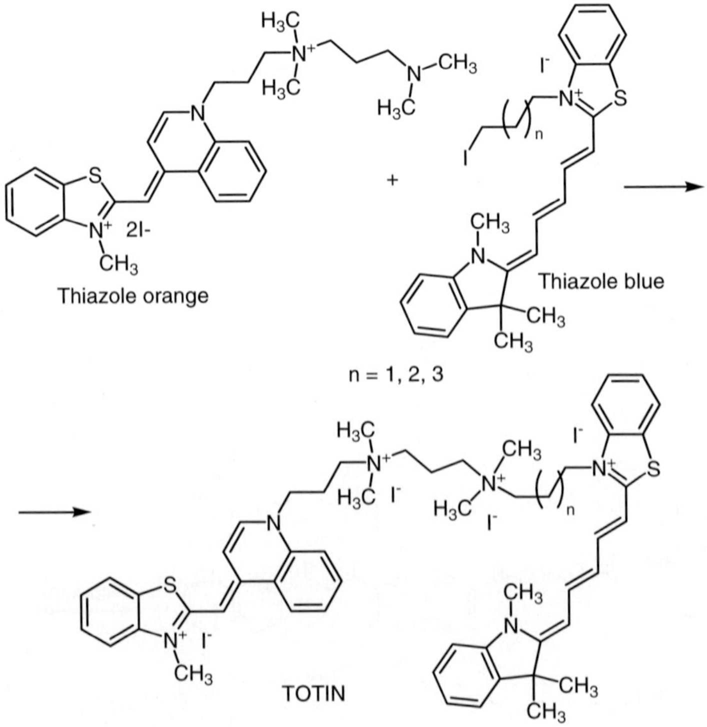

Scheme 45.

chromophore YO-PRO-1. YO-PRO-1 in aqueous solution is virtually non-fluorescent, but upon binding to DNA its fluorescence quantum yield is increased to about 1,000-fold. The authors [81] have observed a similar enhancement of the fluorescence quantum yield of the dye TO-PRO-1 to 0.3 at 4°C in glycerol, compared to the free dye in aqueous solution. The quenching of the fluorescence of the free dye in aqueous solution cannot be explained solely by external quenching, but must be due to some internal quenching mechanism, or twisting or stretching motions of the molecule. The question is whether the internal quenching could be related to a rotation around the internuclear bridge between benzoxazole and quinolinium rings, or the high fluorescence intensity of YO-PRO-1 bound to DNA (as well as in glycerol) is a result of a decreased degree of internal rotation. It is found that the quantum yield of YO-PRO-1 depends on the possibility of rotation around the internuclear bond between the two-ring systems. This explanation is also in agreement with the great enhancement in the quantum yield of the dye upon binding to DNA, because it is shown that YO-PRO-1 binds by intercalation and in the binding mode, where the molecule is inserted between base pairs in DNA, the internal rotation is likely to be

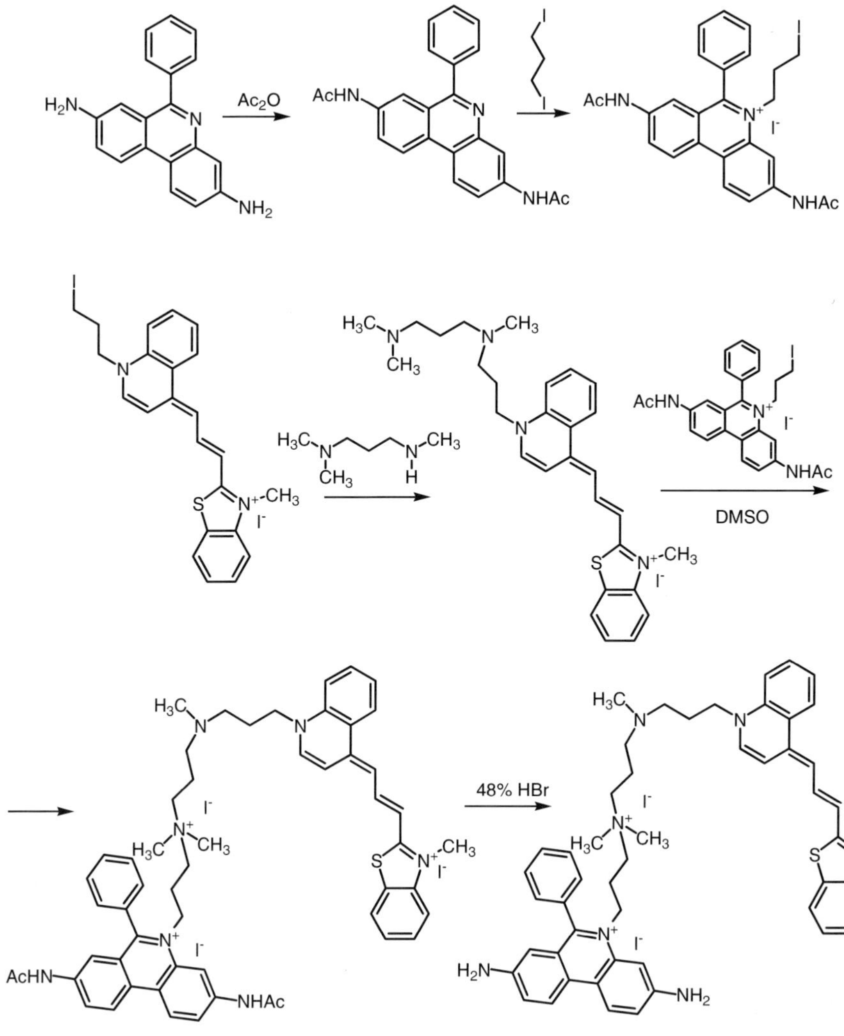

Scheme 46.

strongly hindered. The same explanation is also valid for the other dyes of the TO and YO types.

2.4. Synthesis of styryl dyes as nucleic acid probes

Abott Laboratories patented [82] styryl dyes (Scheme 48) as nucleic acid non-covalently binding probes.

Another representative of these series is a naphthothiazole styryl dye (Scheme 49). The fluorescence intensity in the presence of these dyes is increased 121 and 159-fold for the benzothiazole and naphthothiazole dyes, respectively.

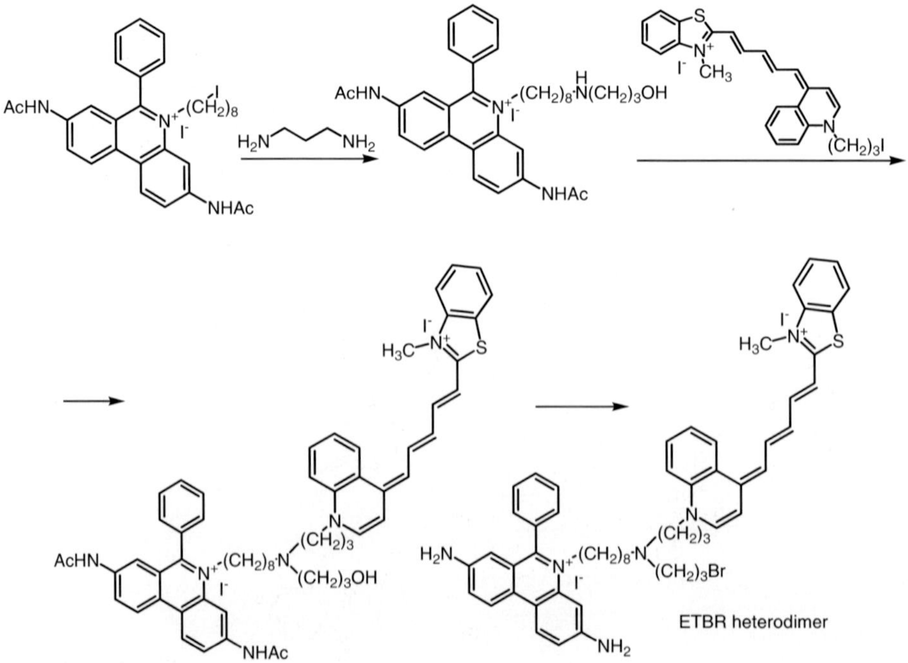

Scheme 47A.

Kumar et al. [83] showed that the styryl dye 4-(4-*N*,*N*-dimethylaminostyryl)-1-methylpyrydinium iodide (Scheme 50) is a non-covalent fluorescent dye for DNA analysis.

Recently, the synthesis (Scheme 51) and spectroluminescent properties of homodimer styrylcyanine dyes were reported [84]. Dyes based on (4-dimethylaminostyryl)pyridinium, (4-dimethylaminostyryl)benzoxazolium, (4-dimethylaminostyryl)benzothiazolium, and (4-dimethylaminostyryl)-1,3,3-trimethyl-3H-indolium heterocycles have been synthesized.

Upon binding to DNA, some of the dyes demonstrated up to 130-fold enhancement of the fluorescence intensity. The dyes have moderate or large Stokes shifts up to 160 nm. Generally, the fluorescence intensity in the free state is insignificant.

A combinatorial approach to the synthesis of styryl dyes has been developed and their subcellular localization (including nuclear, mitochondrial, cytosolic vesicular, granular, and reticular localization) has been investigated [85]. By cell-based screening and fluorescence property measurement upon binding to DNA, a sensitive (according to the same authors [86]) styryl dye was found.

The compound shown above (Scheme 52) is a cell membrane permeable nuclear staining dye and can be useful for live cell imaging purposes [86].

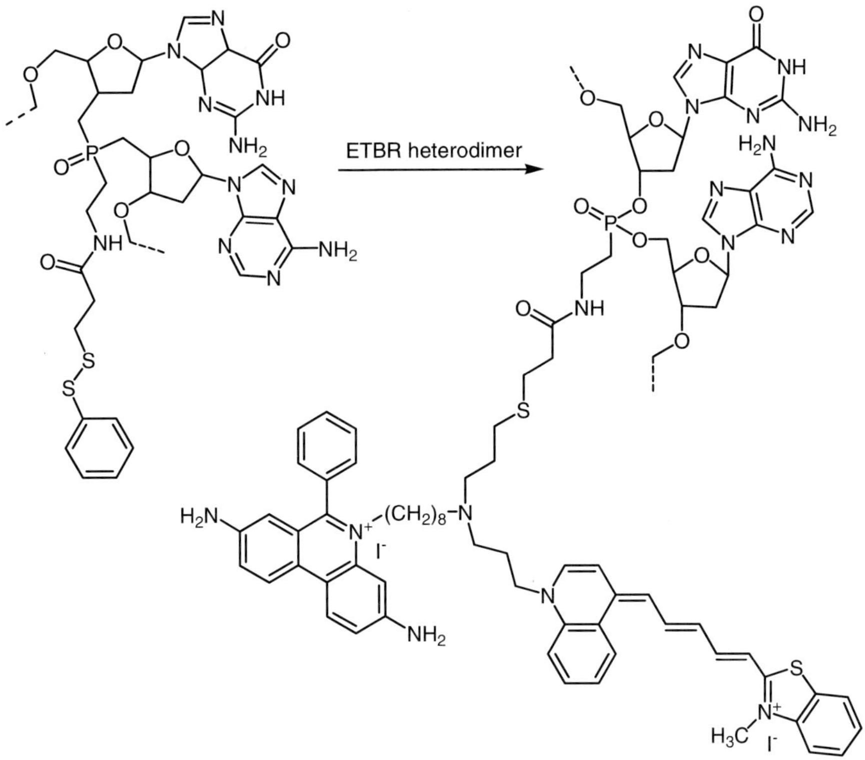

Scheme 47B.

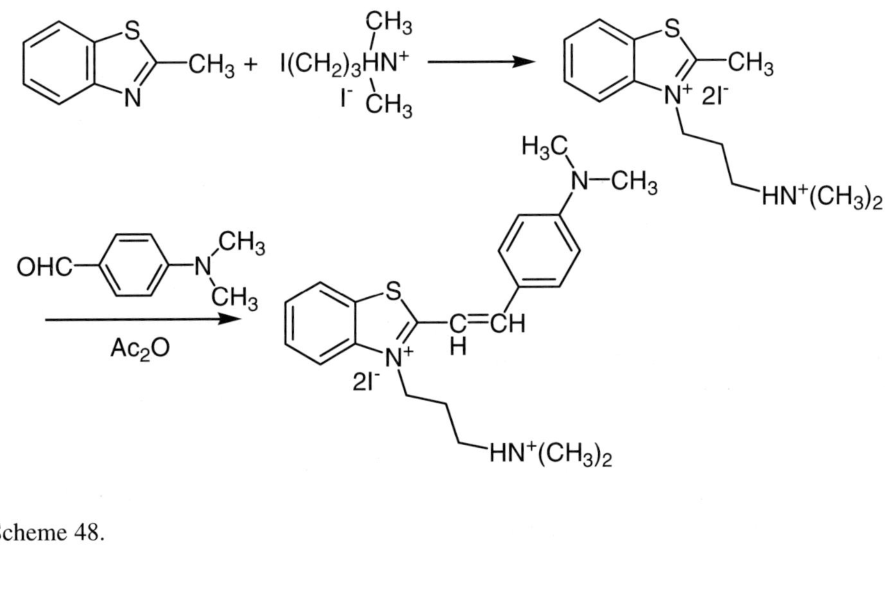

Scheme 48.

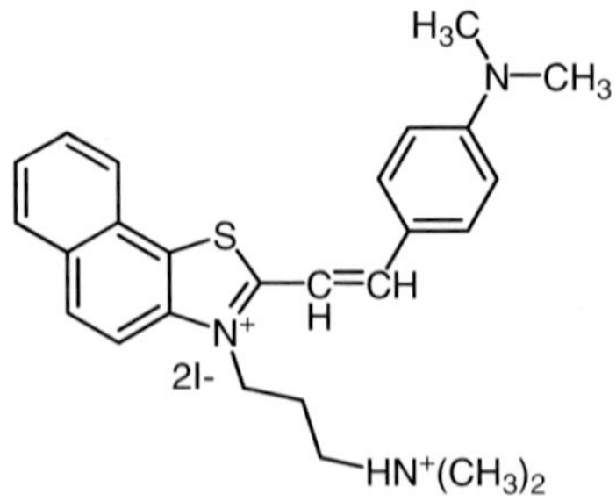

Scheme 49.

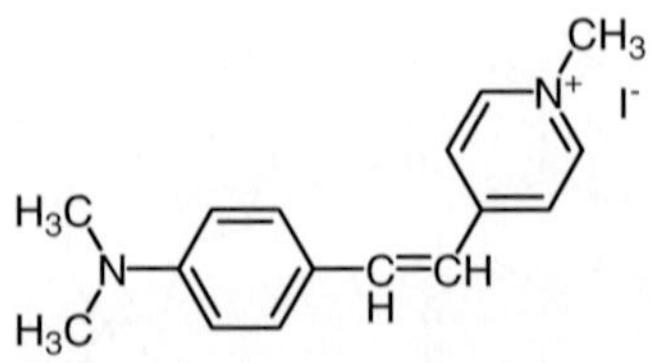

Scheme 50.

A commercial product [19] known as LDS-751 (Scheme 53) is a cell membrane permeable nucleic acid stain that has been used to discriminate intact nucleated cells from non-nucleated and damaged nucleated cells [87], as well as to identify distinct cell types in mixed populations of neutrophiles, leukocytes, and monocytes by flow cytometry [88]. LDS-751 has its peak excitation of around 543 nm when bound to dsDNA. The dye can be excited at 488 nm by the argon laser and is particularly useful in multicolor analysis due to its long-wavelength emission maximum (about 712 nm). Binding of LDS-751 to dsDNA results in 20-fold fluorescence enhancement. When LDS-751 binds to RNA, one observes [19] a significant redshift in its excitation maximum to 590 nm and a blueshift in its emission maximum to 607 nm, suggesting its use to discriminate between DNA and RNA in cells.

A very interesting amplified spontaneous emission [89] is observed from a styryl dye in a complex film composed of a 4-[(4-*N*,*N*-dimethylamino)styryl-1-dodecyl]pyridinium bromide (DMASDPB) dye (Scheme 54), lipid, and DNA.

The dye DMASDPB is known as a non-linear optical material. The dye film shows an amplified spontaneous emission activity under optical pumping, with intensity above the threshold values, when the dye is mixed with DNA–lipid complexes. The preliminary results on the low threshold and long durability indicate a promising possibility for practical applications.

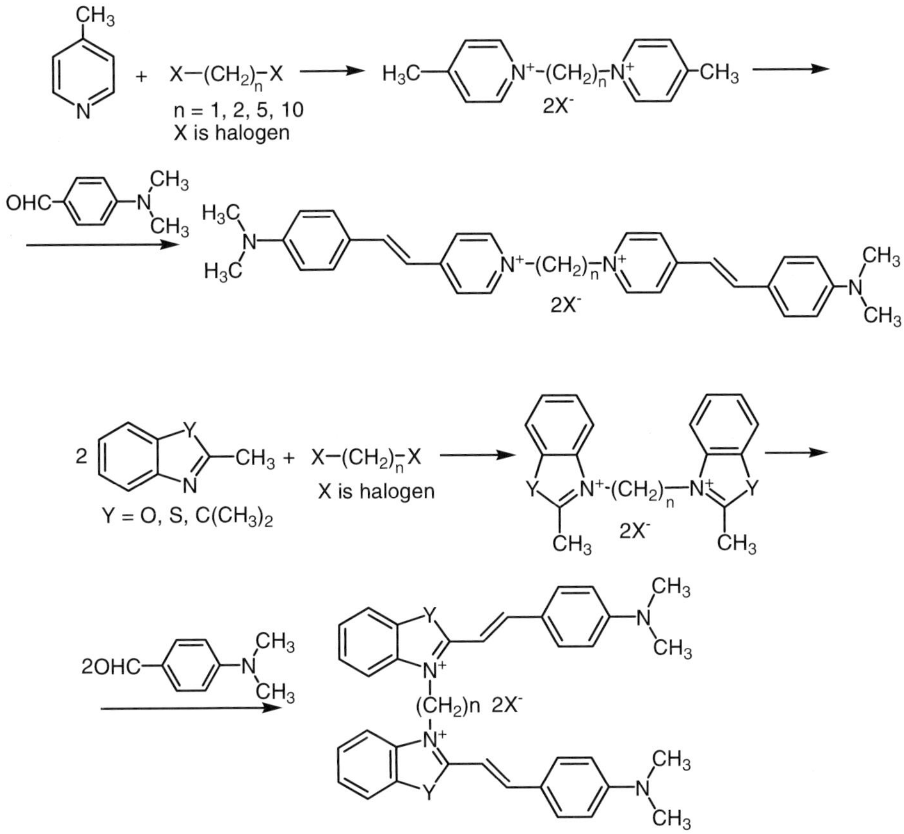

Scheme 51.

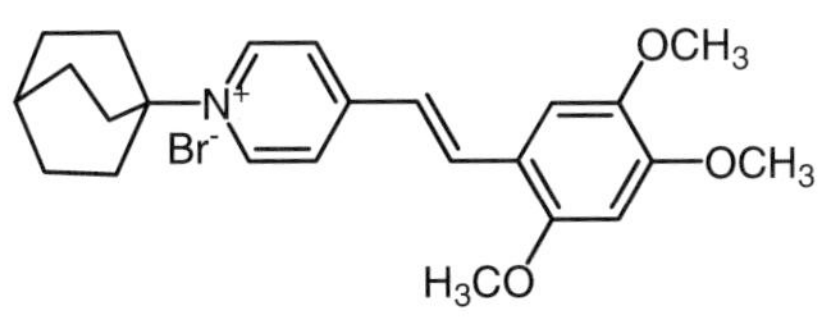

Scheme 52.

2.5. Synthesis of pyrylium dyes as nucleic acid probes

Pyrylium and thiopyrylium compounds (Scheme 55) were first proposed in a patent publication [90] as biological stains with minimal background fluorescence. These compounds are useful as both vital and fixed cell stains and can be used without a wash step. The pyrylium and thiopyrylium salts can be used for differentiation of biological cells and tissues. The dyes are usually metachromic

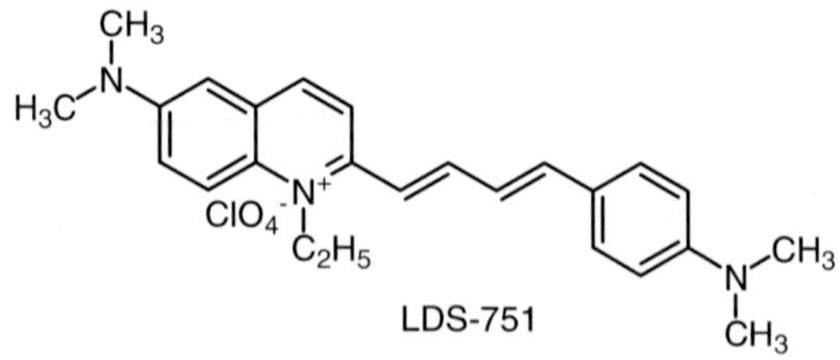

Scheme 53.

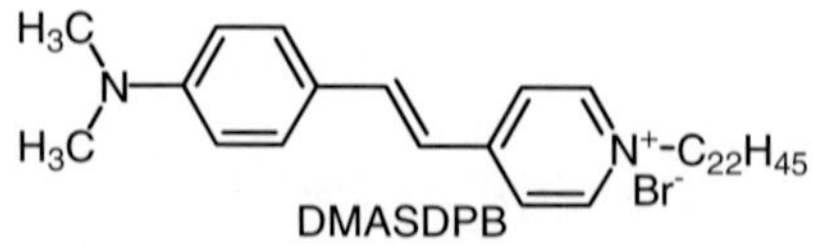

Scheme 54.

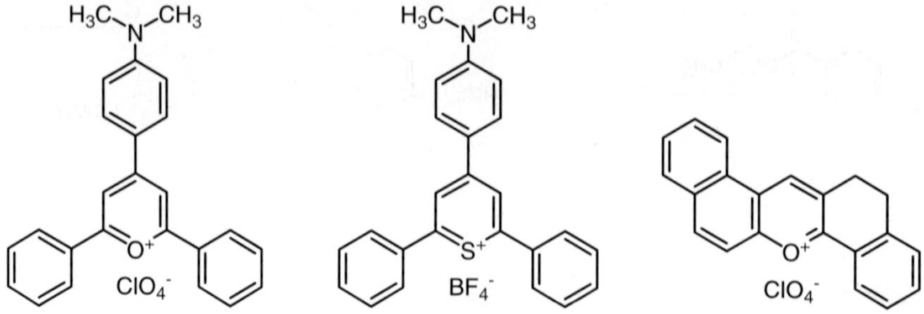

Scheme 55.

fluorochromes and produce unique staining of the cell nucleus and cytoplasm, which stain red and green, respectively.

Japanese researchers reported that certain pyrylium salts are suitable for DNA detection [91, 92] and PCR reaction [93, 94] (Scheme 56).

Such dyes are intercalators with low background and low rate of bleaching, allowing the preparation of the specimens for long-term storage. Pyrylium salts can be used for quantification in solution for gel electrophoresis, for sizing, for live staining of cells and bacteria, for staining of chromosomes, and in PCR reaction [92]. It is possible for a pyrylium dye to be covalently linked to an oligonucleotide for fluorescence *in situ* hybridization (FISH) (Scheme 57).

A comprehensive survey of the synthesis and properties of the pyrylium salts can be found in [95].

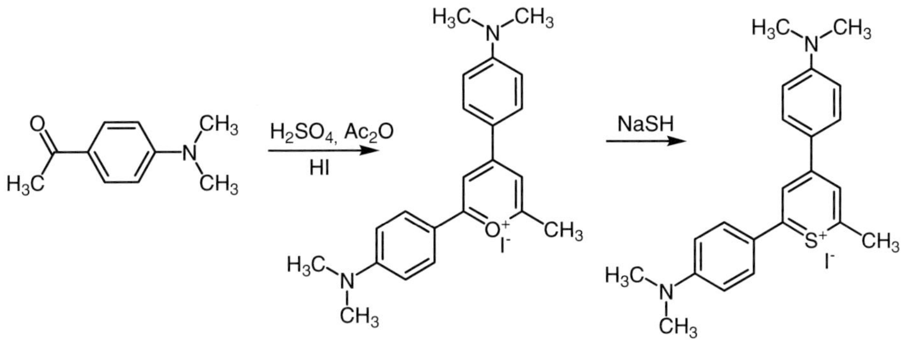

Scheme 56.

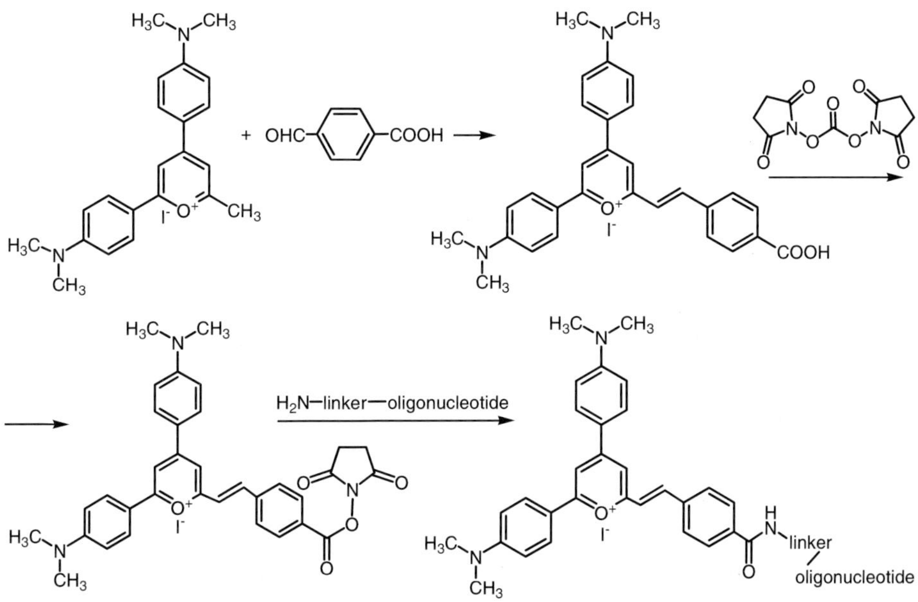

Scheme 57.

3. APPLICATIONS OF CYANINE DYES AS NUCLEIC ACID PROBES

The applications of cyanine dyes as nucleic acid probes in the field of biotechnology are numerous and continuously expanding. We do not claim this to be a comprehensive survey of their applications. For further information, the reader is encouraged to consult www.probes.com where a large number of publications on such dyes, commercially produced by Molecular Probes Inc., are available.

Among the most numerous applications of cyanine dyes is the visualization of DNA by following different biological processes and quantification of DNA in solution. YOYO-1-labeled complexes may be employed as qualitative DNA markers in intracellular delivery studies [96].

The accurate determination of the dsDNA concentration is essential for many processes in molecular biology and physiology. Fluorescence characteristics of variously charged asymmetric monomethine cyanine dyes in the presence of nucleic acids are reported [97]. Highly sensitive fluorescence labels based on a series of newly synthesized homodimeric monomethine cyanine dyes are proposed [98]. Quantification in solution [99], in soils and aquatic sediments [100], and by a cheap custom-made fluorimeter [101] of dsDNA with the dye PG has been described. It was shown [102] that quantitative and sensitive measurement of dsDNA is achieved using a 96-well microplate with SG dye. Ultrasensitive analysis of dsDNA separated by capillary electrophoresis (CE) with POPO-3, YOYO-3, and YOYO-1 is reported [103]. The detection limit of dsDNA is of femtogram amounts. TO-PRO-3 is used in sensitive CE analysis with diode-laser-induced fluorescence [104]. The use is described [105, 106] on-the-fly fluorescence lifetime detection for DNA restriction fragment analysis, in which a dye-labeled digest and a dye-labeled size standard fragment or fragment ladder are simultaneously separated by CE on the same column and distinguished on the basis of fluorescence lifetime of the dye labels. The DNA fragments are labeled with novel monomeric and homodimeric dyes with one, two, three, and four positive charges, respectively [105, 106]. The pyrylium dye [107] proposed by Japanese scientists [91–94] and the SYTOX Green cyanine dye [108] have been used as biosensors for the detection of DNA and live bacteria, respectively.

Enzymatic reaction studies on a single-molecule level aim at the understanding of the functioning of molecular motors, such as DNA polymerases, RNA polymerases, and restriction enzymes at low concentrations. With SG, a novel technique for site-specific enzymatic cleavage of a single stretched DNA molecule is reported [109]. The mismatched heteroduplex DNA protected against exonuclease digestion is proposed to be visualized by direct dyeing of the undigested DNA using sensitive nucleic acid dyes, such as SYBR Gold or SG [110].

A fluorescent viability assay is proposed, using nucleic acid cyanine dyes as cell membrane permeable dyes and other cell membrane impermeable dyes (usually PI) with different spectral characteristics [111]. Chemically reactive cyanine dyes containing a labile chlorine substituent structurally related to the SYTO dyes [19] are proposed as stains for assessing the viability of yeast and fungi [112]. The FUN-1 and FUN-2 stains are freely taken by yeast and fungi and converted from a diffusely distributed pool of yellow-green-fluorescent intercellular stain into compact red-orange or yellow-orange fluorescent intravacuolar structures, respectively. Conversion of the dyes into products with longer emission requires both plasma membrane integrity and metabolic capability. Only metabolically active cells are marked clearly with fluorescent intravacuolar structures, while dead cells exhibit extremely bright diffuse, yellow-green fluorescence [19]. With the combination of a green-fluorescent SYTO 9 dye and red-fluorescent hexidium iodide, it is possible to perform a fluorescent assay for Gram reaction [113]. When mixed populations of live Gram-negative and Gram-positive bacteria are simultaneously stained with the membrane permeable SYTO 9 dye in combination with hexidium iodide, the Gram-negative bacteria fluoresce green, and the Gram-positive bacteria fluoresce red-orange. Dead bacteria do not exhibit predictable staining patterns [19].

Since its introduction [114], the PCR has revolutionized molecular biology and the way of studying, discovering, and understanding fundamental aspects of life. A specific DNA sequence is amplified by an enzymatic reaction using repetitive cycling between three temperatures. The product of the reaction can be diagnostic for the presence of genetic diseases and for the presence of a specific bacterium or virus in a tissue or blood sample and can be used in many other genetic tests, e.g., forensic investigation [115]. Genetic analysis by conventional methods (direct sequencing) is still a time-consuming, manually intensive, and costly procedure in comparison with PCR. The use of the PG dye added after the completion of the reaction and detection by optical fiber-based laser-induced fluorescence allowed an immediate on-line registration of amplification [116]. The same dye has been used for quantitative determination of PCR products in a microplate format, which enables the rapid analysis of large sets of samples, e.g., in connection with diagnostic DNA tests [117]. YOYO-1 has been used for the development of a screening method for a large quantity of PCR products by fluorescence measuring on 98-well polypropylene plates [118]. Generally, all intercalating cyanine dyes are suitable for PCR if they are added after the completion of the reaction, thus giving a fluorescence signal. In a patent publication [119], it is claimed that monomeric dyes, such as BO-PRO-1, TO-PRO-1, YO-PRO-1, and TO-PRO-3, can be used from the beginning of the process to monitor the progress of the amplification. It is stated that homodimeric dyes, such as YOYO-1 and TOTO-3, cannot be used during amplification because they inhibit the PCR process considerably.

In some applications, it is desirable (or even necessary) to use PCR not merely to detect a specific template, but to determine how much of this template is present. The best methods for quantitative PCR are commonly known as real-time PCR, since one can monitor the product accumulation during the reaction, rather than merely assaying it at the end. This has advantages also in diagnostic applications, since a result is obtained much faster.

The most used dye for real-time PCR is SG. Its selectivity for dsDNA is important in many applications, especially in real-time PCR and dsDNA melting-curve analysis [120–122]. Recently, the structures of SG and PG have been determined [27], and they have molar absorptivities of 73,000 l mol^{-1} cm^{-1} at

494 nm and 70,000 l mol^{-1} cm^{-1} at 500 nm, respectively. With respect to the use of SG in real-time PCR, the important property of this dye is the difference in fluorescence intensity when bound to ssDNA or dsDNA. It was shown that SG links efficiently to ssDNA, but the fluorescence intensities are rather low. The data indicate that dye/base pair ratios of more than 0.2 should be applied to distinguish between ssDNA and dsDNA, and dye/base pair ratios of at least 10 appear to be advantageous. Minor groove binding of SG is assumed at dye/base pair ratios higher than 0.15 and this is leading to a significant increase in fluorescence. However, inhibition of PCR [123] as well as potential degradation of dsDNA has to be considered, at least at high dye/base pair ratios. This should also be taken into account in other applications, in particular for the determination of DNase or telomerase activity [124], in fluorescence imaging techniques [125], and for comet assay [126]. However, the dye/base pair ratio is not constant in PCR. It changes with the cycle number as more dsDNA is produced. Therefore, the number of cycles can influence the melting curve analysis [120]. Again, the application of sufficiently high dye/base pair ratios, just below a value where significant inhibition is observed, appears advantageous [27]. There are numerous publications using SG for real-time PCR [127–131].

A sensitive real-time PCR-based telomeric repeat amplification protocol (RQ-TRAP) assay with SG dye to detect telomerase activity for evaluation of the role of telomerase in tumor development and cell immortalization is described [132]. A method reported in [132] is based on the absorptive quantification of genetic elements in a known amount (mass) of genomic DNA by real-time quantitative PCR with SG.

Flow cytometry has been described as automated microscopy that has the advantages of automation objectivity and speed (flow cytometers can analyze many thousands of cells per second) [133]. To do this, the flow cytometer quantitatively measures the optical characteristics of stained cells into a fast flowing fluid stream in front of a focused light beam. The light source used to illuminate the samples is a UV source or laser. As particles pass through the light beam, three parameters are measured using photomultiplier tubes. These are forward scatter, side scatter, and fluorescence [133]. A typical flow cytometer measures the fluorescence in three wavelength ranges. Light of defined wavelengths is channeled to particular detectors, e.g., Channel 1 will measure green fluorescence, Channel 2 – orange fluorescence, and Channel 3 – red fluorescence. Most microorganisms are optically too similar to resolve from each other and from debris without artificially introducing some difference. The microorganisms are modified using fluorescent labeling techniques, such as antibodies, lectins, nucleic acid probes, or fluorescent dyes [133]. The most commonly used fluorescent cell stains are the nucleic acid dyes DAPI, acridine orange, and Hoechst 3334 and in the last 15 years, cyanine dyes have been widely used as nucleic acid probes. Stains are often used in combination in order to gain additional information about a cell or to confirm results.

The major consideration with such dual staining, when using one light source, is that the two or more stains should have different emission spectra [133].

A reproducible method to quantify DNA in DNA and protein solutions at ppb levels with PG using flow cytometry analysis is reported [134]. Different virus families are stained using a variety of fluorescent nucleic acid dyes (SG, SYBR Green II, Oli Green, PG) and examined using a standard flow cytometer equipped with an argon-ion laser [135]. Better results are obtained by flow cytometry for counting viral particles in fresh water and marine environments using SYBR Gold [136]. Homodimeric nucleic acid dyes are also used in flow cytometry to study the viability of bacterial suspensions in milk dairy fermentation starters and probiotic products [137] (TOTO-1). A reliable method to analyze the first steps of DNA-based gene delivery with TOTO-1 and TOTO-3 allows the discrimination between total and internal DNA on a single-cell level [138]. Some patent publications describe flow cytometric methods for the measurement of reticulocytes and leukocytes [139] with asymmetric trimethine cyanine dyes and other cyanine dyes, and for enumeration of blood cells using SYTO dyes [140]. Enumeration of reticulocytes in a blood sample stained with a fluorescent dye (TOTO-3, TO-PRO-3, POPO-3, PO-PRO-3, YOYO-1, YO-PRO-1, YOYO-3, or YO-PRO-3) by flow cytometry is also known [141]. A flow cytometric multiparameter analysis of cells in a body fluid sample, using TO and LDS-751 nucleic acid dyes and at least one fluorescently labeled cell surface marker is described [142]. It is possible to discriminate between intact and damaged cells in a sample with a vital nucleic acid dye LDS-751 (a styryl dye) by selectively staining intact versus damaged cells, which then can be counted and sorted by flow cytometry [143].

A mixture containing chromosomes is stained with a single dye (TOTO-1 or YOYO-1). In a flow cytometer, the fluorescent dye is excited sequentially by a first UV light source to fluoresce at a first intensity and then by visible light to fluoresce at a second intensity. Specific chromosomes may be identified and sorted by intensity relationships between the first and the second fluorescence emission [144].

As already mentioned, the heterodimeric dyes TOTIN and TOTAB are exploiting the FRET [78, 79] phenomenon. These dyes, together with TOTO-1, are useful in multiplex sizing of dsDNA restriction fragments after separation by agarose gel electrophoresis.

FRET is a powerful tool to determine the distance between chromophores bound to macromolecules, since the efficiency of the energy transfer from an initially excited donor to an acceptor depends on the distance between the two dye molecules. By intercalation of two different TOTO dyes, FRET is observed from TOTO-1 as donor to TOTO-3 as acceptor [145]. The results are useful for DNA analysis, e.g., rapid DNA-sizing methods are based on the measurement of the fluorescence intensities of individual, dye intercalated DNA fragments.

Lakowicz and co-workers [146] described an approach for the creation of a new class of luminofores, which display both long wavelength emissions exceeding 600 nm and long lifetimes. These luminofores are based on FRET from a long lifetime donor to a short lifetime but with long wavelength acceptor. The donor is a ruthenium metal–ligand complex, which is known as DNA intercalator. The acceptors are either Nile Blue TOTO-3 or TO-PRO-3. Upon binding of the acceptor to donor-labeled DNA, the fluorescence quantum yield of the acceptor becomes many-fold higher than the intensity of the donor and acceptor alone. The acceptor lifetimes are of 30–80 ns. Similar results were obtained with the donor–acceptor system Ru-complex-BO-PRO-3 [147]. Emission at wavelengths exceeding 600 nm still preserves the long-lifetime component of the Ru donor, retaining average fluorescence lifetimes in the range of 30–50 ns.

The energy transfer between the same dye molecules (YO-PRO-1, YOYO-1, PO-PRO-1, and POPO-1) bound to DNA has been studied by measuring the fluorescent anisotropy of the dye-DNA complexes as a function of the binding density of the dyes [148].

Melting-curve analysis of single-nucleotide polymorphisms (SNP) with SG is reported [149], in which the melting profiles of restriction enzyme-digested PCR products are analyzed. The same SNP genotyping method is described in [150], with a special emphasis on its applicability to forensic science. Scaiano et al. [151, 152] developed a direct method for the determination of single-to-double stranded DNA ratios in solution by measuring the fluorescence lifetimes of SG and PG.

Gel electrophoresis is a broad field of application of such dyes. SYBR Gold is developed as the most sensitive nucleic acid gel stain and can be excited at 300 nm and 495 nm [153]. Pulsed-field gel electrophoresis (PFGE) is widely used to measure DNA double-strand breaks (DSB). The DSB method using SG on PFGE gels is described in [154].

FISH is a sensitive tool for specific DNA and RNA sequences. Its major advantage over other molecular genetic techniques is the possibility to exactly localize the target nucleic acid sequences within individual cells [155]. On paraffin sections, FISH is increasingly used to investigate genetic changes in tumor cells within the histological context. A variety of nucleic acid dyes (YOYO-1, BO-PRO-1, SYTOX Green, TO-PRO-1, PO-PRO-3, TOTO-3, TO-PRO-3, SYTO59, YOYO-3, and YO-PRO-3) were evaluated for specificity of the nuclear staining and the stability of the fluorochromes in sections of paraffin-embedded tissue. Only TO-PRO-3 showed specific staining of nuclei without any staining of the cytoplasm. Because of its high specificity and stability, this dye is suitable for detection of an amplification of a gene by dual-color FISH and confocal laser scanning microscopy [155].

Interactions between short ssDNA oligomers of chromogenous base composition and the probes YO-PRO-1 and homodimeric YOYO-1 are described [156].

Interactions between the dyes and DNA hybrids formed from complementary homogeneous strains of equal length were also investigated. The dyes were found to interact much more strongly with the purine oligomers polydA and polydG than with the pyrimidine oligomers polydT and polydC.

A homogeneous assay for nucleic acid hybridization with the homodimeric dye TOTO-1 has been reported [157]. This assay can detect specific hybridization between ssDNA and non-denaturated dsDNA to form triplexes, thus obviating the need to denature the dsDNA. The analysis can be used to identify accessible regions in folded nucleotide sequences, to determine the number of mismatched pairs in a hybridization complex, and to map genomes.

The sizing of DNA molecules is one of the most widely used analytical methodologies in molecular biology and biochemistry. Measurement of the length distribution of DNA molecules is typically accomplished using slab gel electrophoresis with nucleic acid dyes [158]. However, in recent years, alternative technologies for faster DNA fragment sizing have emerged. CE has been demonstrated to achieve fast separation with better resolution and higher sensitivity, e.g., the detection limit with SG is approximately 80 femtograms dsDNA [159], with a potential for automation. Rapid and reliable staining protocols for DNA fragment sizing by flow cytometry with PG [160] and SYTOX Orange [161] have also been presented.

The use of PG nucleic acid dye made possible a rapid and sensitive assay for DNase activity detection in picogram quantities within an hour [162]. The contamination of the laboratory equipment and solutions by nucleases is highly problematic in experiments involving nucleic acids, because these contaminating enzymes can breakdown the nucleic acids essential to the experiments. To avoid this problem, nuclease-free chemicals must be purchased from vendors, and equipments and solutions must be autoclaved before and during experimentation. Contamination is a particular problem for large manufacturing facilities producing biology-grade chemicals, because of quality control protocols. A patent publication [163] describes an assay to monitor the degradation of DNA and RNA by measuring the decreasing fluorescence (with YOYO-1 and other nucleic acid dyes) with time.

In a patent publication [164], Japanese authors described a chemiluminescent analysis for the detection or quantification with pyrylium salts of dsDNA at an extremely high sensitivity, such as a concentration level of 0.1 femtomoles (in terms of base pairs) or an absolute-quantity level of 0.1 attomoles (in terms of base-pairs). They stated that the chemiluminescent pyrylium salts became luminescent-emissive only when associated with ds nucleic acid and the compound that is not associated with dsDNA does not emit chemiluminescence. Owing to such a mechanism when detecting luminescence, the influence of the background level can be removed and, therefore, the detection for the target nucleic acid can be carried out with high sensitivity. The pyrylium compounds used for the chemiluminescent analysis have one or more aromatic rings as substituents linked to the pyrylium ring by single bonds. For this reason, when the compound is not intercalated into dsDNA (free state), each aryl substituent single-bounded to the pyrylium ring is twisted at an angle of a few tens of degrees, and the resulting structure rarely allows the compound to exhibit luminescence. When the compound is intercalated and inserted between two oligonucleotides of dsDNA, the angle of twisting of the aryl substituents decreases, i.e., the whole structure is in a common plane and allows the compound to readily exhibit luminescence. According to such a mechanism, it is assumed that the pyrylium salts do not have luminescent properties in the absence of dsDNA. The chemiluminescent analysis with pyrylium salts is performed with a combination of an oxalic ester or derivative thereof and a peroxide as a luminescence-inducing reagent [164].

In a series of papers, Jacobsen et al. [165–171] used two-dimensional ^{1}H-NMR spectroscopy and showed that TOTO-1 binds to dsDNA with a considerable preference for a CTAGCTAG site. This site-selectivity is caused by the intercalating chromophores, since the polypropyleneamine linker between the two dye molecules was shown previously of not exhibiting site-selectivity by itself [172]. Most other intercalating compounds show little site-selectivity in the binding to various dsDNA oligonucleotides, and when they do so, the effect is most convincingly explained as due to interactions of bulky side groups or linkers [173, 174]. A characteristic feature of the TOTO-1 chromophore is its ability to adapt to the base pair propeller twist. The benzothiazole moiety of the chromophore is intercalated between the two pyrimidine bases and the quinolinium heterocycle – between the two purine bases. TOTO-1 binds preferentially to 5′-pyrimidine-pyrimidine-purine-purine-3′ sequences with 5′-CTAG-3′ as the preferred binding site. TOTO-1 certainly binds to all the nucleotides, but it has been shown that there is at least a 100-fold preference for *bis* intercalating at the (5′-CTAG-3′)$_2$ site [167]. The elongation of the polyamine flexible linker by adding additional methylene groups leads to a reduction of selectivity [171].

A TOTOBIPY analog of TOTO-1 [171] has a more rigid linker containing the 4,4′-bipyridine (viologen) element. It was shown that TOTOBIPY preferentially spans three base-pair steps as the result of a concerted intercalation and minor groove-binding mode of the linker without a loss of the site-selectivity of the chromophores to the 5′-CTAG-3′ site.

REFERENCES

1. Belmont, Ph., Constant, J.-Fr. & Demeunynck, M., *Chem. Soc. Rev.*, **30** (2001) 70.
2. Wang, Y., Yang, J., Wu, X., Li, L., Sun, S., Su, B. & Zhao, Z., *Anal. Lett.*, **36**(10) (2003) 2063.
3. Underfriend, S. & Zaltzman, P., *Anal. Biochem.*, **3** (1962) 49.
4. Borresen, H. C., *Acta Chem. Scand.*, **17** (1962) 921.
5. Skidmore, W. D. & Duggan, E. L., *Anal. Biochem.*, **14(2)** (1966) 223.
6. Schy, W. E. & Pleva, M. J., *Anal. Biochem.*, **180** (1989) 314.

7. Maxam, A.M. & Gillbert, W., *Enzymology*, **65** (1980) 499.
8. Pasternack, R. F., Gibbs, E. J. & Villafranca, J., *J. Biochem.*, **22**(10) (1983) 2406.
9. Kumar, C. V. & Asuncion, E. H., *J. Am. Chem. Soc.*, **115** (1993) 8547.
10. Carvlin, M. J., Datta-Gupta, N. & Fiel, R., *J. Biochem. Biophys. Res. Commun.*, **108**(1) (1982) 66.
11. Lakowicz, J. R. *Fluorescence Spectroscopy, Biochemical Applications*, Plenum Press, New York, Vol. 3 (1992) p. 137.
12. Lankeker, M. V. & Gheyssens, L. C., *Anal. Lett.*, **19** (5,6) (1986) 615.
13. Lown, J. W., *Anticancer Drug Des.*, **3**(1) (1988) 25.
14. Singh, M. P., Joseph, T., Kumar, S., Yadagiri, B. J. & Lown W., *Chem. Res. Toxicol.*, **5** (1992) 597.
15. Le Pecq, J. B. & Paoletti, C. J., *J. Mol. Biol.*, **27** (1967) 87.
16. Tang, H. W., Cheng, B. Chen, G. Q. & Zeng, Y. E., *Chem. J. Chinese Univ.*, **18**(12) (1997) 1960.
17. US Patent 4 357, 774 (1981).
18. Hanafi-Bagby, D., Piunno, P. A. E., Wust, C. C. & Krull, U. *J., Anal. Chim. Acta*, **411** (2000) 19.
19 Haugland R. P., *Handbook of Fluorescent Probes and Research Chemicals*, 9th ed., Molecular Probes Inc., Eugene, OR, (2002).
20. US Patent 5437980, (1995).
21. Lee, L.G., Chen, Ch.-H., & Chiu, L.A., *Cytometry*, **7** (1986) 508.
22. US Patent 4 883 887 (1989).
23. Rye, H.S., Quesada, M.A., Peck, K., Mathies, R.A., & Glazer, A.N., *Nucleic Acids Res.*, **19**(2) (1991) 327.
24. US Patent 5 321 130 (1994).
25. US Patent 5 436 134 (1995).
26. US Patent 5 658 751 (1997).
27. Zipper, H., Brunner, H., Bernhogen, J. & Vitzhum, F., *Nucleic Acids Res.*, **32(12)** (2004) e103
28. US Patent 5 783 687 (1998).
29. Deligeorgiev T. G., Gadjev, N. I., Drexhage, K-H. & Sabnis, R. W., *Dyes Pigments*, **29** (1995) 315.
30. Deligeorgiev T. G., Zaneva, D., Katerinopulos, H. & Kolev, V., *Dyes Pigments*, **41** (1999) 49.
31. Deligeorgiev T. G., Zaneva, D. & Kim, S. M., *Dyes Pigments*, **17** (1998) 205.
32. Gadjev, N., Deligeorgiev, T. & Kim S. H., *Dyes Pigments*, **40** (1999) 181.
33. Yarmoluk, S. M. & Kryvorotenko, D. B., *Dyes Pigments*, **48** (2001) 165.
34. Isaesson, J. & Westamann, G., *Tetrahedron Lett.*, **42** (2001) 3207.
35. Deligeorgiev T. G., Gadjev, N., Vasilev, A. & Drexhage, K-H., *Dyes Pigments*, In press.
36. WO 0066664 (2000).
37. Chu-Moger, M. Y. & Berger, R. J., *Org. Chem.*, **60** (1995) 5721.
38. Smith, K., Lindsoy, C., Movis, I. K., Mathews, I. & Pritchard, G. J., *Sulfur Lett.*, **17** (1994). 197
39. Vasilev, A., Deligeorgiev, T., Gadjev, N. & Drexhage, K.-H., *Dyes Pigments*, **66** (2005) 135.
40. Kovalska, V. B., Tokar, V. P., Losytskyy, M. Yu., Deligeorgiev, T., Vasilev, A., Gadjev, N., Drexhage, K.-H. & Yarmoluk, S. M., *J. Biochem. Biophys. Methods*, In press.
41. US Patent 6 015 902 (2000).
42. Hamer, F.M., *The Cyanine Dyes and Related Compounds*, Interscience Publisher, New York (1964) pp. 58–59.
43. Giglio, S., Monis P. T. & Saint C. P., *Nucleic Acids Res.*, **31** (2003) e136.
44. Rizie, K. M., Rasmussen, R. P. & Wittwer, C. T., *Anal. Biochem.*, **245** (1997) 154.
45. WO 02090443 (2002).
46. Karlsson, H. J., Lincoln, P. & Westman, G., *Bioorg. Med. Chem.*, **11** (2003) 1035.

47. Bengtsson, M., Karlsson, H. J., Westman, G. & Kubista, K., *Nucleic Acids Res.*, **31**(8) (2003) e45
48. Eriksson, M., Karlsson, H. J., Westman, G. & Hermann, B. *Nucleic Acids Res.*, **31** (2003) e235.
49. Karlsson, H. J., Eriksson, M., Perzon, E., Akerman, B., Lincoln, P. & Westman, G., *Nucleic Acids Res.*, **31** (2003) 6227.
50. Karlsson, H. J., Bergqvist, M. H., Lincoln, P. & Westman, G., *Bioorg. Med. Chem.*, **12** (2004) 2369.
51. Rock, D., Prosad, K., Repic, O. & Blacklock, T. J., *Tetrahedron Lett.*, **41** (2000) 2083.
52. Chaundhuri, N. C., *Synth. Commun.*, (1996) 3783.
53. Singh, B., Pennock, P., Lesher, G. Y., Bacon, E. R. & Paye, D., *Heterocycles*, **36** (1993) 133.
54. Svanvik, N., Westman, G., Wang, D. & Kubista, M., *Anal. Biochem.*, **281** (2000) 26.
55. Engholm, M., Buchardt, O., Chrisiansen, L., Behreus, K., Freier, S. M., Driver, D. A., Bery, R. H., Kim, S. K., Norden, B. & Nielsen, P. E., *Nature*, **365** (1993) 566.
56. Svanvik, N., Nygren, J., Westman, G. & Kubista, M., *J. Am. Chem. Soc.*, **123** (2001) 803.
57. Svanvik, N., Stahlberg, A., Sehstedt, U., Sjoback, R. & Kubista, M., *Anal. Biochem.*, **287** (2000) 179.
58. US Patent 5 863 753 (1999).
59. Mason, S. J. & Balasubramanian, S., *Org. Lett.*, **4** (2002) 4261.
60. WO 0186264 (2001).
61. US Patent 5 826 127 (1998).
62. US Patent 5 658 735 (1997).
63. Gaugain, B., Barbet, J., Oberlin, R., Roques, B.P. & Le Pecq, J.-B., *Biochemistry*, **24** (1978) 5071.
64. Rye, H.S., Yue, S., Wemmer, D.E., Quesada, M.A., Haugland, R.P., Maties, R.A. & Glazer, A.N., *Nucleic Acids Res.*, **20** (1992) 2803.
65. WO 93106482 (1993).
66. US Patent 5410030 (1993).
67. Timcheva, I., Maximova, V., Deligeorgiev, T., Gadjev, N., Sabnis, R. & Ivanov, I., *FEBS Lett.*, **405** (1997) 141.
68. Benson, S.C., Mathies, R.A. & Glazer, A.N., *Nucleic Acids Res.*, **21** (1993) 5720.
69. Quesada, M.A., Rye, H.S., Gingrich, J.C., Glazer, A.N. & Mathies, R.A., *Biotechniques*, **10** (1991) 616.
70 Netzel, T., Nafisi, K., Zhao, M., Lenhard, J. & Johnson, I., *J. Phys. Chem.*, **99** (1996) 17936.
71. Carlsson, C., Jonsson, M. & Akerman, B., *Nucleic Acids Res.*, **23** (1995) 2413.
72. Deligeorgiev, T., Gadjev, N., Timchiva, I., Maximova, V., Katerinopoulos, H. & Fukaraki, E., *Dyes Pigments*, **44** (2000) 131.
73. Gadjev, N., Deligeorgiev, T., Timchiva, I. & Maximova, V., *Dyes Pigments*, **57** (2003) 161.
74. Deligeorgiev, T., Timtcheva, I., Maximova, V., Gadjev, N., Vasilev, A., Jacobsen J.-P. & Drexhage, K.-H., *Dyes Pigments*, **61** (2004) 79.
75. Benson, S., Mathies, R. & Glazer, A., *Nucleic Acids Res.*, **21** (1993) 5720.
76. Benson, S., Mathies, R. & Glazer, A., *Nucleic Acids Res.*, **21** (1993) 5727.
77. US Patent 5646264 (1997).
78. Benson, S., Zheng, Zh. & Glazer, A., *Anal. Biochem.*, **231** (1995) 247.
79. Zheng, Zh., Benson, S. & Glazer, A., *Anal. Biochem.*, **231** (1995) 256.
80. EP 1223226 (2002).
81. Carlsson, C., Larsson, A., Jonsson, M., Albinsson, B. & Norden, B., *J. Phys. Chem.*, **98** (1994) 10313.

82. US Patent 4544546 (1994).
83. Kumar, C., Turner, R. & Ascunsion, E., *J. Photochem. Photobiol. A*, **74** (1993) 231.
84. Kovalska, V., Kryvorotenko, D., Balanda, A., Losytskyy, M., Tokar, V. & Yarmoluk, S., *Dyes Pigments*, **67**, (2005) 47.
85. Rosania, G., Lee, J., Ding, L., Yoon, H. & Chang, Y., *J. Am. Chem. Soc.*, **125** (2003) 1130.
86. Lee, J., Jung, M., Rosania, G. & Chang, Y., *Chem. Commun.*, (2003) 1852.
87. Lokeshwar B. & Lin H.-S., *J Immunol. Methods*, **123** (1989) 103.
88. McCarthy D. & Macey M., *J Immunol. Methods*, **163** (1993) 155.
89. Kawabe, Y., Wang, L., Nakamura, T. & Ogata, N., *Appl. Phys. Lett.*, **81** (2002) 1372.
90. US Patent 4555396 (1985).
91. Yamamoto, N., Okamoto, T. & Kawaguchi, M., *Nucleic Acids Symp. Ser.*, **29** (1993) 83.
92. US Patent 6022961 (2000).
93. Yamamoto, N. & Okamoto, T., *Nucleic Acids Res.*, **23** (1995) 1445.
94. US Patent 5670315 (1997).
95. Balaban, A., Dinculescu, A., Dorofeenko, G., Fischer, G., Koblik, A., Mezheritskii, V. & Schroth, W., Pyrylium salts: Syntheses, reactions and physical properties *Adv. Hetrocycl. Chem.*, (Suppl. 2), Academic Press, New York, London, Paris, (1982).
96. Wong, M., Kong, S., Dragowska, W. & Bally, M., *Biochim. Biophys. Acta*, **1527** (2001) 61.
97. Deligeorgiev, T. Timtcheva, Il., Maximova, V., Gadjev, N. & Drexhage, K.-H., *J. Fluoresc.*, **12** (2002) 225.
98. Timcheva, I., Maximova, V., Deligeorgiev, T., Zaneva, D. & Ivanov, I., *J. Photochem. Photobiol. A*, **130** (2000) 7.
99. Singer, V., Jones, L., Yue, S. & Haugland, R., *Anal. Biochem.*, **249** (1997) 228.
100. Zipper, H., Buta, Ch., Laemmle, K., Brunner, H., Bernhagen, J. & Vitzhum, F., *Nucleic Acids Res.*, **31** (2003) e39.
101. Schofield, G., *Biotechniques*, **37** (2004) 778.
102. Vitzhum, F., Georg, G., Bisswanger, H., Brunner, H. & Brunner, H., *Anal. Biochem.*, **276**, (1999) 59.
103. Figeys, D., Arriaga, E., Renborg, A. & Dovichi, N., *J. Chromatogr. A*, **669** (1994) 205.
104. Rampal, S., Liu, M.-S. & Chen, F-T., *J. Chromatogr. A*, **781** (1997) 357.
105. McIntosh, S., Deligeorgiev, T., Gadjev, N. & McGown. L., *Electrophoresis*, **23** (2002) 1473.
106. McIntosh, S., Nunnally, B., Nesbit, A., Deligeorgiev, T., Gadjev, N. & McGown. L., *Anal. Chem.*, **72** (2000) 5444.
107. Jakeway, S. & Krull, U., *Can. J. Chem.*, **77** (1999) 2083.
108. Cheng, A-Ch., Gillespie, J. & Tabacco, M., *Anal. Chem.*, **73** (2001) 467.
109. Namasivayam, V., Larson, R., Burke, D. & Burns, M., *Anal. Chem.*, **73** (2001) 4188.
110. Sachadyn, P., Stanislawska, A. & Kur, J., *Nucleic Acids Res.*, **28** (2000) e36.
111. US Patent 5534416 (1996)
112. US Patent 5445946 (1995)
113. US Patent 5545535 (1996)
114. Mullis, K. & Faloona, F., *Methods Enzymol.*, **155** (1987) 335.
115. Brinkmann, B., *Methods Mol. Biol.*, **98** (1998) 105.
116. Curcio, M. & Roeraade, J., *Anal. Chem.*, **75** (2003) 1.
117. Romppanen, E.-L., Savolainen, K., & Mononen, I., *Anal. Biochem.*, **279** (2000) 111.
118. Ogura, M. & Mitsuhashi, M., *Anal. Biochem.*, **218** (1994) 458.
119. EP 684316 (1995).
120. Giglio. S., Monis, P. & Saint Ch., *Nucleic Acids Res.*, **31** (2003) e136.
121. Ririe, K., Rasmussen, R. & Wittwer, C., *Anal. Biochem.*, **245** (1997) 154.
122. Wittwer, C., Hermann, M., Moss, A. & Rasmussen, R., *Biotechniques*, **22** (1997) 130.

123. Nath, K., Sarosy, J., Hahn, J. & Di Como, Ch., *J. Biochim. Biophys. Methods*, **42** (2000) 15.
124. Iida, R., Yasuda, T., Tsubota, E., Nakashima, Y., Sawazaki, K., Aoyama, M., Matsuki, T. & Kishi, K., *Electrophoresis*, **19** (1998) 2416.
125. Iwabuchi, S., Muramatsu, H., Chiba, N., Kinjo, Y., Murakami, Y., Sakaguchi T., Yokoyama, K. & Tamiya, E., *Nucleic Acids Res.*, **25** (1997) 1662.
126. Wart, T., & Murples, B., *Int. J. Radiat. Biol.*, **76** (2000) 61.
127. Karsai, A., Muller, S., Platz, S. & Hauser, M., *Biotechniques*, **32** (2002) 790.
128. Baek, K-H. & Skinner, D., *Electron. J. Biotechnol.*, **7** (2004) 55.
129. Burgos, J., Ramirez, C., Tenorio, R., Sastre, I. & Bullido, M., *Mol. Cell. Probes*, **16** (2002) 257.
130. Vanderbroucke, I., Vandesompele, J., De Paepe, A.& Messianen, L., *Nucleic Acids Res.*, **26** (2001) e68.
131. Jung, M., Muche, M., Lukowski, A., Jung, K. & Loeninig, St., *Anal. Biochem.*, **289** (2001) 292.
132. Wilhelm, J., Pingoud, A. & Hahn, M., *Nucleic Acids Res.*, **31** (2003) e56.
133. Veal, D., Deere, D., Ferrari, B., Piper, J. & Attfield, P., *J. Immunol. Methods*, **243** (2000) 191.
134. Higuchi, A., Murayama, Y., Tsukamoto, Y., Kojima, T. & Kitamura, H., *Bull. Chem. Soc. Jpn.*, **77** (2004) 2021.
135. Brussaard, C., Marie, D. & Bratbak, G., *J. Virol. Methods*, **85** (2000) 175.
136. Cheng, F., Lu, J.-R., Binder, B., Liu, Y.-Ch. & Hodson, R., *Appl. Environ. Microbiol.*, **67** (2001) 539.
137. Bunthof, Ch. & Abee, T., *Appl. Environ. Microbiol.*, **68** (2002) 2934.
138. Ogris, M., Wagner, E. & Steinlein, P., *Biochim. Biophys. Acta*, **1474** (2000) 237.
139. US Patent 5891731 (1999).
140. US Patent 6197593 (2001).
141. US Patent 5563070 (1996).
142. US Patent 5047321 (1991).
143. US Patent 5057413 (1991).
144. US Patent 5418169 (1995).
145. Laib, S. & Seeger, S., *J. Fluoresc.*, **14** (2004) 187.
146. Lakowicz, J., Piszczek, G. & Kang, J., *J. Fluoresc.*, **12** (2002) 97.
147. Kang, J., Piszczek, G. & Lakowicz, J., *J. Fluoresc.*, **12** (2002) 17.
148. Carlsson, Ch., Larsson, A., Bjoerkman, M., Jonsson, M. & Albinsson, B., *Biopolymers*, **41** (1997) 481.
149. Akey, J., Sosnoski, D., Parra, E., Dios, S., Hiester, K., Su, B., Bonilla, C., Jin, L. & Shriver, M., *Biotechniques*, **30** (2001) 358.
150. Ye, J., Parra, E., Sosnoski, D., Hiester, K., Underhill, P. & Shriver, M., *J. Forensic Sci.*, **47** (2002) 593.
151. Cosa, G., Focsaneanu, K.-S., McLean, J. & Scaiano, J., *Chem. Commun.*, (2000) 689.
152. Cosa, G., Focsaneanu, K.-S., McLean, J., MacNamee, J. & Scaiano, J., *Photochem. Photobiol.*, **73** (2001) 585.
153. Tuma, R., Beaudet, M., Jin, X., Jones, L., Cheung, Ch.-Y., Yue, S. & Singer, V., *Anal. Biochem.*, **268** (1999) 278.
154. Kiltie, A. & Ryan, A., *Nucleic Acids Res.*, **25** (1997) 2945.
155. Bink, K., Walch, A., Feuchtinger, A., Eisenmann, H., Hutzler, P., Hoefler, H. & Werner, M., *Histochem. Cell Biol.*, **115** (2001) 293.
156. Simon, L., Abramo, K., Sell, J. & McGown, L., *Biospectroscopy*, **4** (1998) 17.
157. WO 0146467 (2001).
158. Carlsson, Ch., Jonsson, M. & Akermann, B., *Nucleic Acids Res.*, **23** (1995) 2413.

159. Skeidsvoll, J. & Ueland, P., *Anal. Biochem.*, **231** (1995) 359.
160. Yan, X., Grace, W., Yoshida, T., Habbersett, R., Vellapan, N., Jett, J., Keller, R. & Marrone, B., *Anal. Chem.*, **71** (1999) 5470.
161. Yan, X., Habbersett, R., Cordek, J., Nolan, J., Habbersett, R., Jett, J. & Marrone, B., *Anal. Biochem.*, **286** (2000) 138.
162. Choi, S-J. & Szoka, F., *Anal. Biochem.*, **281** (2000) 95.
163. US Patent 5554502 (1996).
164. EP 837141 (1998).
165. Spielmann, H., Wemmer, D. & Jacobsen, J-P., *Biochemistry*, **34** (1995) 8542.
166. Petersen, M. & Jacobsen, J-P., *Bioconjugate Chem.*, **9** (1999) 331.
167. Bunkenborg, J., Stidsen, M. & Jacobsen, J-P., *Bioconjugate Chem.*, **10** (1999) 824.
168. Petersen, M., Hamed, A., Pedersen, E. & Jacobsen, J-P., *Bioconjugate Chem.*, **10** (1999) 17066.
170. Bondensgaard, K. & Jacobsen, J-P., *Bioconjugate Chem.*, **10** (1999) 735.
171. Staerk, D., Hamed, A., Pedersen, E. & Jacobsen, J-P., *Bioconjugate Chem.*, **8** (1997) 869.
172. Bunkenborg, J., Gadjev, N., Deligeorgiev, T. & Jacobsen, J-P., *Bioconjugate Chem.*, **11** (2000) 861.
173. Wemmer, D., Srivenugopal, K., Reid, B. & Morris, D., *J. Mol. Biol.*, **185** (1985) 457.
174. Searle, M., *Progr. NMR Spectrosc.*, **25** (1993) 403.

Functional Dyes
Sung-Hoon Kim (Editor)

Chapter 5

Functional dyes for surface plasmon resonance-based sensing system

Sung-Hoon Kim[a] and Kwangnak Koh[b]

[a]Department of Textile System Engineering, Kyungpook National University, Daegu, Korea
[b]College of Pharmacy, Pusan National University, Pusan, Korea

1. INTRODUCTION

Molecular recognition is the informatics involved in the binding and selection of substrate by a given receptor molecule. It thus implies reading-out of molecular information. The transfer of information from molecular domain to human knowledge requires transducer principles and technologies. There are several kinds of transducers getting a signal from a molecular interaction event. Surface plasmon resonance (SPR) is a unique optical transducer with numerous applications in a variety of molecular interactions. SPR can be used to probe refractive index changes that occur at the interface.

Surface plasmon waves (SPWs) are excited at the interface between a thin, highly reflecting metal layer (e.g. silver or gold) and a sample, by coupling through a substrate with a high refractive index, as illustrated in Fig. 1a (Kretschmann configuration). SPWs are transverse waves with an oscillating electric field normal to the surface. Since surface plasmons only have an electric field component, which is normal to the surface, p-polarized light is required to satisfy the boundary conditions necessary to excite SPR. At the SPR angle, the energy of the metal surface coincides with the incident photon and the charge density wave. The photon energy is then transferred to the SPW, which results in energy loss from reflected intensity. This phenomenon can be observed as a sharp dip in the reflected light intensity (Fig. 1b). Outside the metal, an evanescent electric field exists. This evanescent wave is part of the internally reflected light beam and partially penetrates into the lower refractive index medium. The evanescent wave is the 'sensing' component and can interact optically with compounds close to or at the surface. Changes in the optical properties of this

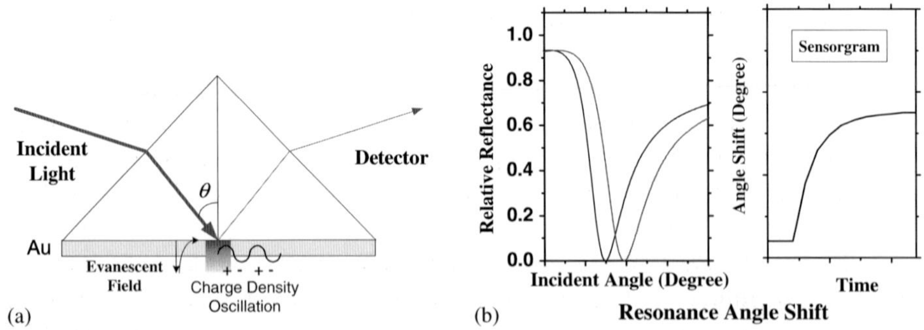

Fig. 1(a) The Kretschmann configuration and (b) SPR-reflectance waves. The light reflectance is recorded as a function of the angle of incidence, θ.

region will influence the SPR angle, and accounts for the use of SPR for sensing purposes.

Thus, any phenomenon of molecular interaction at the surface that alters the refractive index will elicit a signal from molecular interaction.

The measurement of refractive index can be used to determine the concentration of an analyte in solution, which can be explained by the fact that the refractive index of a solution depends on analyte concentration. This kind of study can be performed effectively by the SPR technique with selective sensing membranes containing recognition-functional dyes such as squarylium and spiroxazine derivatives as sensing molecules. A sensing membrane consists of a thin film (e.g. polymeric membrane or self-assembled monolayer (SAM)) having optical properties that vary with the concentration of the analyte to be sensed.

2. Ag^+ ION SENSING MEMBRANE CONTAINING DITHIOSQUARYLIUM DYE

The design of chemically sensitive interfaces requires an interdisciplinary study such as design and synthesis of molecular recognition system, the characterization of materials and interfaces, and the strategies for signal transduction. Furthermore, a dual function (such as the molecular recognition by selective interaction occurring at the interfaces and simultaneous signal amplification) is a key criterion for the design of the sensing interfaces and materials [1]. In particular, the development of those sensing interfaces related to the metal cation detection is an ongoing research topic in biochemical analysis and in analytical chemistry [2].

From this perspective, we have previously reported the synthesis and detection characteristics of squarylium (SQ) and dithiosquarylium (DTSQ) dyes as metal ion-recognition materials [3–4]. However, despite the implementation of those studies, the need for further improvement in detection limits and sensitivity is significant.

The SPR sensor has been widely used for the precise detection of biochemical molecules [5]. Most of those biochemical applications are based on the highly sensitive refractive index change usually caused by the interaction between specific analytes and their selectively capturing molecules immobilized in the sensing membrane of the sensor [6]. In addition, because the refractive index change is related to the light absorption change, the absorbance change of the sensing membrane caused by the selective recognition of the analyte gives rise to more effective refractive index change [7]. Subsequently, the metal ion detection based on the absorbance-changeable sensing membrane can also be useful for the precise metal ion detection in the SPR sensor system.

Accordingly, one can investigate the Ag^+ ion-selective sensing polymeric membrane containing a DTSQ dye by the use of the SPR technique, and also interpret the relation between absorbance change of the sensing membrane according to Ag^+ ion concentrations and the SPR angle shift related to refractive index change (by Fresnel equations and Kramers–Kronig relation) [8].

2.1. Formation of membrane

DTSQ dye can be designed and synthesized according to our previously reported results [9] (Fig. 2). SQ dyes are generally prepared by condensing one equivalent of squaric acid with two equivalents of various types of electron-donating carbocyclic and heterocyclic compounds such as azulenes, [10] pyrroles [11] or heterocyclic methylene bases [12] in an azeotropic solvent.

We have found for the first time that thio-analogs of SQ dyes can be synthesized by the reaction of SQ with Lawesson's reagent. DTSQ dyes were also

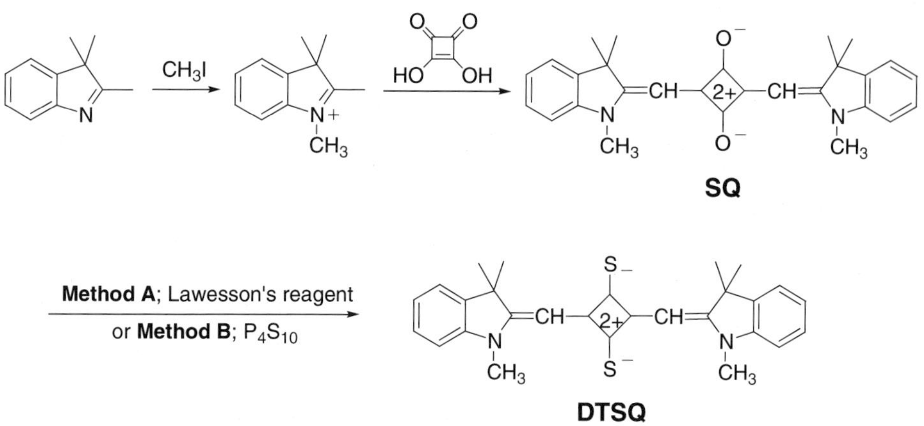

Fig. 2. Synthesis of SQ and DTSQ dyes.

prepared from SQ dyes, by using phosphorous pentasulfide, P_4S_{10}, as the thionation agent. The general synthetic routes to the DTSQ dyes are outlined in Fig. 2. The reactivity of Lawesson's reagent (Method A) against P_4S_{10} (Method B) was very poor. DTSQ dyes generally absorb at much longer wavelengths than the corresponding SQ dyes.

The Ag^+ ion-sensing polymeric membrane can be constructed according to the following procedure. For substrate preparation, after immersing the glass slide (18 × 18 mm) in a cleaning detergent for 10 h, it is cleaned with deionized water in an ultrasonic cleaner for 10 min and dried for 30 min at 80°C. Au is then coated by electron beam evaporation to a thickness of 50 nm after 3 nm nickel–chromium deposition on the glass slide (Fig. 3). The casting solution of the polymeric sensing membrane of 20 nm thickness for SPR measurement is composed of PVC–PVAc–PVA matrix copolymer (M.W. = 27,000 Da, 14.82 wt%), plasticizer (dioctyl phthalate, 64.66 wt%), anionic site (potassium *Tetrakis*(4-chlorophenylborate), 6.26 wt%) and DTSQ dye (14.26 wt%) in tetrahydrofuran (THF). This mixture is cast at 5000 rpm for 30 s on the surface of the Au chip by a conventional spin-coating method.

Target metal ions (Ag^+, Cu^{2+}, K^+, Na^+, Zn^{2+}, Mg^{2+}, Ca^{2+}, Co^{2+}, Li^+, Cd^{2+} and Hg^{2+} = 10^{-4}–10^{-12} M, except Pb^{2+} = 10^{-5}–10^{-12} M as it precipitates at above >10^{-5} M in 0.05 M Tris–H_2SO_4 buffer solution, pH 7.0) of various concentrations can be used to investigate the metal ion–dye selective interaction. All test solutions have to be freshly prepared just before each measurement with nitrate salts under N_2 gas environment by bubbling for 30 min to avoid metal carbonate complex formation. Ag^+ ion-selective behavior of the DTSQ dye to the

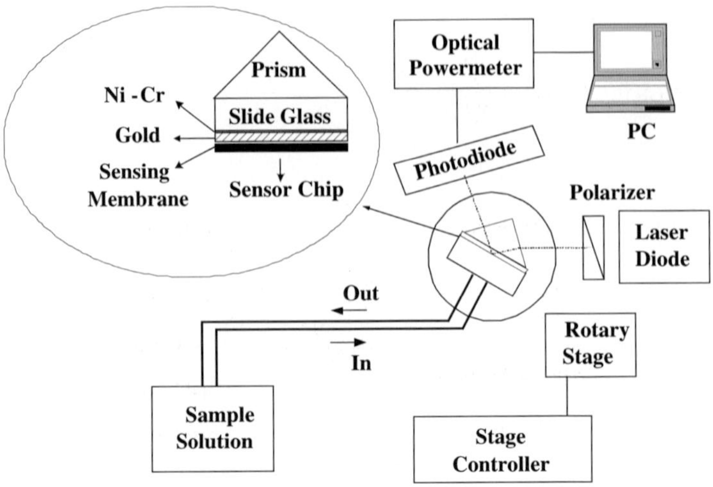

Fig. 3. Schematic diagram of the SPR measurement system and sensor chip.

other interfering ion is investigated usually by means of the separate solution method (SSM) that determines separate calibration plots for the Ag^+ ion and the interfering ion at constant conditions [13].

2.2. SPR measurement

The most widely used methods of exciting surface plasmons are based on the evanescent wave excitation using a Kretschmann prism configuration (Fig. 3) [14].

In these methods, a p-polarized laser beam (4.5 mW, 675 nm) is incident upon the metal–prism interface and the reflected beam is detected by a photodiode. Its signal is recorded with an optical power meter. When the incident angle reaches an appropriate value, the reflection decreases sharply to a minimum (i.e. the surface plasmon is resonantly excited at this angle). The incident angle is controlled using a stepping motor with a 0.004° of minimum resolution. Each sample solution is pumped into the sample cell from low to high concentrations.

2.3. Determination of refractive index change in sensing membrane

The four-layer system composed of a prism, metal film (gold), a dielectric sensing membrane and an external dielectric environment is shown in Fig. 4.

This diagram has been used as a model of a typical Kretschmann configuration. Particularly, in order to estimate the refractive index changes of a DTSQ membrane formed on a gold surface, the following equations can be applied to the four-layer model [15].

First, the theoretical reflectance of the four-layer system is described from the Fresnel equations for p-polarized light as follows:

$$|r_{1234}|^2 = \left| \frac{r_{01} + r_{12}s_1^2 + r_{23}s_1^2 s_2^2 + r_{01}r_{12}r_{23}s_2^2}{1 + r_{01}r_{12}s_1^2 + r_{12}r_{23}s_2^2 + r_{01}r_{23}s_1^2 s_2^2} \right|^2$$

where $s_i = \exp(ik_{ki}d_i)$, $r_{ij} = \dfrac{\varepsilon_i k_{xj} - \varepsilon_j k_{xi}}{\varepsilon_i k_{xj} + \varepsilon_j k_{xi}}$, $k_{xi} = \dfrac{2\pi}{\lambda}(\varepsilon_i - \varepsilon_0 \sin^2\phi)^{1/2}$

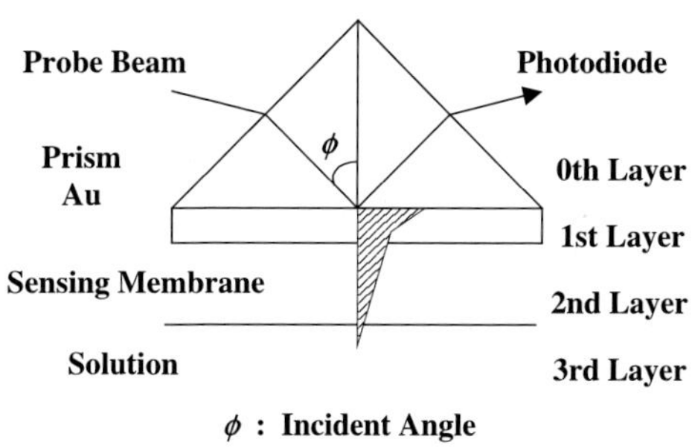

Fig. 4. Four-layer model for Kretschmann configuration.

ε_i is the dielectric constant of the ith layer, d_i the thickness of the ith layer, ϕ the incident angle at the prism–metal interface and λ the wavelength of the incident radiation.

After obtaining the optical constants of bare gold with three layers (without sensing membrane), fitting the equation to the measured reflectance curves of the sensing membrane results in estimating the thickness and the refractive index of the sensing membrane. However, an ambiguity still exists in their separate determinations. Thus, to overcome this problem, various multiple SPR measurement techniques have been proposed [16]. Here, the multiple measurements in different environments (air and buffer solution) are implemented to separately determine the thickness and the refractive index of the sensing membrane. Based on these results, the resonance angle changes by the increase of Ag^+ ion concentration can be easily evaluated by the above equation.

2.4. Ag^+ ion sensing

According to our previously reported results of UV–visible spectroscopy [3], the ion-sensing membrane containing a DTSQ dye on a glass plate showed that the intensity at maximum absorption (λ_{max} = 665 nm) decreased when increasing the concentration of Ag^+ ions (Fig. 5).

On the other hand, the effect of the other interfering metal ions was not found in the absorption spectra. This Ag^+ ion-selective detection of DTSQ dye can also be reproduced by the SPR measurement in more detail (Fig. 6). As the concentration of Ag^+ ions becomes higher, the resonance angle (reflectance minimum) shifts to a higher angle. The re-measured resonance angles did not return

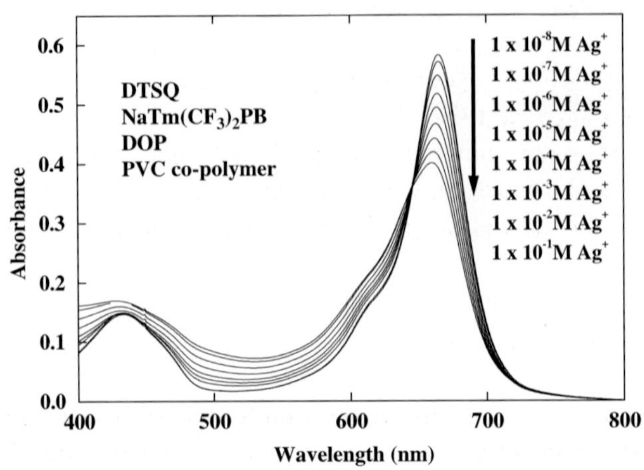

Fig. 5. Absorbance spectra of Ag^+-selectivity optode film (4.5 μm thick) with various Ag^+ concentrations in pH buffered solutions (pH 7.0).

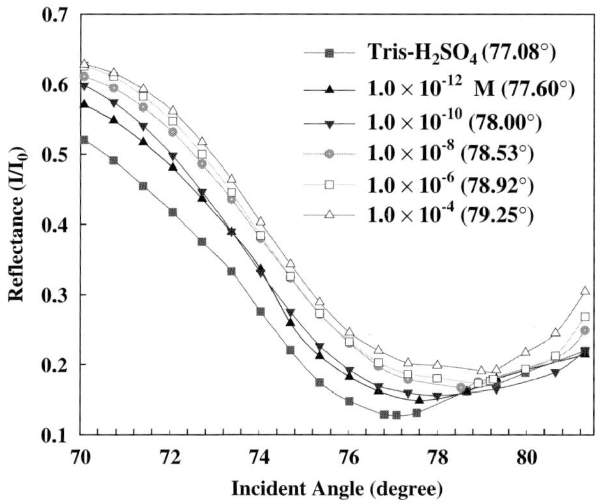

Fig. 6. Experimental reflectance curves corresponding to various Ag^+ ion concentrations.

to its original value in buffer solution. It mostly keeps a half value of the original resonance angle. It is considered that there is a memory effect from the remaining metal ions in the sensing membrane.

In particular, the resonance angle shift in case of Ag^+ ions with the concentration of 1.0×10^{-4} M appears as the value of 2.17° from the resonance angle of the buffer solution. This value is over four times that of the other metal ions (about 0.5° or below) (Fig. 7). Here, each plot corresponds to the shift from the resonance angle of the buffer solution as a reference angle. Furthermore, Fig. 7 depicts that this molecular system shows the selective detection for Ag^+ ion concentrations over a wide range of from 1.0×10^{-4} to 1.0×10^{-12} M (except Hg^{2+}, maximum angle shift is 1.72°).

A method for the numerical fitting of each plot is obtained by using the Boltzmann sigmoidal function, which has been applied as the typical analysis of response characteristics to the optode-type sensor [17]. It is given by the following equation:

$$y = (A_1 - A_2)/(1 + e^{(x-x_0)/\mathrm{d}x}) + A_2$$

where A_1 is the lower limit, A_2 the upper limit for the resonance angle, x_0 the center value for the concentration and dx the slope at each point. The regression coefficient about this non-linear function appears in Table 1.

Additionally, the offset of the y-axis in Fig. 7 arises from the difference of refractive index caused by the content of the metal ion in the solution. Above all,

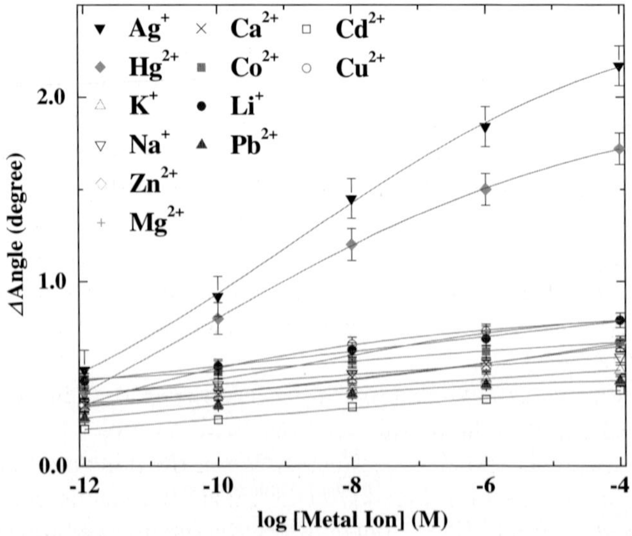

Fig. 7. Calibration plots of the resonance angle shifts for various metal ions and concentrations ranging from 10^{-4} to 10^{-12} M.

Table 1
Coefficients of the different metal ions fitted from the Boltzmann sigmoidal function

Target metal ion	A_1	A_2	X_0	d_x
Ag^+	−0.09963	2.52099	−8.87464	2.64398
Cu^{2+}	−11.90719	0.86912	−25.05836	4.18113
K^+	−2.4566	1.46476	−39.7456	31.20397
Na^+	−0.3778	0.92896	−16.05239	11.53026
Zn^{2+}	0.12445	7.92422	17.53424	8.28288
Mg^{2+}	0.33484	0.83411	−8.21539	1.81079
Ca^{2+}	−0.005063	2.25166	4.36943	10.08111
Co^{2+}	0.39955	0.74991	−7.71512	3.07531
Li^+	−1.59293	4.95331	16.28317	36.15431
Pb^{2+}	0.14979	0.48225	−10.35709	2.35754
Cd^{2+}	0.04714	0.49625	−9.41448	3.83434
Hg^{2+}	−0.54631	2.01085	−10.34564	3.1019

there is a considerable different offset between the other metal ion solutions and buffer solution. In order to investigate the resonance angle shift in terms of refractive index change, the Fresnel calculation by equation is performed after obtaining the optical constants of bare gold. As a result of multiple measurements in different environments, the thickness and refractive index of the sensing membrane are estimated at about 20 nm and 1.455, respectively.

Each uncertainty results from an observational and numerical fitting error. This uncertainty value for the thickness is ± 1 nm and the refractive index change is about 5×10^{-3}. These values are calculated by using the equation considering the uncertainty range for theoretical and experimental data (resonance angle and minimum reflectivity).

In a recent research work [18], it has been proposed that a thin aqueous layer is formed between the liquid membrane and the substrate. From this point of view, theoretical and experimental evaluation of the thin aqueous layer can be performed. Though the thin aqueous layer exists, it remains constant and the results obtained from the calculation and experiments showed little difference in the refractive index value irrespective of whether it exists or not. Therefore, it is reasonable to suppose that the sensing membrane is considered as a single layer in disregard of the presence of the thin aqueous layer. Furthermore, the fitting results to the measurement curves of Ag^+ ion sensing reveal that the resonance angle shifts by Ag^+ ion sensing correspond to the increase of the refractive index of the sensing membrane containing a DTSQ dye (Fig. 8).

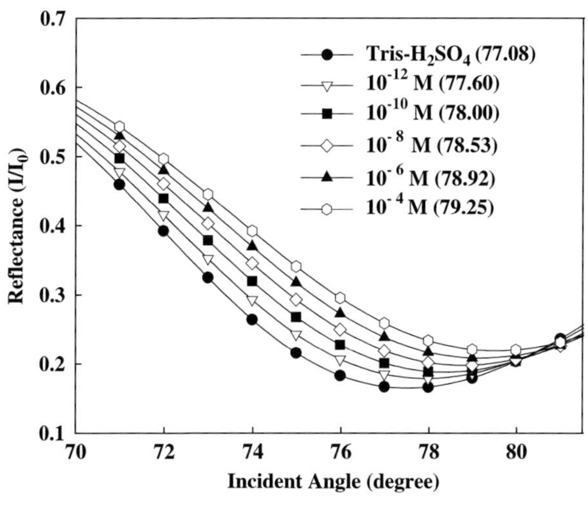

Fig. 8. Calculated reflectance curves corresponding to various Ag^+ ion concentrations for the sensing membrane containing DTSQ dye.

2.5. Refractive index change of sensing membrane

The refractive index increases in the sensing membrane are significant, considering the Kramers–Kronig calculation for determining the relation between refractive index and the absorption coefficient [8]. According to a previous report [7], refractive index on the shoulders of sharp and narrow peaks of absorbance curves abruptly changes in contrast with no change in refractive index at the maximum absorbance. Thus, the wavelength of the SPR probe beam can strongly affect the behavior of the resonant angle shift corresponding to the refractive index change [19].

The absorption curves in Fig. 9 (a) are based on our previously reported results of UV–visible spectra. The maximum absorbance wavelength of DTSQ (665 nm) is located on the left of that of the SPR probe beam (675 nm, as characterized by a fluorescence spectrophotometer F-4500 (Hitachi)). Thereby, the decrease of absorption coefficients in accordance with the increased concentration of Ag^+ ions results in the increase of the refractive index of the sensing membrane.

Although the maximum absorption wavelength of the sensing membrane is not coincident with that of the maximum refractive index changing, it shows significant changes in the refractive index of the sensing membrane through the Ag^+ ion-detection process. This means that the signal from the interaction between an Ag^+ ion and the DTSQ dye can be amplified effectively through the sensitive property of the surface plasmon. This amplified refractive index change results in a greater SPR angle shift. These results indicate that enhanced sensitivity can be obtained through the choice of the SPR probe beam wavelength or through the appropriate molecular system design of the sensing membrane. Therefore, chromogenic sensing with recognition-functional dyes approaches related to SPR can offer a novel strategy for highly sensitive metal ion detection. In other words, as described in the introduction, the dual function of the molecular recognition and signal amplification can be realized more effectively through the novel molecular system design strategy related to the SPR phenomena. Moreover, in comparison with a common Ag^+ ion-selective electrode system [20], the SPR method has some advantages such as simple manufacturing, easy measuring, a low cost for maintenance and a detection ability of low concentration metal ions.

An Ag^+ ion sensor can be fabricated by a novel DTSQ dye containing a polymeric film. The selective sensing signal of the Ag^+ ion sensor appeared as the refractive index change corresponding to the shifts of the SPR angle. In addition, selective Ag^+-ion detection is possible over a wide range from 10^{-4} to 10^{-12} M. In the light of the calculation by Fresnel equations and the Kramers–Kronig relation, the refractive index increase in a sensing membrane appears to have been caused by the decrease of absorption coefficient around the wavelength of the SPR probe beam. These results suggest that λ_{max} control of DTSQ dye related

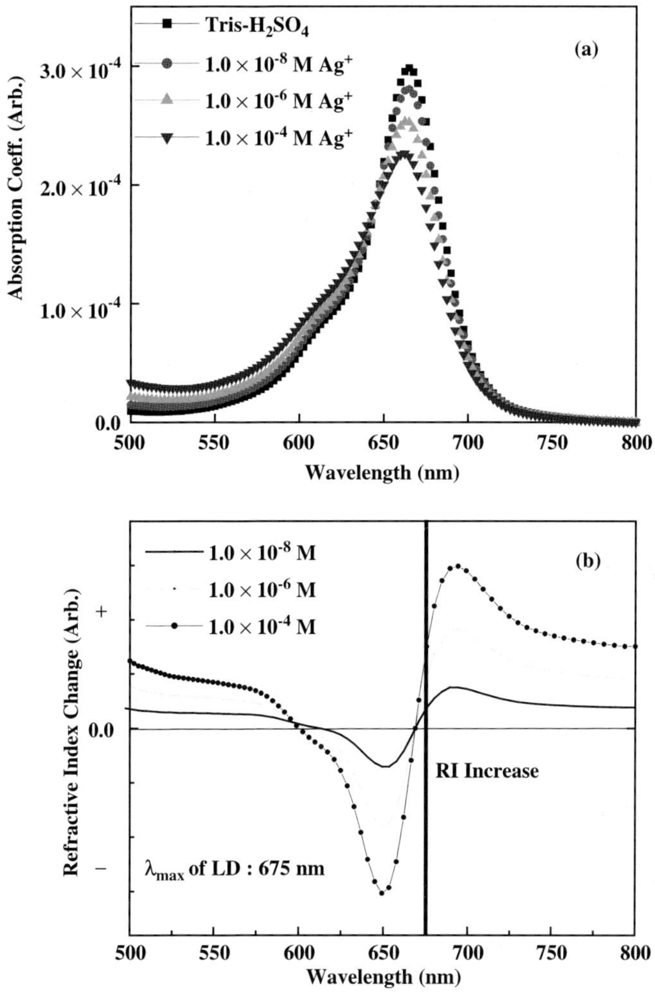

Fig. 9. The calculated absorption coefficient (a) and refractive index change (b) of the sensing membrane containing DTSQ dye for various Ag^+ ion concentrations.

to SPR phenomena can offer a novel strategy to the highly sensitive metal ion detection, i. e. a dye derivative system used as a recognition material guarantees the high sensitivity in the SPR due to enhanced refractive index change.

3. DETECTION OF Cu^{2+} ION USING AN SQ DYE

Soil and water pollution by massive amount of industrial waste, caused by rapid industrial development and growth of population, have become one of the major social problems. Especially, the pollution caused by harmful heavy metal ions discharged from the factories is the most serious one [21–22]. It is well known that many heavy metal ions among the pollutants are not decomposed and accumulate continuously and in the end, it is impossible to extract them out fully. Furthermore, the metal ion absorbed into a body not only reduces body metabolism but also causes harmful effects when the ion is accumulated over a certain level. Copper, among various heavy metals, is an important element in many industrial processes and is an essential inorganic substance for the human body. But if it is taken in excess, it damages liver cells and causes various diseases. It also obstructs the growth of many crops [23, 24].

Thus, it is necessary to resort to a new approach in order to facilitate more selective and sensitive detection of Cu^{2+} ion in solution rather than using conventional methods [3, 25, 26]. According to our previous UV—visible spectroscopic study on the SQ dye, the ion-sensing thick film containing SQ dye on a glass plate (optode) showed that the maximum absorption peak at 640 nm was decreased by increasing the concentration of the Cu^{2+} ion (Fig. 10). On the other hand, the absorption changes by the other co-existing metal ions were not detectable [3]. Thus, naturally, these preliminary results on the superior characteristics of the SQ dye require the effort to combine with these characteristics a promising optical technique that can maximize the Cu^{2+}-ion sensing property of the SQ dye.

The SPR method is a suitable optical technique, which can effectively monitor the interaction between SQ dye and Cu^{2+} ion. The SPR technique is a well-known signal-transducing principle that can convert the chemical signal caused by the interaction between the analyte and the recognition material to a useful

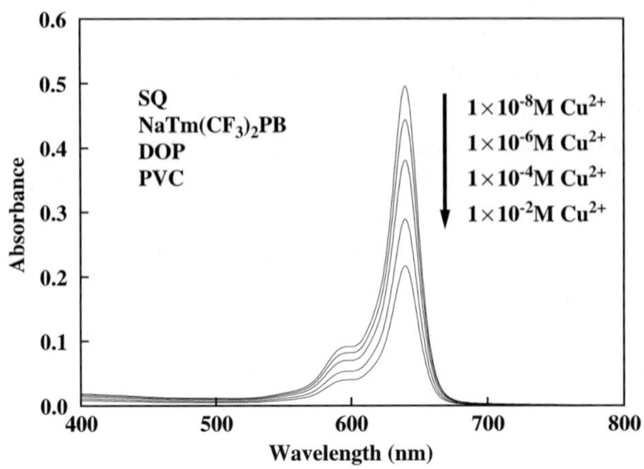

Fig. 10. Absorption spectra of Cu^{2+}-selective optode film (3.7 μm thick) with various Cu^{2+} concentrations in pH buffered solution (pH 7.0).

optical signal [5]. Its optical signal may appear as the refractive index change and/or as the thickness change [7]. In the case of the polymeric ion-sensing film, the refractive index change can be the dominant optical signal. Especially, a polymeric film containing a dye that absorbs light in the visible range can produce more improved refractive index changes as a result of the sensing mechanism. It can be explained easily by the anomalous dispersion near the absorption maximum [7, 19, 27].

3.1. Formation of polymeric sensing thin film

The SQ dye can be synthesized according to a previously reported method as follows [28]. The casting solution to make a polymeric sensing thin film on Au surface is composed of PVC–PVAc–PVA copolymer (14.96 wt%), plasticizer (dioctylphthalate 65.30 wt%), anionic site (potassium *Tetrakis* (4-chlorophenylborate), 6.32 wt%), SQ dye (13.42wt%) and THF as a solvent. This solution mixture is spin coated on the surface of the Au layer (membrane thickness: about 12 nm, Au thickness: 50 nm and Ni–Cr thickness: about 3 nm).

3.2. SPR measurement

The SPR system is based on the traditional Kretschmann configuration (Fig. 3, [14]). Each metal ion-containing sample solution was allowed to flow into the sample cell from a low concentration to a higher one by pumping. Metal ions (Cu^{2+}, Ag^{+}, K^{+}, Na^{+}, Zn^{2+}, Mg^{2+}, Ca^{2+}, Co^{2+}, Li^{+}, Cd^{2+} and Hg^{2+} = $1.0 \times 10^{-12}-1.0 \times 10^{-4}$ M, except Pb^{2+} = $1.0 \times 10^{-12}-1.0 \times 10^{-5}$ M, because this ion precipitates at above $>10^{-5}$ M) were dissolved in 0.05 M Tris–H_2SO_4 buffer solution of pH 7.0 to make various concentrations of sample solutions. Tris–H_2SO_4 buffer solution can be used as a reference solution (background). All counter anions of metal ions are nitrates. Cu^{2+} ion-selective behavior of the SQ dye with the other interfering ions can be investigated by means of SSM that determines separate calibration plots for the Cu^{2+} ion and the interfering ions under constant conditions [17].

Cu^{2+}-ion molecular sensing system using SQ dye shows excellent selectivity and remarkable sensitivity in the measurement of ion concentrations over a wide range from 1.0×10^{-12} to 1.0×10^{-4} M linearly. Even though the reproducible results greatly depend on the film deposition conditions, especially on the sensing film thickness, if the thickness could preserve constantly at each measurement, reproduction of the selectivity and sensitivity is guaranteed.

Fig. 11 indicates typical SPR curves related to the sample solutions with different Cu^{2+} ion concentrations. The resonance angle difference between Tris–H_2SO_4 buffer solution (as a reference) and the highest concentration of Cu^{2+} ion (1.0×10^{-4} M) is 2.30° in the SPR sensing system.

Moreover, even at the lowest Cu^{2+} ion concentration, 1.0×10^{-12} M, the resonance angle is significantly changed to 0.70°. This is a distinguishable value

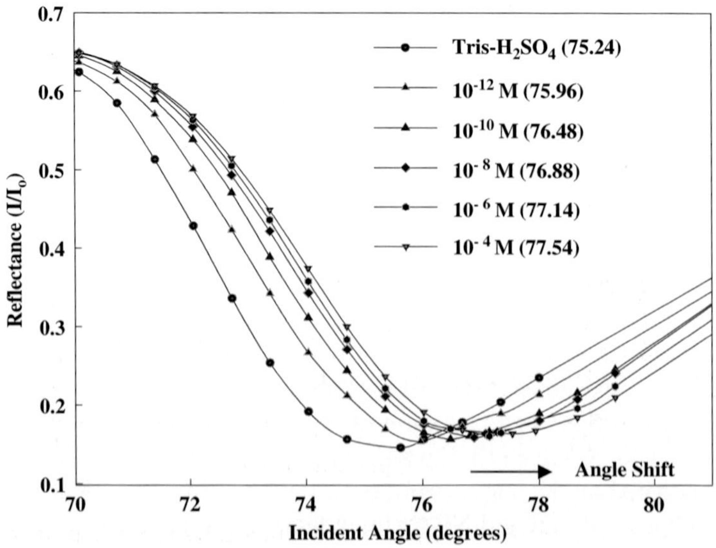

Fig. 11. Resonance angle shifts according to various Cu^{2+} ion concentrations.

in comparison with angle shifts of the other metal ions at high concentration, which stay at around 0.50°. These results describe that the good selectivity and sensitivity of the SQ-dye sensing to the Cu^{2+} ion in the polymeric thin film can be obtained through the SPR sensing method, although the SQ-dye system used as a molecular sensing interface is not a well-designed molecule for selective Cu^{2+}-ion sensing. It is considered that both cooperative "hard–soft" metal ion–ligand interaction and size-selective recognition of Cu^{2+} ion to SQ dye may contribute to high selectivity [29]. Furthermore, the signal related to the effective electrostatic interaction between SQ dye and Cu^{2+} ion may be amplified through the SPR sensing system with high sensitivity in the wider linear dynamic range.

As described previously, the light absorption property of SQ dye can be regarded as an important factor for constructing the highly sensitive Cu^{2+} ion sensor using SPR method. Because the SQ dye has the maximum absorption peak at 640 nm, the Cu^{2+}-ion sensing polymeric film containing the SQ dye has the highest refractive index at the wavelength of laser diode (675 nm) according to the anomalous dispersion relation [7, 19, 27]. Thus, the variation of absorbance near 675 nm by the increase of ion concentration can give rise to a change in the refractive index of the ion sensing film greatly. This can strongly affect the SPR phenomenon such as the reflectance variation, resonance angle shift and so on. Consequently, it can offer a more sensitive Cu^{2+}-ion detection

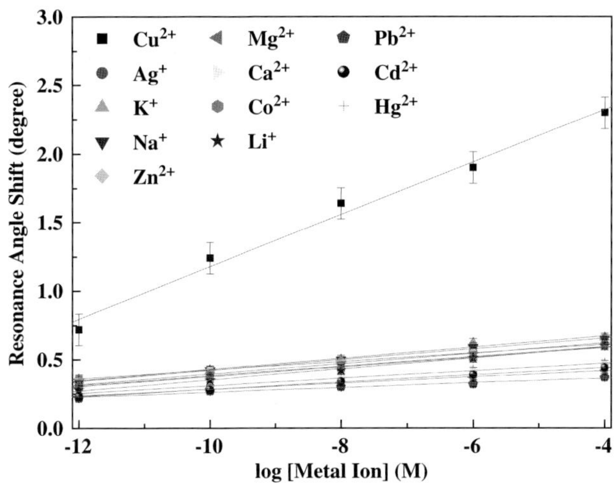

Fig. 12. Resonance angle shifts corresponding to different concentrations of several metal ions.

mechanism related to the improved resonance angle shifts. More details of physical analysis on these interesting light absorption properties related with SPR are being investigated.

From the above results, the possibility of constructing a more accurate Cu^{2+}-ion sensor system with a recognition-functional dye can be established. The highly selective detection of the Cu^{2+} ion is achieved by the SQ-dye molecular system that produces the effective electrostatic interaction between SQ dye and Cu^{2+} ion (Fig. 12). When combining this molecular sensing system with the SPR method, the sensitivity of Cu^{2+}-ion detection is maximized. The maximum resonance angle shift is about 2.30° at 1.0×10^{-4} M Cu^{2+}-ion concentration. The dynamic detection range of this chemical sensor varies widely from 1.0×10^{-12} to 1.0×10^{-4} M. Furthermore, the absorption property of the SQ dye can improve the sensitivity to Cu^{2+} ion detection. It can be maximized through matching the wavelength of the probing radiation of SPR (675 nm) with the absorption maximum of the SQ dye, exactly the shoulder of the absorption maximum. Thus, this Cu^{2+} ion sensor may be useful in the medical, biomedical, and environmental fields.

4. SPIROXAZINE MONOLAYERS

Recently, photochromic materials have attracted attention because of their importance and potential application in the area of optics, which includes optical switching, display, optical memory and non-linear optical devices [30–35]. Spiroxazines are photochromic compounds analogous to spiropyrans. These two classes of compounds are similar in many respects. However, the replacement of the benzopyran ring by a naphthoxazine ring results in spiroxazine having the advantage of greatly improved resistance to prolonged UV irradiation, which confers a much more commercial importance on them [36]. Upon UV irradiation, the C–O bond of the colorless spiroxazine is cleaved and the colored merocyanine form is obtained. Thus, the interconversion of spiroxazine (SP)–merocyanine (MC) systems (SP–MC), has been extensively investigated because of their potential applications in molecular devices and uses in biotechnology [37]. The SAM of photochromic material is of particular interest in view of its high efficiency.

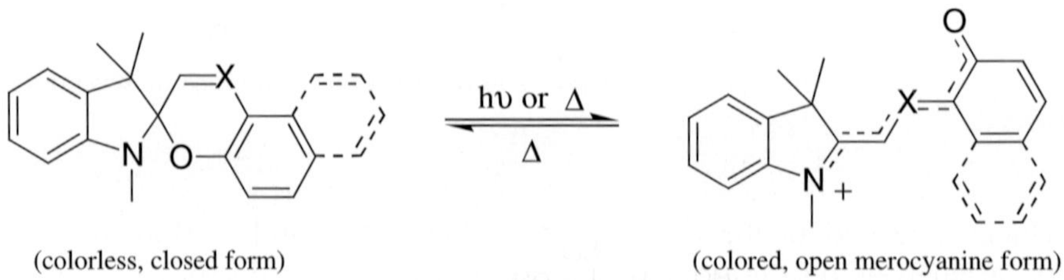

X = CH : spiropyran
X = N : spiroxazine

Photochromism is the change in color under light irradiation, accompanied by changes in color, optical constants, ε and thickness, d. The ε of a photochromic material is one of the main parameters in selecting the material for an appropriate application and it is calculated from optical constants [35]. Therefore, precise measurement of optical constants of a photochromic monolayer is very important.

To characterize the optical property of SAM, many surface analytical techniques have been applied. However, a few of these are able to characterize the properties of SAM. Moreover, SPR is an optical method that is widely gaining recognition as a valuable tool for determining the optical constants (refractive index, n and extinction coefficient, k) and d [5, 38, 39, 40]. In addition, ε of the SAM is derived from optical constants [38, 40, 41]. That is to say, SPR is a very sensitive tool for determining the ε and d values of the SAM.

The UV-induced ε and d values of spiroxazine monolayer can be determined both in the open and closed form by using SPR, i.e. the changes in the ε and d values due to the transformation from the open MC form to the closed SP form can be determined from the theoretical fitting of measured SPR curves [35, 42, 43]

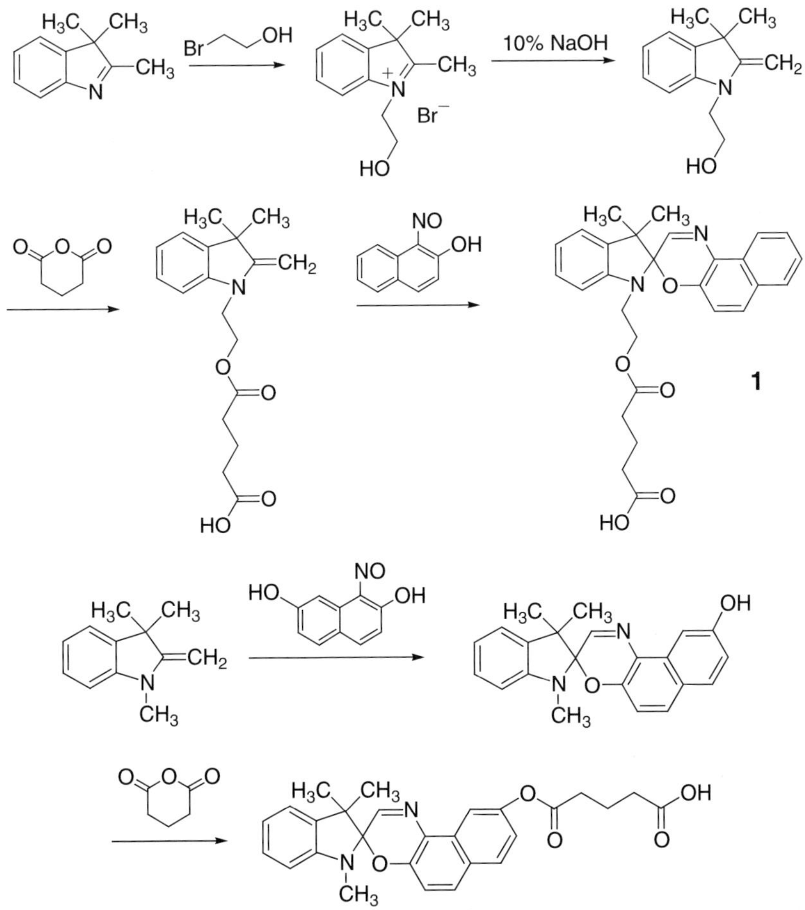

2

For comparing the UV-induced change in the ε of spiroxazine SAMs having different structures, spiroxazines **1**, **2** and **3** SAMs are recommendable (Fig. 13) [42–45]. The synthetic route to spiroxazines **1**, **2** and **3** is outlined in Fig. 13. The ε and d values of the spiroxazine SAMs can be calculated from the theoretical fitting of experimental SPR curves that describe the angle-dependent reflectance for each measurement.

4.1. Formation and characterization of monolayers

Fig. 14 shows spiroxazines **1**, **2** and **3** with different structures that can be used for the formation of spiroxazine SAM [46–49]. Characterization by Fourier

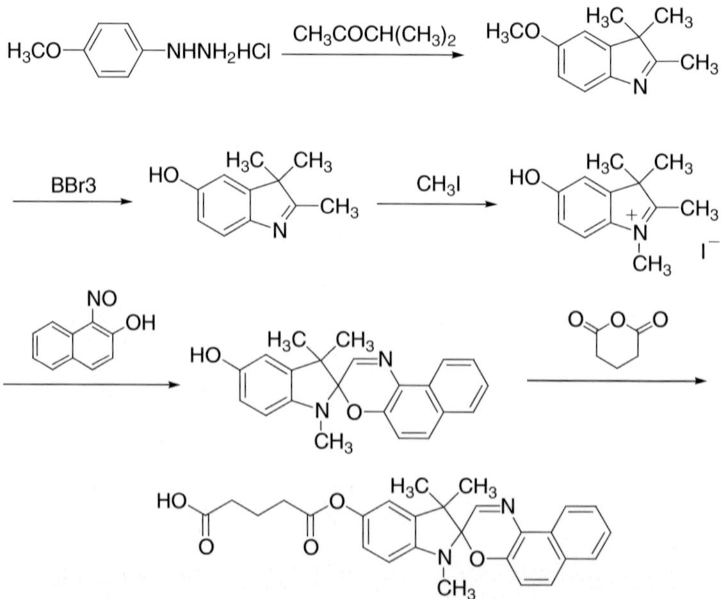

Fig. 13. Synthesis of spiroxazines **1**, **2** and **3**.

transfer infrared reflectance absorption spectroscopy (FT-IRRAS), element analysis (EA) and atomic force microscopy (AFM) are usual.

To fabricate the spiroxazine monolayer, the deposited gold on cover glass is used as a substrate. Gold (50 nm) is deposited on a clean cover glass by direct current (DC) sputter. First, cystamine solution is used to fabricate the cystamine SAM. After that, SAMs of spiroxazine derivatives are formed by immersing the gold-deposited cover glass into an ethanol solution of spiroxazine derivatives for 2 h. For spiroxazine SAM, the solution consists of a spiroxazine derivative, 1-[3-(dimethylamino)propyl]-3-ethylcarbo diimidehydrochloride (EDC) and ethanol as a solvent. Kim and Koh [46–48] reported the synthesis and self-assembly of the photochromic spiroxazine-containing alkanethiol with an amide group incorporated into the backbone.

ε and d values of the spiroxazine monolayer can be investigated using an SPR system based on the Kretschmann configuration [41]. The SPR apparatus consisted of a laser diode, a polarizer, a prism, a detector, a motorized stage, a stage controller, a UV light source, a peristaltic pump, a temperature controller and a computer (Fig. 3). The laser was a laser diode (4.5 mW) of wavelength 675 nm. A cell with BK7 prism (n = 1.514 at 675 nm) and detector (photodiode) are placed on a motorized stage, which was driven by a stepping motor making steps of 0.004°. Signal processing and motor control were done with

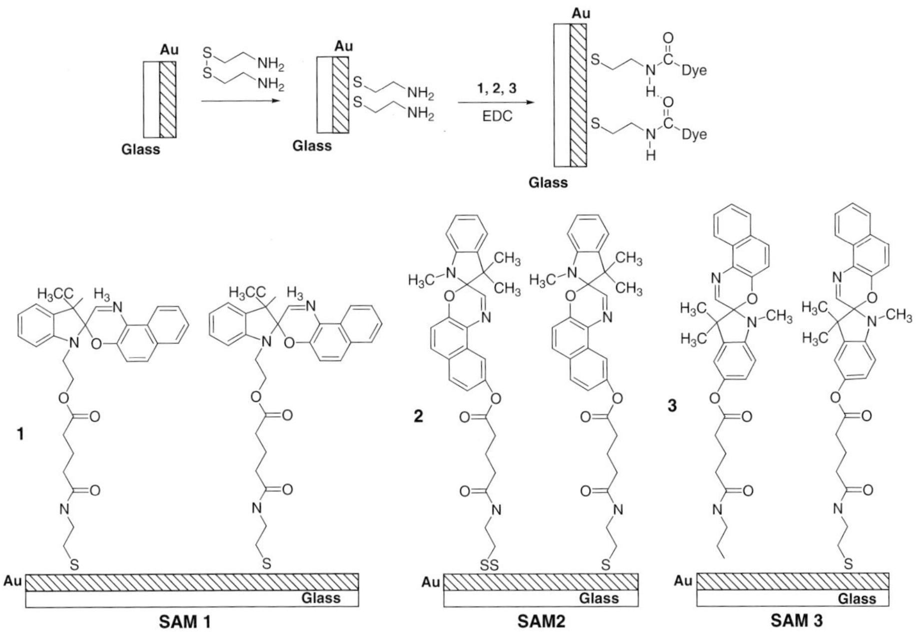

Fig. 14. Stepwise organization of the spiroxazine dyes' monolayers on gold.

a computer and motor controller. To obtain SPR curves, the reflectance of polarized laser light from metal–prism interface is measured as a function of the incident angle, θ.

To investigate the UV-induced ε, SPR curves of spiroxazine **1**, **2,** and **3** SAMs have to be measured both in the open (MC form under UV irradiation) and closed (SP form in the dark) form by using SPR [44–45] (Fig. 15). UV irradiations are performed by a UV light source composed of a UV lamp with a filter. Optical constants (n and k) and d are determined by the theoretically fitted SPR curve using the Fresnel equation and ε is calculated from the obtained optical constants using the equation $\varepsilon = n^2 - k^2$ [38].

The system displayed in Fig. 16 can be represented with the characteristic matrix below

$$\begin{pmatrix} E_0^+ \\ E_0^- \end{pmatrix} = \prod_{m=1}^{3} \frac{C_m}{t_m} \begin{pmatrix} E_3^+ \\ E_3^- \end{pmatrix}$$

where

$$C_m = \begin{pmatrix} e^{i\delta_{m-1}} & r_m e^{-i\delta_{m-1}} \\ r_m e^{-i\delta_{m-1}} & e^{i\delta_{m-1}} \end{pmatrix}$$

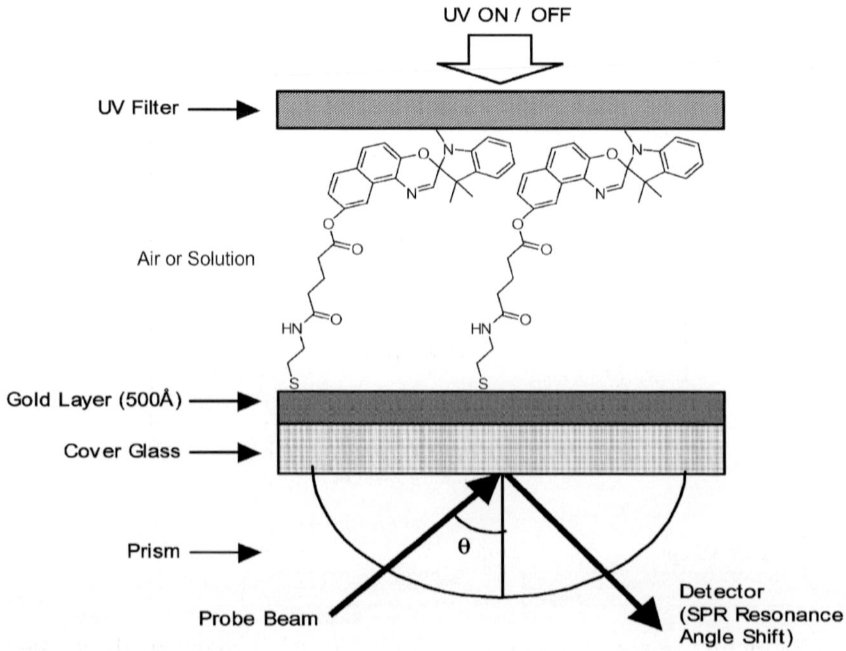

Fig. 15. Schematic diagrams of four-layer structure based on Kretschmann configuration.

Here, E_0^+ and E_0^- represent the positive going incident wave and negative going reflected wave in the 0th medium (air), respectively. Also, E_3^+ indicates the positive going transmitted wave in the third medium (spiroxazine SAMs). Since there is no negative going wave in the third medium, one can set $E_3^- = 0$. To solve the equation , the phase term δ_m and Fresnel coefficient r_m, t_m can be considered to be the complex value described below in the equations under the conditions of arbitrary incidence angle, absorbing media and p-polarization, respectively.

$$\delta_{m-1} = \frac{2\pi}{\lambda}(n_{m-1} - jk_{m-1})\cdot d_{m-1}\cdot\cos\theta_{m-1}$$

$$r_m = g_m + jh_m$$

$$\equiv \frac{\boldsymbol{n}_{m-1}\cos\theta_m - \boldsymbol{n}_m\cos\theta_{m-1}}{\boldsymbol{n}_{m-1}\cos\theta_m + \boldsymbol{n}_m\cos\theta_{m-1}}, (\boldsymbol{n}_m = \boldsymbol{n}_m - jk_m)$$

$$t_m = 1 + g_m + jh_m$$

where λ, d_m, $\boldsymbol{n}_m$, n_m, k_m and θ_m are the wavelength of incident wave, thickness of layer, complex refractive index, real part of the refractive index, imaginary part

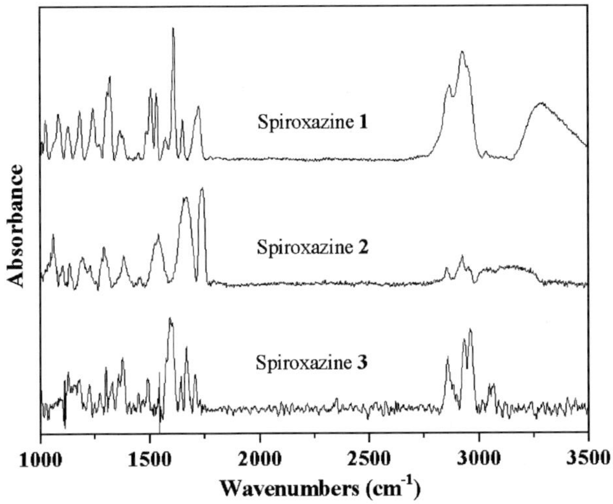

Fig. 16. FT-IRRAS spectra of spiroxazine **1**, **2** and **3** monolayers.

of the refractive index (extinction coefficient) and incident angle at the *m*th medium, respectively. Furthermore, the incident angles in each interface determined by Snell's law are also complex if an angle is over the critical value. As a result, the manipulation of complex angles must be included in the evaluation of the equation.

Writing the matrix product

$$\prod_{m=1}^{3} C_m = \begin{pmatrix} a & b \\ c & d \end{pmatrix}$$

in the equation, we obtain the reflectance that is given by

$$R = \frac{(E_0^-)(E_0^-)^*}{(E_0^+)(E_0^+)^*} = \frac{cc^*}{aa^*}$$

In order to estimate the optical constants of each layer, first, the previous equations should be applied to fit the reflectance curve of bare gold (second medium). Then, the best-fit values can be used to estimate the thickness and the refractive index of the spiroxazine SAM. However, if one of these parameters cannot be determined before, an ambiguity exists in the separate determination of them. Since the optical thickness (refractive index ($\boldsymbol{n}_m$)$\times$geometrical thickness (d_m)) in the phase term, δ_m, has many pairs of values ($\boldsymbol{n}_m$, d_m) to the given reflectance curve, one of them must be fixed prior to Fresnel calculation to remove any uncertainty.

Based on the above procedures, the photochromic change of spiroxazine SAM can be investigated in terms of the resonance angle shift using the SPR method. Using the experimental results and calculations from Fresnel equations, the physical meaning of the photochromic change in terms of optical parameters such as the refractive index and thickness can be analyzed.

In Fig. 16, the FT-IRRAS result shows strong N–H stretching (amide) at 3292 (spiroxazine **1**) and 3163 (spiroxazine **2**) cm^{-1}. C=O stretching (amide) appears at 1649 (spiroxazine **1**), 1653 (spiroxazine **2**), and 1666 (spiroxazine **3**) cm^{-1}. Peaks of 2928 (spiroxazine **1**), 2926 (spiroxazine **2**) and 2933 (spiroxazine **3**) cm^{-1} are assigned as the asymmetric C–H stretching mode of a methylene group. Peaks of 2863 (spiroxazine **1**), 2855 (spiroxazine **2**) and 2858 (spiroxazine **3**) cm^{-1} are assigned as the symmetric C–H stretching mode of a methylene group. A C–H asymmetric stretching band of the CH_3 group is observed at 2951 (spiroxazine **1**), 2953 (spiroxazine **2**) and 2960 (spiroxazine **3**) cm^{-1}.

AFM images of gold, cystamine and Spiroxazine **1**, **2** and **3** SAMs are shown in Fig. 17. Root mean square roughnesses of AFM images obtained are 10.4, 12.5, 15.9, 21.0 and 12.9 Å, respectively. Those results confirm that the cystamine and spiroxazine SAMs are successfully formed on gold surface.

UV irradiation causes a decrease of 0.079°, 0.082° and 0.077° in the resonance angles of spiroxazine **1**, **2** and **3** SAMs, respectively (Fig. 18). As shown

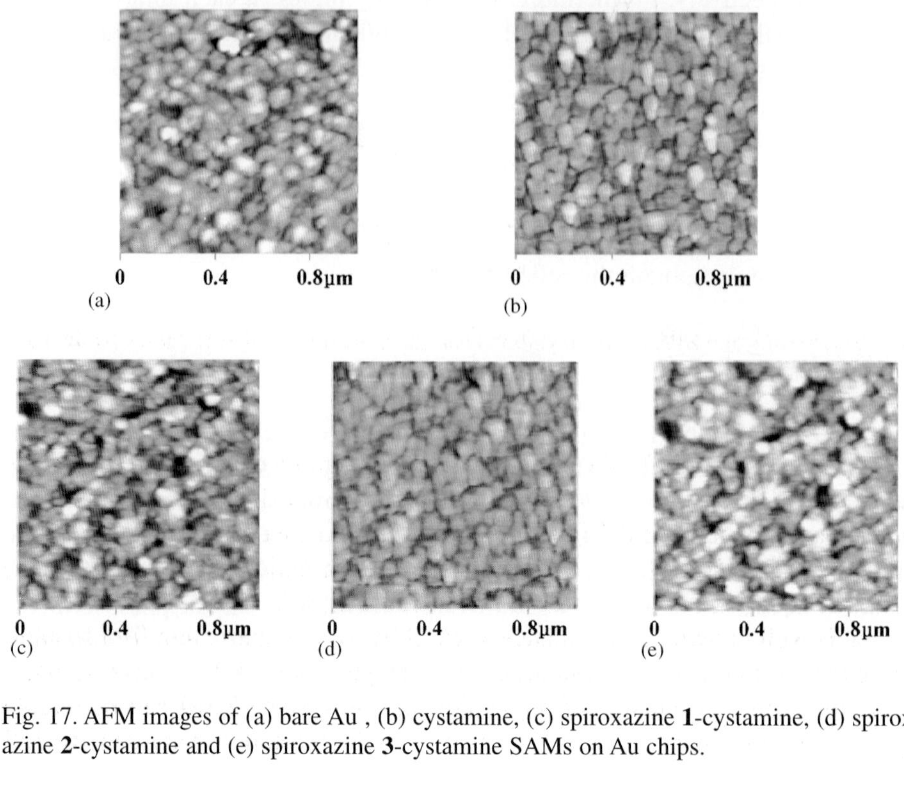

Fig. 17. AFM images of (a) bare Au , (b) cystamine, (c) spiroxazine **1**-cystamine, (d) spiroxazine **2**-cystamine and (e) spiroxazine **3**-cystamine SAMs on Au chips.

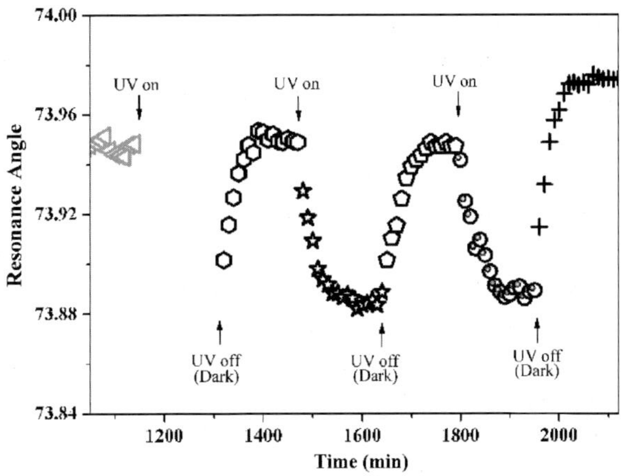

Fig. 18. UV-induced SPR angle shift of spiroxazine **1** monolayers.

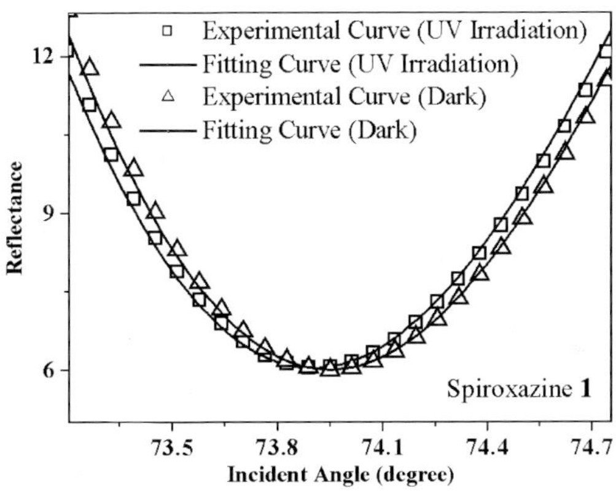

Fig. 19. Experimental SPR curves (symbols) and theoretically fitted SPR curves (solid lines) of spiroxazine **1** monolayer.

in Fig. 19, experimental SPR curves are in good agreement with the theoretical curve [5, 38]. Therefore, the *n*, *k* and *d* values of spiroxazine SAMs can be determined from the theoretical fitting of the experimental data using Fresnel equation (four-layer model) [41]. ε can be determined from optical constants (*n* and *k*) using the equation $\varepsilon = n^2 - k^2$, as shown in Table 2 [38].

It is well known that resonance angle shifts of polymeric thin film containing photochromic dye are influenced by a change of *n* [49]. In contrast to polymeric thin film, the change of angle shift of spiroxazine SAM can be mainly described by thickness change. As shown in Table 2, ε and *d* of spiroxazine **1, 2**

Table 2
Calculated refractive index (n), extinction coefficient (k), thickness (d, nm), dielectric constant (ε), coloration (r_c, min^{-1}) and decoloration (r_d, min^{-1}) rate constants of spiroxazine **1**, **2** and **3** SAMs

	Spiroxazine **1**		Spiroxazine **2**		Spiroxazine **3**	
	UV on	UV off	UV on	UV off	UV on	UV off
n	1.4510	1.4506	1.4512	1.4523	1.4505	1.4532
k	0.197	0.135	0.115	0.107	0.059	0.044
d	1.3	1.7	1.5	1.9	1.5	1.8
ε	2.066	2.086	2.092	2.098	2.100	2.110
r_c	2.20×10^{-2}		2.51×10^{-2}		2.32×10^{-2}	
r_d	2.70×10^{-2}		2.98×10^{-2}		2.64×10^{-2}	

and **3** SAMs decreased by UV irradiation. These results show that ring opening of photochromic spiroxazine derivatives can lead to a decrease in ε and d. In addition, a difference of the spiroxazine structure results in a difference of UV-induced ε and d changes. In the determination of ε, the above-mentioned method did not consider the influence of magnetic permeability on ε, which needs to be further investigated [49].

The ring opening of spiroxazine and the ring closing of MC are described by the following equation: $\varepsilon_t - \varepsilon_\infty = \varepsilon_0 \exp(-rt)$ [50, 51], where r, t, ε_t, ε_0 and ε_∞ are the coloration (r_c) or decoloration (r_d) rate constants, time (s) and dielectric constant at t, 0 and ∞, respectively. For more stable spiroxazine forms, r_d of spiroxazine **1, 2** and **3** SAMs are faster than their r_c. Absorption of UV light by spiroxazine causes opening of the carbon–oxygen bond with the formation of an MC structure. The MC structure reverts to the SP one by a ring closure reaction when the UV light source is removed.

With an assumption that most spiroxazine molecules tilt in the vertical plane (*X–Y* plane) of the Au layer (*X–Z* plane) due to an interaction between spiroxazine molecules and cystamine linked with spiroxazine, inclination angle of each spiroxazine SAM from MOPAC AM1 approximation and theoretical fitting of SPR curves are obtained (Fig. 20). Those results demonstrate that UV-induced structural changes of spiroxazine **1**, **2** and **3** SAMs result in a change in the value of d.

SPR systems can determine UV-induced structural changes of spiroxazine SAMs and which have considerable potential. This method has an advantage for the determination of precise optical parameters. n, k, d, ε, r_c, r_d and other properties of the photochromic film can be easily determined from the SPR curve.

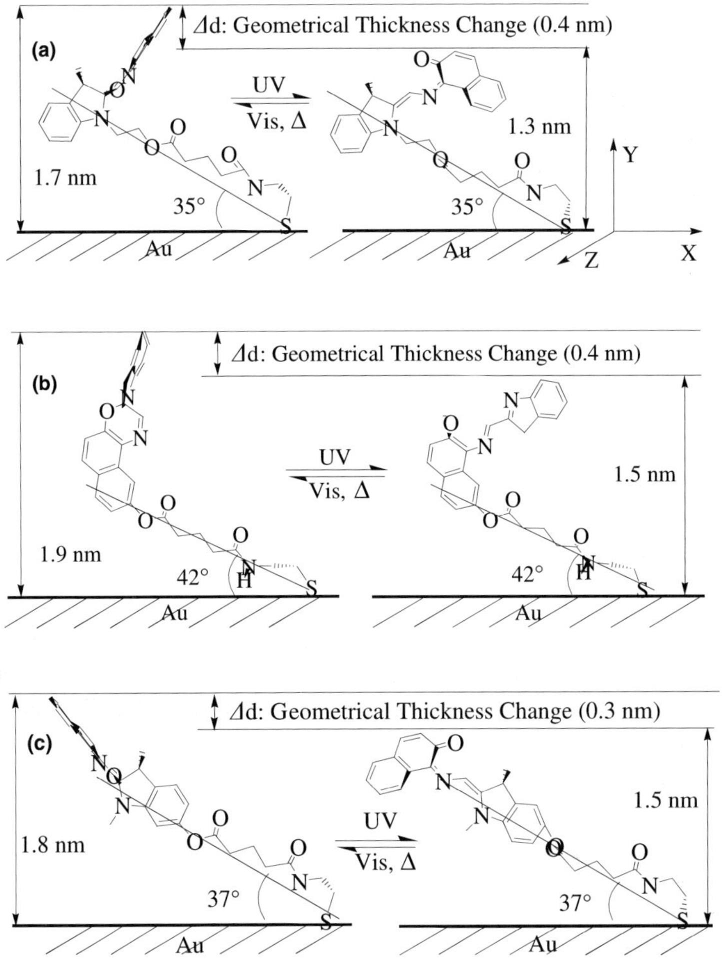

Fig. 20. Structural changes of spiroxazine (a) **1**, (b) **2** and (c) **3** obtained by MOPAC AM1 approximation and theoretical fitting of SPR curves.

4.2. UV-addressable phenylalanine sensing

Spiroxazines, which are similar to spiropyrans, are easily converted to the ring-opened form upon UV irradiation and have excellent fatigue-resistance property to light [52]. A spiropyran derivative was reported that was involved in the transportation of phenylalanine across liposomal bilayers mediated by a photoresponsive carrier [53]. Based on these facts, the ring-opened structure of a zwitterionic spiroxazine is expected to form an ionic complex with zwitterionic phenylalanine in an aqueous buffered solution (Fig. 21). Therefore, the study of UV-addressable phenylalanine sensing with spiroxazine **3** is interesting.

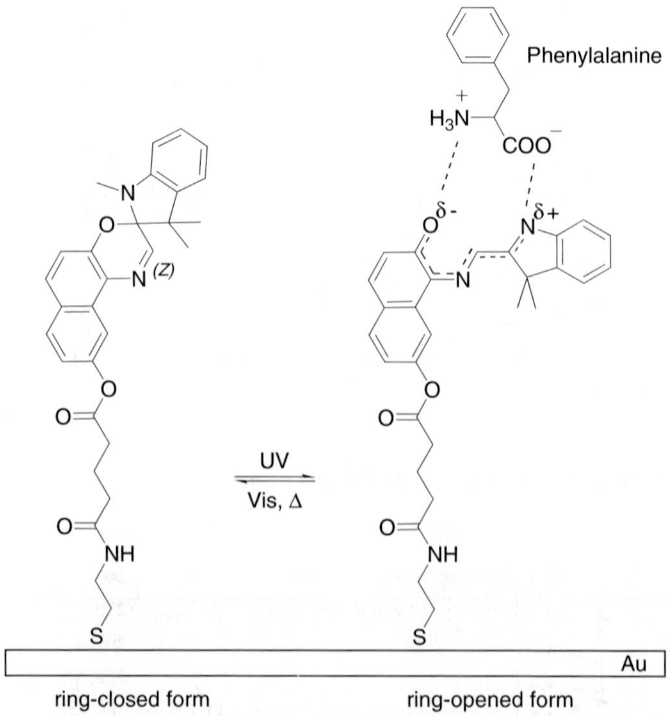

Fig. 21. Reversible photoconversion of ring-closed spiroxazine into ring-opened spiroxazine and the interaction between ring-opened spiroxazine and phenylalanine.

Fabrication of phenylalanine-sensing spiroxazine **3** on Au surface is easily done by the self-assembly technique and monitored by the SPR technique (Fig. 22). SPR spectroscopy is a relatively simple optical technique that is capable of real-time monitoring for molecular interaction at the surface. Investigation of the interaction between spiroxazine **3** self-assembled monolayer (SAM) and phenylalanine under UV irradiation can be performed by SPR spectroscopy (Fig. 22, [5]).

Confirmation of the UV-addressable behavior of the spiroxazine **3** SAM in buffer solution can be done by measuring the SPR angle shift for spiroxazine **3** SAM upon UV irradiation. The SPR angle for spiroxazine **3** SAM decreased under UV irradiation and returned to its initial value when UV was switched off. These SPR angle shifts are caused mainly by the thickness changes of the monolayer by spiroxazine photoisomerization of spirooxazine **3** [5, 43].

The relative SPR angle shift is expressed as $(\theta_0 - \theta)/\theta_0$, where θ_0 is the SPR angle for spiroxazine **3** SAM in phenylalanine solution before UV irradiation and θ is the SPR angle for spiroxazine **3** SAM at the photoisomerization equilibrium state. The relative SPR angle shifts increased with an increase in the concentration

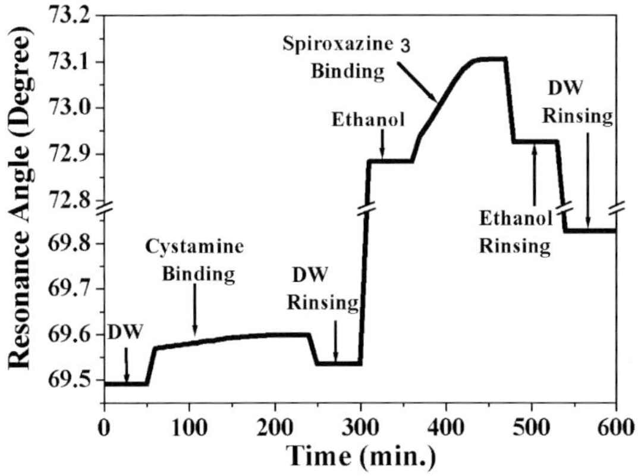

Fig. 22. SPR sensogram of binding cystamine and spiroxazine **3**.

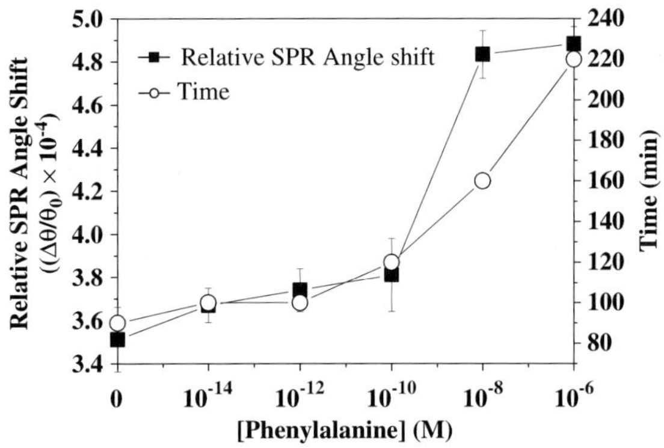

Fig. 23. Concentration dependence of the relative angle shift and the time to reach photoisomerization equilibrium of spiroxazine **1** SAM.

of phenylalanine solution (Fig. 23). In addition, with an increase in the concentration of phenylalanine, more time is required to reach the photoisomerization equilibrium of spiroxazine **3**. These results can be explained by the fact that the zwitterionic phenylalanine induces the formation of the more ring-opened spiroxazine. Thus, it is demonstrated that there is interaction between phenylalanine and ring-opened spiroxazine.

The interaction between phenylalanine and spiroxazine **3** SAM upon UV irradiation can be confirmed by SPR study. The different SPR angle shifts caused by the concentration difference of phenylalanine means that spiroxazine **3** SAM has a potential application in UV-addressable phenylalanine sensing.

REFERENCES

1. Mallouk, T.E. & Harrison, D.J., *Interfacial Design and Chemical Sensing*, American Chemical Society, Washington, DC (1994) p. 2.
2. Bargossi, C., Fiorini, M.C., Montalti, M., Prodi, L. & Zaccheroni, N., *Coord. Chem. Rev.*, **208** (2000) 17.
3. Kim, S.H., Han, S.K., Park, S.H., Lee, S.M. & Koh, K.N., *Dyes Pigments*, **41** (1999) 221.
4. Kim, S.H., Han, S.K., Jang, G.S., Koh, K.N., Keum, S.R. & Yoon, C.M., *Dyes Pigments*, **44** (2000) 169.
5. Homola, J., Yee, S.S. & Gauglitz, G., *Sensors Actuators B*, **54** (1999) 3.
6. de Bruijn, H.E., Lenferink, A.T.M., Kooyman, R.P.H. & Greve, J., *Opt. Comm.*, **86** (1991) 444.
7. Hanning, A., Roeraade, J., Delrow, J.J. & Jorgenson, R.C., *Sensors Actuators B*, **54** (1999) 25.
8. Pankove, J., *Optical Processes in Semiconductors*, Prentice-Hall, Englewood Cliffs, NJ (1971) p. 89.
9. Kim, S.H. & Hwang, S.H., *Dyes Pigments*, **35** (1997) 111.
10. Ziegenbein, W. & Sprenger, H.E., *Angew. Chem.*, **78** (1996) 937.
11. Treibs, A. & Jacob, K., *Angew. Chem.*, **77** (1965) 680.
12. Spreger, H.E. & Ziegenbein, W., *Angew. Chem.*, **79** (1967) 581.
13. Seiler, K., *Ion-Selective Optode Membranes*, Fluka Chemie AG, Buchs (1993) p. 33.
14. Kretschmann, E., *Zh. Phys.*, **241** (1971) 313.
15. Phelps, J.M. & Taylor, D.M., *J. Phys. D: Appl. Phys.*, **29** (1996) 1080.
16. Chinowsky, T.M. & Yee, S.S., *Sensors Actuators B*, **51** (1998) 321.
17. Seiler, K., *Ion-Selective Optode Membranes*, Fluka Chemie AG, Buchs (1993) p. 22.
18. Darmani, H., Coakley, W.T., *Biochim. Biophys. Acta*, **1021** (1990) 182.
19. Pockrand, I., Swalen, J.D., Santo, R., Brillant, A. & Philpott, M.R., *Chem. Phys.*, **69** (1978) 4001.
20. Vesely, J., *Analytica Chim. Acta*, **62** (1972) 1.
21. Brown, P.A., Gill, S.A. & Allen, S.J., *Water Res.*, **34** (2000) 3907.
22. Gan, Q., *Waste Manage.*, **20** (2000) 695.
23. Song, Y. & Chen, M., *Clin. Biochem.*, **33** (2000) 589.
24. Brun, L.A., Maillet, J., Hinsinger, P. & Pepin, M., *Environ. Pollut.*, **111** (2001), 293.
25. Cleij, M.C., Scrimin, P., Tecilla, P. & Tonellato, U., *J. Org. Chem.*, **62**(1997) 5592.
26. Seleznev, B.L., Legin, A.V. & Vlasov, Y.G., *J. Anal. Chem.*, **51** (1996) 882.
27. Boussaad, S., Pean, J. & Tao, N., *J. Anal. Chem.*, **72** (2000) 222.
28. Kim, S.H. & Hwang, S.H., *Dyes Pigments,* **36** (1998) 139–148.
29. Koh, K.N., Imada, T., Nagasaki, T. & Shinkai, S., *Tetrahedron Lett.*, **35** (1994) 4157.
30. Christie, R.M., Agyako, C.K. & Mitchell, K., *Dyes Pigments,* **29** (1995) 241.
31. Mitchell, R.H., Ward, T.R., Chen, Y., Wang, Y., Weerawarna, S.A., Dibble, P.W., Marsella, M.J., Almutairi A., & Wang, Z. Q., *J. Am. Chem. Soc.*, **125** (2003) 2974.
32. Berkovic, G., Krongauz, V. & Weiss, V., *Chem. Rev.*, **100** (2000) 1741.
33. Sekkat, Z., Wood, J., Geerts, Y. & Knoll, W., *Langmuir,* **12** (1996) 2976.
34. Evans, S.D., Johnson, S.R., Ringsdorf, H., Williams, L.M. & Wolf, H., *Langmuir*, **14** (1988) 6436.
35. Biteau, J., Chaput, F. & Boilot, J.P., *J. Phys. Chem.*, **100** (1996) 9024.
36. Chu, N.Y.C., Durr, H. & Bouas-Laurent, H., *Photochromism – Molecule and Systems*, Elsevier, Amsterdam (1990).
37. (a) Berkovic, G., Krongauz, V. & Weiss, V., *Chem. Rev.*, **100** (2000) 1741;(b) Kawata, S. & Kawata, Y., *Chem. Rev.*, **100** (2000) 1777;(c) Collins, G.E., Choi, L.S., Edwing, K.J., Michelet, V., Bowen, C.M. & Winkler, J.D., *Chem. Commun.* (1999) 321.

38. Kazuyoshi, K. & Koji, S., *Anal. Chem.*, **74** (2002) 696.
39. Pockrand, I., Swalen, J.D., Gordon, J.G. & Philpott, M.R., *Surface Sci.*, **74** (1978) 237.
40. Peterlinz, K.A. & Georgiadis, R., *Opt. Comm.*, **130** (1996) 260.
41. Heavens, O.S., *Optical Properties of Thin Solid Films*, Dover Publications Inc., New York (1991).
42. Kim, S.H., Ock, K.S., Kim, J.H. & Koh, K.N., *Mol. Cryst. Liq. Cryst.*, **349** (2000) 39.
43. Kim, S.H., Ock, K.S., Im, J.H., Kim, J.H. & Koh, K.N., *Dyes Pigments*, **46** (2000) 55.
44. Suh, H.J., Jin, S.H., Gal, Y.S., Koh, K. & Kim, S.H., *Dyes Pigments*, **58** (2003) 127.
45. Biteau, J., Chaput, F., Lahlil, K. & Boilot, J.P., *Chem. Mater.*, **10** (1998) 1945.
46. Kim, S.H., Lee, S.M., Park, J.H., Kim, J.H. & Koh, K.N., *Dyes Pigments*, **45** (2000) 51.
47. Kim, S.H., Choi, S.W., Kim, J.H., Jin, S.H., Gal, Y.S., Ryu, J.H., Cui, J.Q. & Koh, K., *Dyes Pigments*, **50** (2001) 109.
48. Kim, S.H., Choi, S.W., Suh, H.J., Jin, S.H., Gal, Y.S. & Koh, K., *Dyes Pigments*, **55** (2002)17.
49. Sasaki, K. & Nagamura, T., *J. Appl. Phys.*, **83** (1998) 2894.
50. Levitus, M. & Aramendia, P.. F., *J. Phys. Chem. B.*, **103** (1999) 1864.
51. Tang, X.. C., Jia, D.Z., Liang, K., Zahng, X.G., Xia, X. & Zhaou, Z.Y., *J. Photochem. Photobiol. A*, **134** (2000) 23.
52. Willner, I., Doron, A., Katz, E. & Levi, S. *Langmuir*, **12** (1996) 946.
53. Sunamoto, J., Iwamoto, K., Mohri, Y. & Kominato, T., *J. Am. Chem. Soc.*, **104** (1982) 503.

Functional Dyes
Sung-Hoon Kim (Editor)

Chapter 6

Syntheses and application of squarylium dyes

Shigeyuki Yagi and Hiroyuki Nakazumi

Department of Applied Chemistry, Graduate School of Engineering, Osaka Prefecture University, Sakai, Osaka, Japan

1. INTRODUCTION

Squarylium dyes, often called squaraines, involve two aromatic or heterocyclic moieties at both the ends of an oxocyclobutenoate core. They are often classified into polymethyne dyes like cyanine dyes [1], because their localized π-conjugation structures are quite similar to those of cyanines, as shown in Fig. 1. Both cyanines and squaryliums exhibit sharp and intense electronic absorption in the region from visible to near-IR, and sometimes show fluorescence emission properties. However, from the viewpoint of the electronic conjugation structure there are mainly two differences between them. One is that typical cyanine dyes are cationic although squaryliums are neutral showing zwitter-ionic structures. The other is the difference in electronic distribution in dye molecules. As has been discussed so far, the intramolecular electronic resonance in the cyanine π-conjugation system extended from the heterocyclic component at one end to that at the other end. On the other hand, squarylium dyes consist of the central cyclobutene and the heterocycles at both ends, yielding an electron donor–acceptor–donor (D–A–D) charge transfer structure, and the electronic structures are also represented by the cyclobutene diylium (Scheme 1).

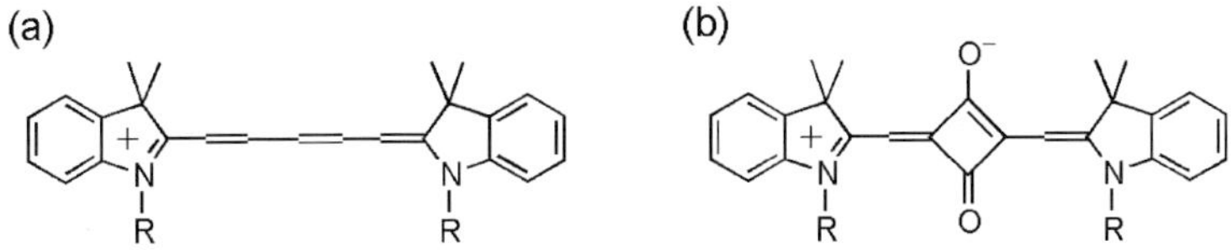

Fig. 1. Structures of typical cyanine and squarylium dyes: (a) indolinocyanine dye and (b) indolinosquarylium dye.

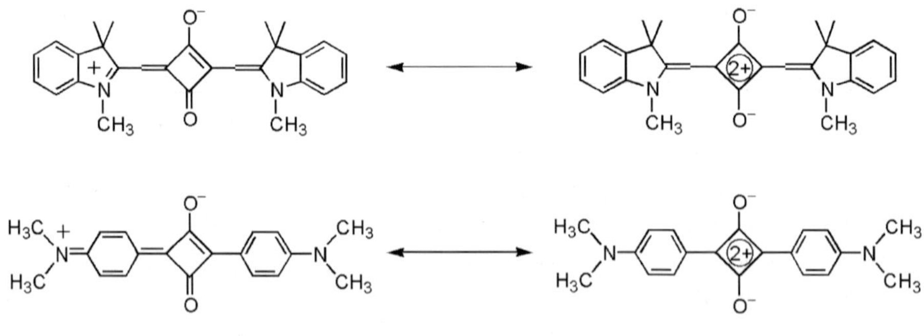

Scheme 1.

At the early stage of the squarylium chemistry, interest of the chemists has been poured into the synthesis of a new series of squarylium dye skeletons [2–6]. Especially, the representation of the π-conjugation structures has been lying at the center of squarylium studies and attracting interest of theoretical chemists as well as X-ray crystallographers. Thus, a variety of squarylium dyes with classical D–A–D structures have been eagerly synthesized and they have established their position as one of important organic functional dyes. Along with increasing applicability of organic functional dyes to electric industries, squarylium dyes have also been receiving much attention as materials for xerographic imagings, organic semiconductors, photosensitizers in organic solar cells, and so on [7]. From such demands, researches not only on the basic squarylium chemistry but also on the application of squarylium dyes have been facilitated since the late 1970s. The utilization of diode laser for optical recording has also required intense light absorption properties of squarylium dyes in the near-IR region [8,9]. Nowadays, squarylium dyes are also applied to the clinical and bioanalytical fields such as photodynamic therapy and chemosensory systems for biologically and environmentally important ions and molecules. However, there was a serious problem in squarylium researches: the classical synthetic procedures could yield only symmetrical dyes with highly electron donating aromatic or heterocyclic components, limiting the variability of the electronic structures of squarylium dyes. Therefore, it is of quite urgency to develop the methods to construct novel squarylium-based chromophoric systems.

In the present chapter, the authors review synthetic procedures of squarylium dyes and their homologues, involving their original works. The topics start from the classical squarylium synthesis, and then, move to the synthesis of unsymmetrical squarylium dyes. The syntheses of new types of squarylium homologues are also mentioned with their optical properties covered. In the last part, application of squarylium dyes to industrial fields and future aspects to advanced materials are discussed.

2. SYNTHESIS AND OPTICAL PROPERTIES OF SQUARYLIUM DYES

2.1. Symmetrical squarylium dyes

The typical squarylium synthesis is carried out by condensation of 1 mol equivalent of squaric acid (3,4-hydroxy-3-cyclobutene-1,2-dione) with 2 mol equivalents of aromatic or heterocyclic compounds. The first synthesis of squarylium dyes was reported by Treibs and Jacob in 1965 (Scheme 2) [2]. Therein, condensation reactions of squaric acid with α-unsubstituted pyrrole and 1,3,5-trihydroxybenzene were examined to afford the dyes **1** and **2** in 65 and 67% yields, respectively. The solvents used for the synthesis were ethanol with a catalytic amount of perchrolic acid for **1** and acetic acid for **2**. Just after the first report by Treibs and Jacob, Ziegenbein and Sprenger reported azuleno dye **3** [4] and anilino dye **4** [5], colored blue-to-green for **3** and blue for **4**. The condensation reactions efficiently underwent (yields; 90 and 60% for **3a** and **4a**, respectively) using an azeotropic solvent mixture of butanol–benzene, where water produced during the condensation was removed by azeotropic distillation. Since then, a mixture of butanol and benzene has been employed as the most common solvent system in the squarylium synthesis. The reaction mechanism for the squarylium formation was proposed by Sprenger and Ziegenbein in their review as shown in Scheme 3, [3]. The nucleophilic attack of the aromatic compound (Ar–H) against the carbonyl carbon in the half-ester of squaric acid followed by the removal of the alcohol is expected to yield the monoaryl squaric acid as an intermediate. The subsequent attack of another Ar–H followed by dehydration should produce the squarylium dye.

In addition to electron-rich aromatics described above, heterocyclic compounds with active methyl groups, transformed to the enamine structures under basic conditions, were also converted to the corresponding squarylium dyes. The heterocycles include 2-methylindolium (**5**), 2-methylbenzothiazolium (**6**), 2-methylbenzoselenazolium (**7**), 2-methylquinolinium (**8**), and so on. Squarylium dyes of this class were first reported by Sprenger and Ziegenbein [10], where the butanol–benzene solvent mixture contained a small amount of quinoline to convert the quaternary heterocyclic salts to the corresponding enamine form. The typical examples and their electronic absorption maxima are summarized below (**9–12**). The squarylium dyes formed from the heterocyclic enamines possess the vinylogous structures to the dyes formed from the aromatic precursors. It is worthy to note that employing dialkyl squarate, instead of squaric acid, as a starting material did not yield any squarylium dyes, but 3,4-disubstituted 1,2-dioxocyclobut-3-enes [10]. Compounds of this type are sometimes obtained as byproducts in the squarylium synthesis, probably due to esterification of squaric acid by butanol solvent. Although squarylium dyes are cyanine-like polymethynes, these 1,2-substitutetd dyes adopt merocyanine-like structures, showing blue-shifted absorption spectra, because the intramolecular

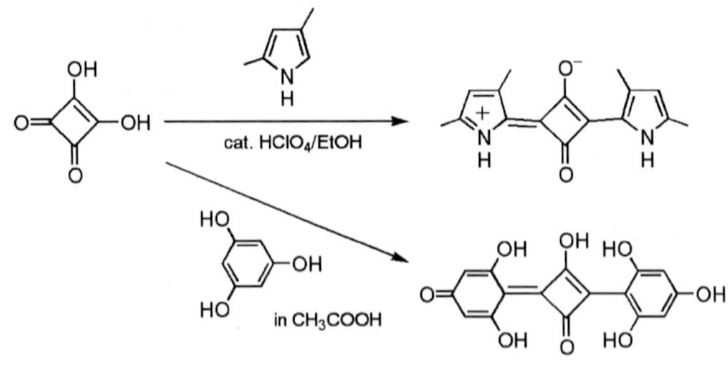

1: λ_{max}; 550 nm in chloroform

2

Scheme 2.

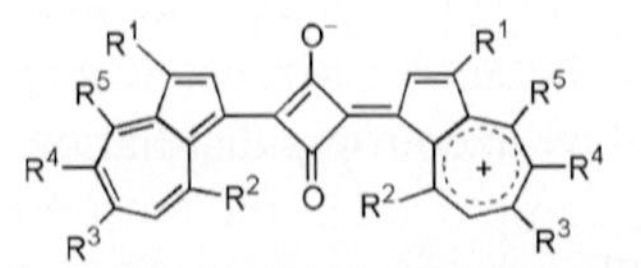

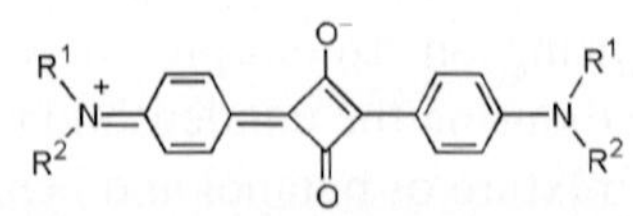

3a: R^1-R^5 = -H λ_{max}; 680 nm ($CHCl_3$)
3b: R^1, R^5 = -CH_3; R^4 = -*i*-Pr; R^3, R^5 = -H λ_{max}; 770 nm ($CHCl_3$)
3c: R^1, R^4 = -H; R^2, R^3, R^5 = -CH_3 λ_{max}; 720 nm ($CHCl_3$)

4a: R^1, R^2 = -CH_3 λ_{max}; 628 nm ($CHCl_3$)
4b: R^1, R^2 = -$(CH_2)_2OH$ λ_{max}; 640 nm (MeOH)
4c: R^1, R^2 = -CH_2Ph λ_{max}; 624 nm ($CHCl_3$)
4d: R^1, R^2 = -$(CH_2)_2O(CH_2)_2$- λ_{max}; 621 nm (MeOH)

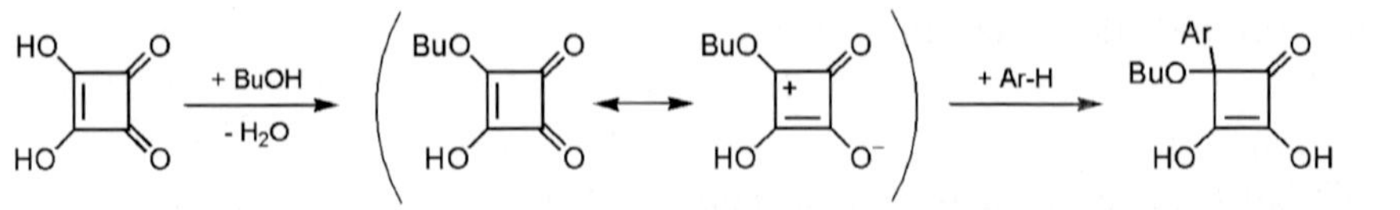

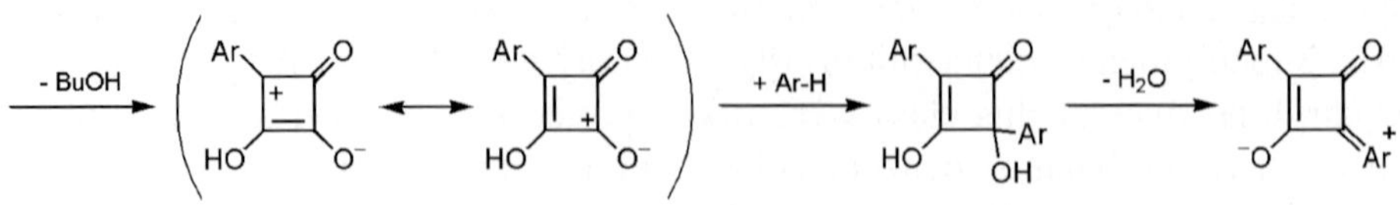

Scheme 3.

cross-conjugation from the aromatic (or heterocyclic) moieties to the central cyclobutenedione disturbs extensive π-electron resonance over a whole molecule (Fig. 2).

2.2. Unsymmetrical squarylium dyes

As mentioned above, the traditional synthetic method of squarylium dyes, i.e., condensation of squaric acid with electron-rich aromatics or enamine-type

5 **6** **7** **8**

9: R = -CH_3 λ_{max}; 630 nm ($CHCl_3$)

10: R = -CH_2CH_3 λ_{max}; 670 nm ($CHCl_3$)

11: R = -CH_2CH_3 λ_{max}; 678 nm (DMF)

12: R = -CH_2CH_3 λ_{max}; 730 nm ($CHCl_3$)

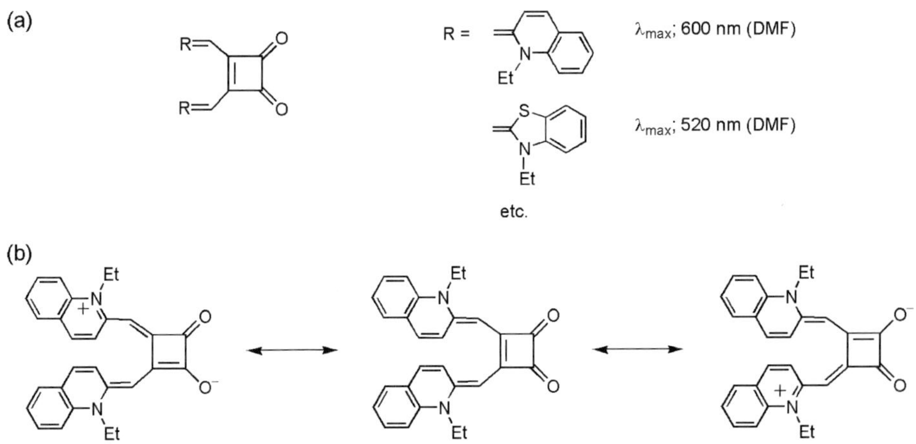

Fig. 2. (a) Structures of 1,2-substituted squarylium homologues and (b) the possible resonance structures showing merocyanine-like cross-conjugation.

heterocycles, yields only the symmetrical structures. This is one of most significant problems in squarylium chemistry and its application, because the limitation of the variability in the obtained electronic structures and physicochemical properties gives rise to restricted applicability and utility of the dyes in any fields. In order to extend the variability of squarylium dyes and to make their electronic structures tunable on demand, the establishment of synthetic procedures for unsymmetrical squarylium dyes has been intensely required.

Taking into consideration the reaction mechanism shown in Scheme 3, the rational strategy to unsymmetrical squarylium dye is to obtain a squaric acid substituted with an aromatic or a heterocyclic component followed by the condensation reaction with another component. That is, mono-substituted squaric acids, namely semi-squaryliums, have been intensely required for this purpose. Several research groups have established the semi-squarylium synthesis and opened the door to the synthesis of unsymmetrical squarylium dyes. Green and Neuse first reported the Friedel–Crafts-type reaction of benzene with 3,4-dichloro-3-cyclobutene-1,2-dione in the presence of $AlCl_3$ [11]. The dichlorinated cyclobutenedione was available from squaric acid by chlorination with thionyl chloride [12]. Thereafter, it was found that such a reaction employing electron-rich aromatics underwent under milder conditions, i.e., in the absence of $AlCl_3$ (Scheme 4) [13]. The obtained mono-substituted squaryl chlorides were converted to the semi-squaryliums such as **13** and **14** by hydrolysis under acidic conditions. It is also applicable to enamine-type heterocycles as represented by **15**. Thus, aromatic–aromatic, aromatic–heterocylic and heterocyclic–heterocyclic types of unsymmetrical squarylium dyes have become available by condensation of semi-squaryliums with aromatics or heterocyclic enamines. Nowadays, this procedure has been one of convenient synthetic routes to unsymmetrical squarylium dyes. For example, Yagi and Nakazumi prepared aromatic–heterocyclic as well as heterocyclic–heterocyclic types of dyes absorbing near-IR light (**16**–**21**; λ_{max}, 739–821 nm) (Table 1) [13]. These are quite unique near-IR dyes in terms of their relatively short π-conjugations. The key essence is the introduction of strongly electron-donating heterocyclic components at one end of the cyclobutene ring to induce intense intramolecular charge transfer. It is obvious that the development of the synthesis of unsymmetrical dyes make optical (electronic absorption and fluorescence emission) and electrochemical (redox potential) properties more tunable. Besides the dyes described above, lots of unsymmetrical squarylium dyes have been synthesized to append functionality toward chemosensors, protein-labeling materials, and so on. Detailed discussions on the application of squarylium dyes are described in the following section.

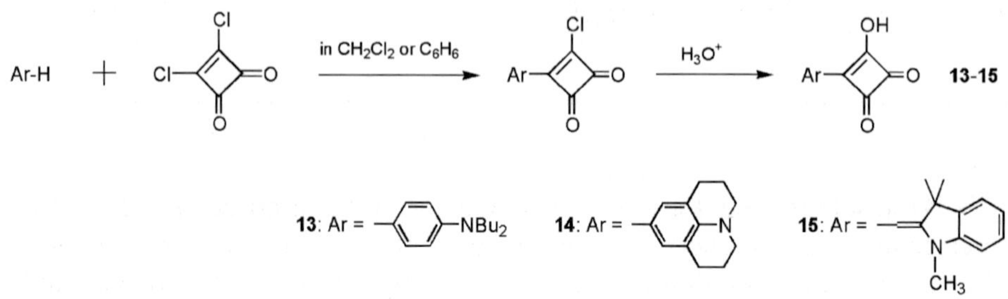

Scheme 4.

Table 1
Structures, yields, and electronic absorption data for unsymmetrical squarylium dyes **16–21**

Compd.	R^1	R^{2+}	Yield/%	λ_{max}/nm (log ε) in $CHCl_3$
16	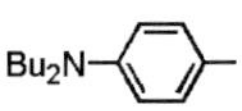	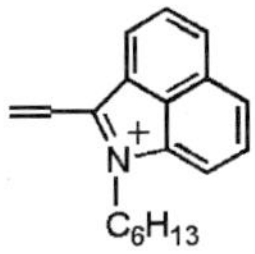	58	729 (4.92) 794 (5.03)
17	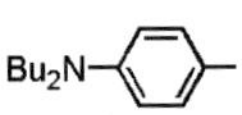	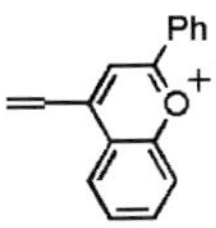	20	712 (4.99) 782 (5.18)
18	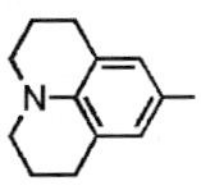	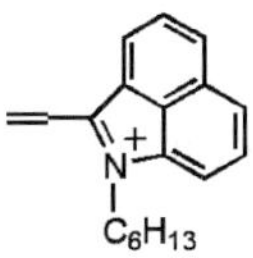	71	751 (4.92) 821 (4.96)
19	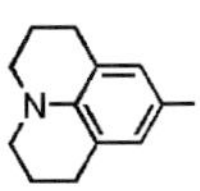	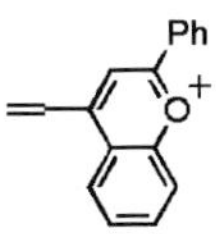	19	725 (5.01) 800 (5.10)
20	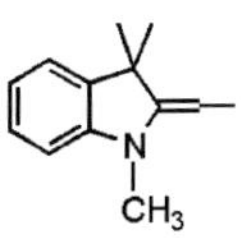	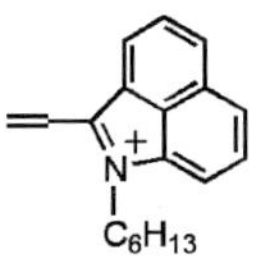	55	720 (4.97) 774 (5.00)
21	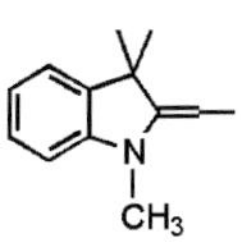	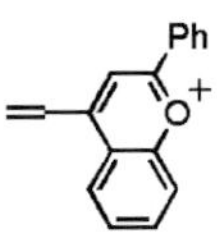	73	682 (5.02) 739 (4.98)

Terpetschnig and Lacowicz reported a similar preparation of a semi-squarylium where dialkyl squarate reacted with an equimolar amount of 1,2,3,3-tetramethylindolium salt followed by hydrolysis to afford indolinylidenemethyl-substituted squaric acid [14]. This procedure was applicable to benzothiazolium and benzoselenazolium salts with an active methyl group at their

2-position to produce indolenine–benzothiazole (**22**), indolenine–benzoselenazole (**23**), and benzothiazole–benzoselenazole (**24**) types of unsymmetrical dyes (Scheme 5) [15]. Without hydrolysis of alkylsquarate, quaternary heterocyclic compounds reacted with the esters to achieve the stepwise synthesis of unsymmetrical squarylium dyes. In this respect, a recent report has pointed out the problem of the reactivity, focusing on the electronic factor of the reactant in the case of quinaldine-based unsymmetrical dye formation [16]. As shown in Scheme 6, the reaction of a quinaldinium salt with an electron deficient or electron-withdrawing group at its 6-position (X = H, Br, I, NO_2, CN) with squaric acid smoothly produced the corresponding symmetrical squarylium. On the other hand, the reaction of an electron-rich quinaldinium (X = OH, OEt) was terminated at the stage producing the semi-squarylium. The isolated semi-squarylium (X = OH) reacted with an iodo-substituted quinaldinium salt to produce the unsymmetrical dye, but did not react with the hydroxy-substituted salt to yield any squarylium dyes (Scheme 7). This problem was explained by the acidity of the 2-methyl proton of the quinaldinium: the formation of enamine in the reaction system, rather than the nucleophilicity, seriously affects the yield of the dye.

Law and Bailey reported the synthesis of unsymmetrical squarylium dyes without employing squaric acid, in order to investigate and improve photoconductivity of the dye for xerographic application [17,18]. They chose the [2 + 2] cycloaddition reaction of tetraethoxyethane with arylacetylchloride reported by

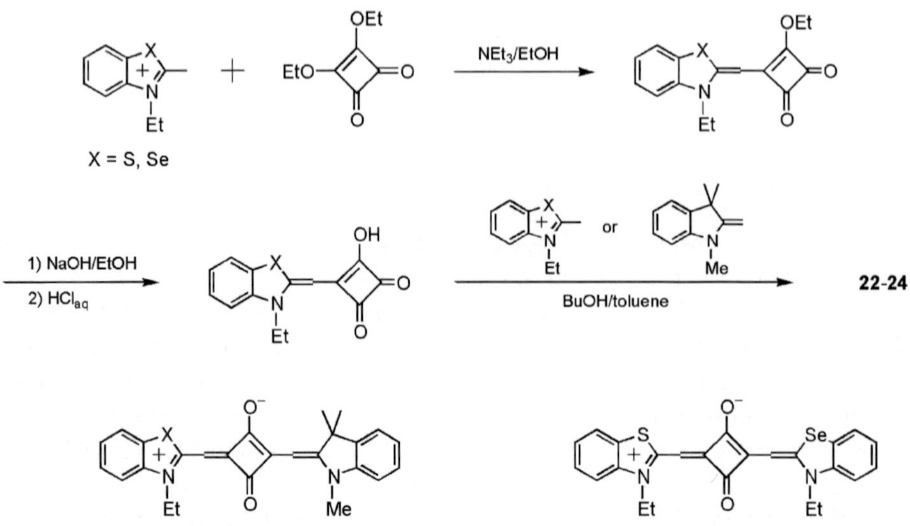

22: X = S, λ_{max}; 649 nm (log ε; 5.20) in $CHCl_3$
23: X = Se, λ_{max}; 657 nm (log ε; 5.03) in $CHCl_3$
24: λ_{max}; 677 nm (log ε; 5.03) in $CHCl_3$

Scheme 5.

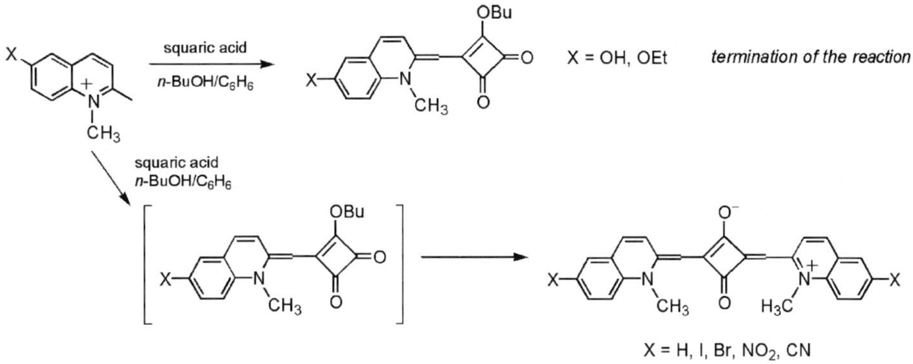

Scheme 6.

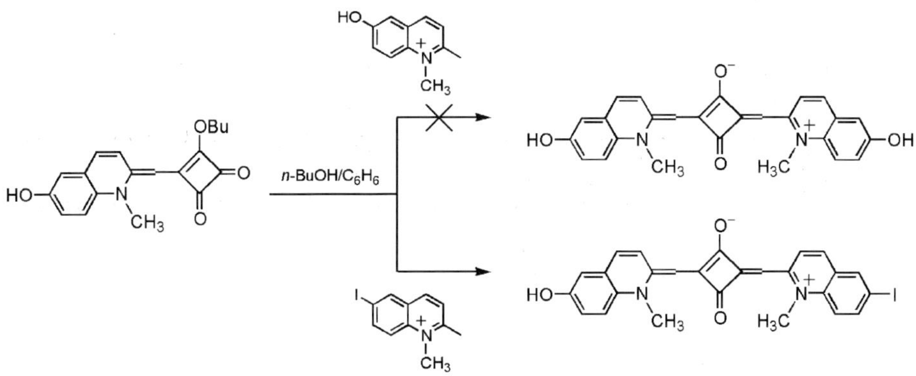

Scheme 7.

Belluš [19] to obtain semi-squaryliums. Three types of semi-squaryliums were reported, which were converted to the unsymmetrical squarylium dyes by condensation with a series of aniline derivatives via the salt formation of both components (Scheme 8). Two methods were examined; the condensations of the organic salts in 2-propanol containing a small amount of tributyl orthoformate (method A) and in a butanol–benzene solvent mixture (method B). In most of the cases reported here, the former method afforded unsymmetrical dyes in better yields than the latter, as shown in Table 2, and suppressed the side reaction producing unwanted polymer. The absorption maxima of the prepared dyes ranged from 562 to 592 nm in chloroform, red- and blue-shifted in comparison with the symmetrical methoxyphenyl-type (λ_{max}; 536 nm) and dialkylaminophenyl-type (λ_{max}; 624 nm) squarylium dyes, respectively. This cycloaddition–condensation reaction sequence was the first example of the preparation of unsymmetrical squarylium dyes without using squaric acid as a synthetic precursor.

2.3. Squarylium dyes derived from bisquaryl

As described above, by choosing aromatic and/or heterocyclic enamine moieties at both ends of a squarylium skeleton, a variety of squarylium dyes have been synthesized. Of course, the preparation of unsymmetrical dyes has extended the variation of the π-conjugated structures of the squarylium skeleton. On the other hand, quite a small number of examples of squarylium homologues has so far been reported except for the dyes with amino- and dithio-substituted cyclobutene rings [20–23]. Besides these dye skeletons, Nakazumi *et al.* reported the preparation of new squarylium homologues with a bisquaryl core **25–28** [24]. In this term, the Liebeskind *et al.*'s bisquaryl synthesis [25], as shown in Scheme 9, made possible the preparation of these dyes: the reaction of dialkyl squarate with tri(*n*-butyl)trimethylsilylstannane ($Bu_3SnSiMe_3$) in the presence of a catalytic amount of cyanide anion yielded tributylstannylated squarate [26], which was coupled with chlorosquarate to yield a bisquaryl ester. Although the reaction of bisquaric acid with enamines afforded too many unknown byproducts, the condensation of bisquarate with 2 eq of quaternary salts such as 2-methylindolium,

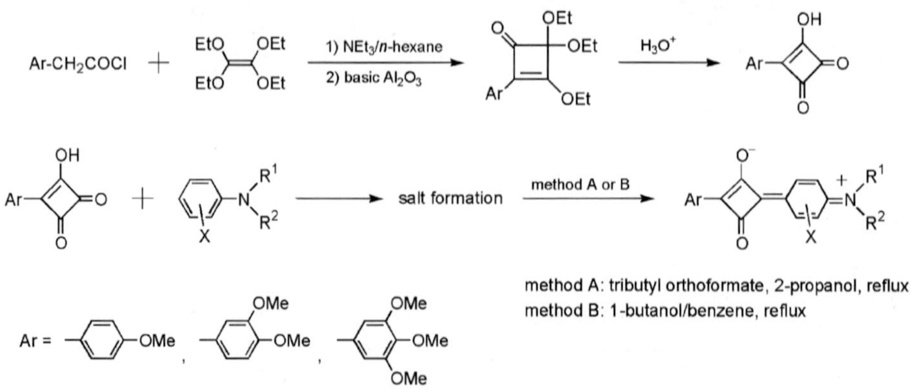

Scheme 8.

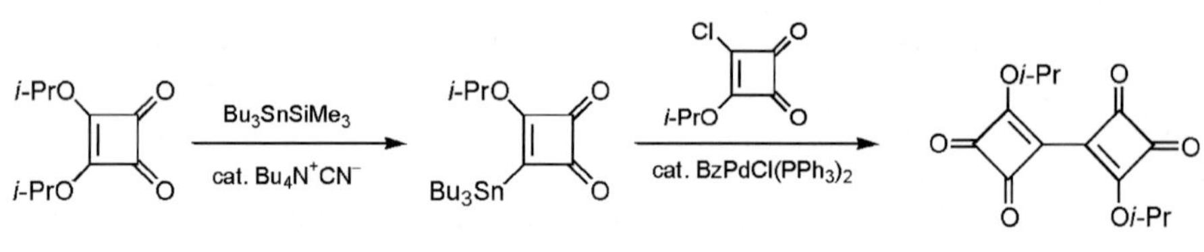

Scheme 9.

Table 2
Structures, yields, and electronic absorption data for Law's unsymmetrical squarylium dyes

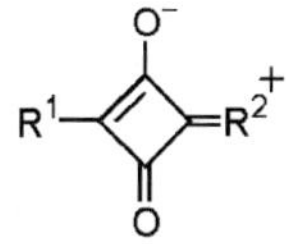

R^1	R^{2+}	Yield (%)		λ_{max}/nm (log ε) in $CHCl_3$
		Method A	Method B	
	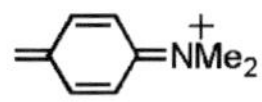	67	16	578.8 (5.37)
	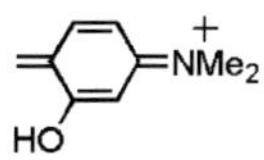	86	75	563.6 (5.20)
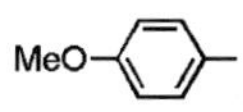	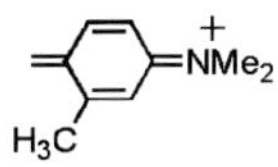	73	41	583.5 (5.32)
	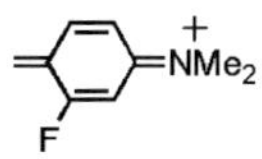	32	4.5	581.1 (5.40)
	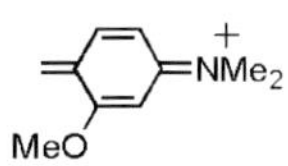	43	3.0	583.6 (5.23)
	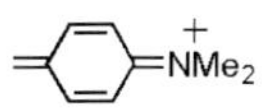	77	33	587.0 (5.34)
	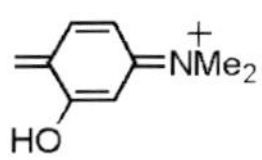	87	78	572.1 (5.20)
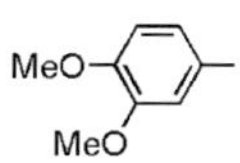	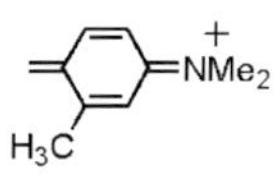	83	37	592.4 (5.35)
	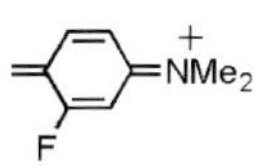	59	7.7	590.6 (5.32)
	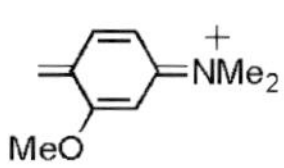	52	32	582.4 (5.08)
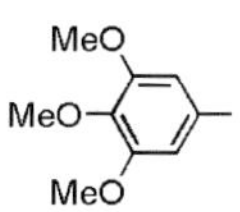	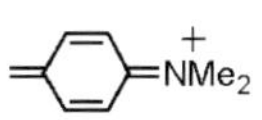	57	----	583.1 (5.39)
	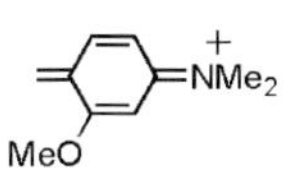	----	51	562.4 (5.12)

2-methyl benzothiazolium, and 2- and 4-methylquinaldiniums in the presence of triethylamine afforded the bisquarylium dyes in 36–67% yields (Table 3). These dyes show broadened light-absorption spectra splitting into a few peaks, and the absorption maxima of the dyes, ranging from 653 to 757 nm in chloroform, are red-shifted in the order of **26** < **25** < **27** < **28**.

Similar to the preparation of unsymmetrical squarylium dyes, the isolation of semi-bisquaryliums offered the opportunity to synthesize unsymmetrical bisquarylium dyes [24]. The stepwise synthesis of the semi-bisquaryliums is shown in Scheme 10. The reaction of dichlorocyclobutenedione with one equivalent of an aniline derivative or a heterocyclic quaternary salt followed by the Libeskind's palladium-catalyzed cross-coupling reaction with tributylstannylcyclobutenedione afforded a semi-bisquarylium. It is worthy to note that the reaction of bisquarate with any aromatics or heterocyclic salts does not yield the synthetic precursors. The subsequent reaction with one equivalent of a series of heterocyclic quaternary salts in the presence of triethylamine yielded unsymmetrical bisquarylium dyes **29**–**37** (Scheme 10, showing the case of the indolenine-benzothiazole-type bisquarylium dye as an example). The structures and light-absorption data of a series of obtained bisquarylium dyes are listed in Table 4. The

Table 3
Structures, yield, and electronic absorption data for bisquarylium dyes **25**–**28**

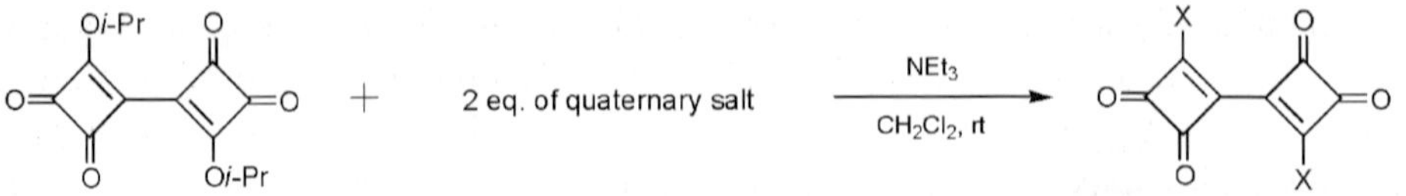

Compd.	X	Yield (%)	λ_{max} (nm)(log ε) in $CHCl_3$
25	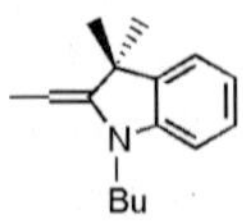	39	657 (4.87) 601 (4.73)
26	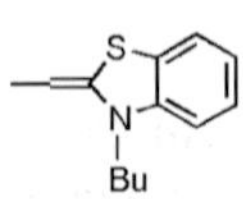	67	653 (4.92) 595 (4.73)
27	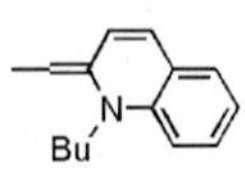	41	692 (5.04) 628 (4.79)
28	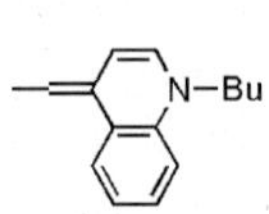	36	757 (5.14) 682 (4.76)

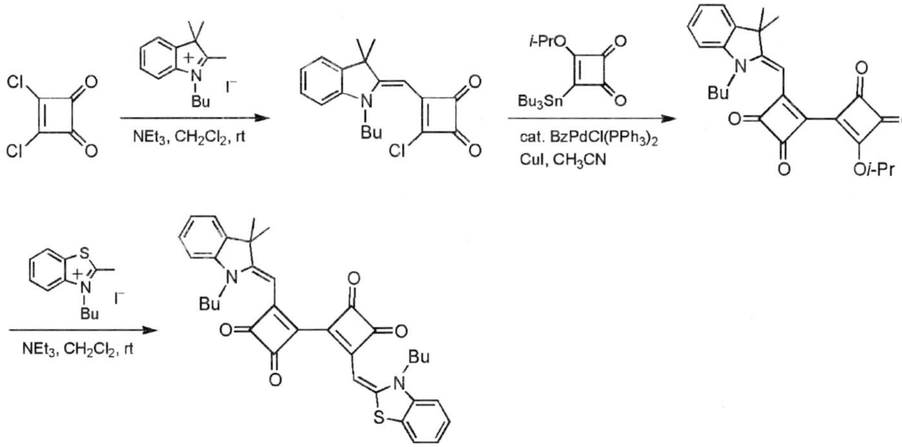

Scheme 10.

absorption maxima of these unsymmetrical dyes varied from 640 to 705 nm in chloroform at 298 K, showing splitting absorption bands similar to symmetrical bisquaryliums. Among the symmetrical and unsymmetrical bisquarylium dyes, the dye derived from the 4-methylquinolinium-4-methylquinolinium pair exhibited most red-shifted absorption maximum (λ_{max}; 757 nm). It is interesting that the absorption maximum of the unsymmetrical dye is an averaged value of λ_{max} of the symmetrical dyes possessing the corresponding aromatic or heterocyclic components: for instance, the value of λ_{max} of **31** (λ_{max}; 705 nm) is the half of the sum of the values of λ_{max} of **25** (657 nm) and **28** (757 nm). Thus, it is possible to expect the absorption maxima of bisquarylium dyes without theoretical methods such as molecular orbital calculations.

The π-conjugation systems of the bisquarylium dyes are quite complicated, and obviously, they can not adopt a cyanine-like resonance structure, contrary to typical squarylium dyes: cross-conjugation should occur in the central bisquarate component, and merocyanine-like resonance structures should be involved. Therefore, the dyes of this series are regarded as homologues of 1,2-squaraine dyes, as discussed in Fig. 2.

2.4. Squarylium dyes with extended π-conjugation structures

Along with the development of squarylium chemistry in the synthesis as well as the material application, increasing attention have been poured into the exploration of novel π-conjugation systems based on the squarylium chromophores. Especially, the extension of the π-conjugation systems affords unique physicochemical properties such as near-IR light absorption, enlarged light-absorbing ability, conductivity due to low band-gapping, and so on. In this regard, novel reactions and synthetic pathways have been enthusiastically studied over the past decade.

Table 4
Structures, yields, and electronic absorption data for unsymmetrical bisquarylium dyes **29–37**

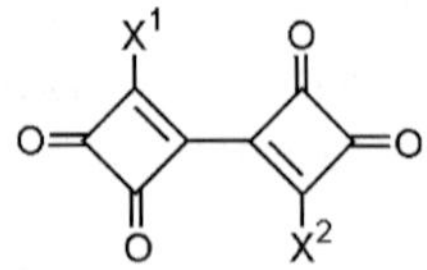

Compd.	X^1	X^2	Yield/%	λ_{max}/nm (log ε) in $CHCl_3$
29		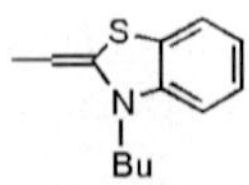	70	653 (5.01) 598 (4.85)
30	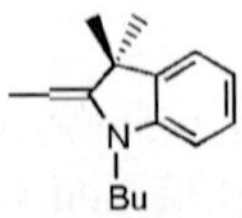	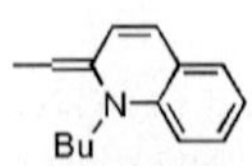	50	672 (5.02) 613 (4.82)
31		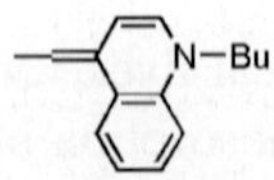	50	705 (5.10) 641 (4.82)
32	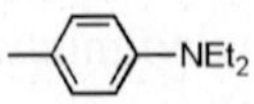	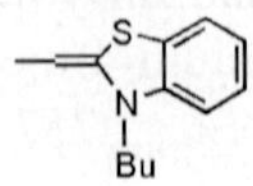	38	640 (4.71) 600 (4.68)
33		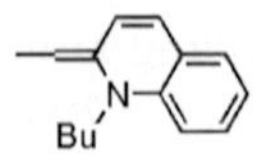	70	648 (4.73) 620 (4.72)
34	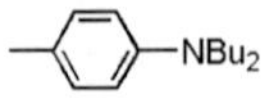	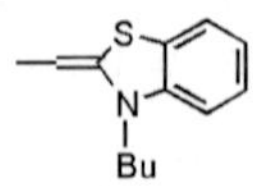	33	642 (4.70) 604 (4.67)
35		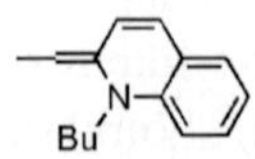	30	649 (4.73) 620 (4.72)
36	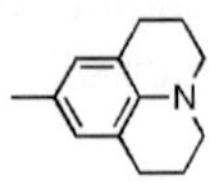	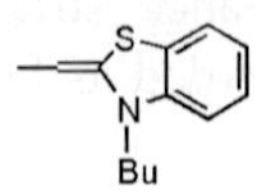	35	656 (4.70) 620 (4.61)
37		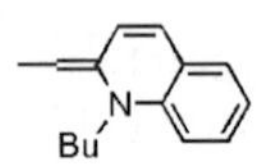	30	673 (4.73) 630 (4.67)

In 1994, Nakazumi *et al.* reported a unique squarylium-based cationic dye **38** with two semi-squarylium and one benzothiazolium components connecting to the same carbon atom [27]. This dye was produced by the reaction of 2-methylbenzothiazolium iodide with squaric acid in 1-butanol/benzene under azeotropic conditions in the absence of a base catalyst such as quinoline. Taking it into consideration that the normal squarylium dye was produced in the same mixed solvent containing quinoline, the proton generating upon the enamine formation from the benzothiazolium promoted the formation of the cationic dye. The reaction mechanism was proposed as shown in Scheme 11: the benzothiazolinosquarylium was once formed followed by the addition of the protonated semi-squarylium to the methyne carbon to produce the triarylmethane-like cationic skeleton. The structure was confirmed for the tetraphenylborate salt by X-ray crystallographic analysis, where the π-conjugation is extended from one benzothiazolino half-squarylium to the other via the methyne bridge, stabilized by intramolecular hydrogen bonding between the two squarate components (discussed in the following section). The exo-benzothiazolium ring was placed in the pseudo-perpendicular manner, hardly conjugating to the main chromophoric system.

The electrophilic attack of a series of semi-squarylium onto the methyne carbon of the benzothiazolinosquarylium dye was also successful, and various types of unsymmetrically substituted cationic squarylium homologues **39**–**43** were prepared by the acid-promoted reaction of five types of semi-squaryliums with the benzothiazolinosquarylium, as shown in Scheme 12 [28]. The semi-squaryliums **44** and **45** directly reacted with the squarylium dye to yield **39** and **40** in 72 and 35%, respectively (method A). The ethyl squarates **46**–**48** were converted *in situ* to the corresponding semi-squaryliums, which were subjected to protonation on

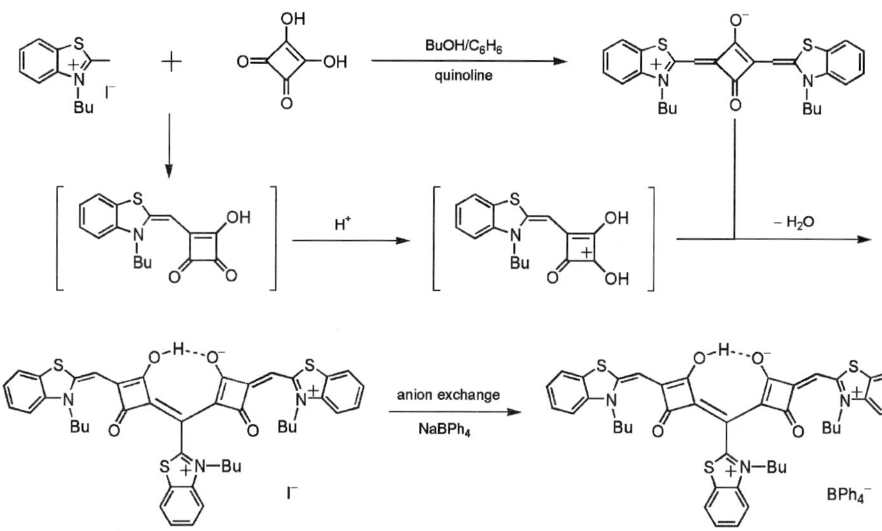

38: λ_{max}; 797 nm (log ε; 5.55) in $CHCl_3$

Scheme 11.

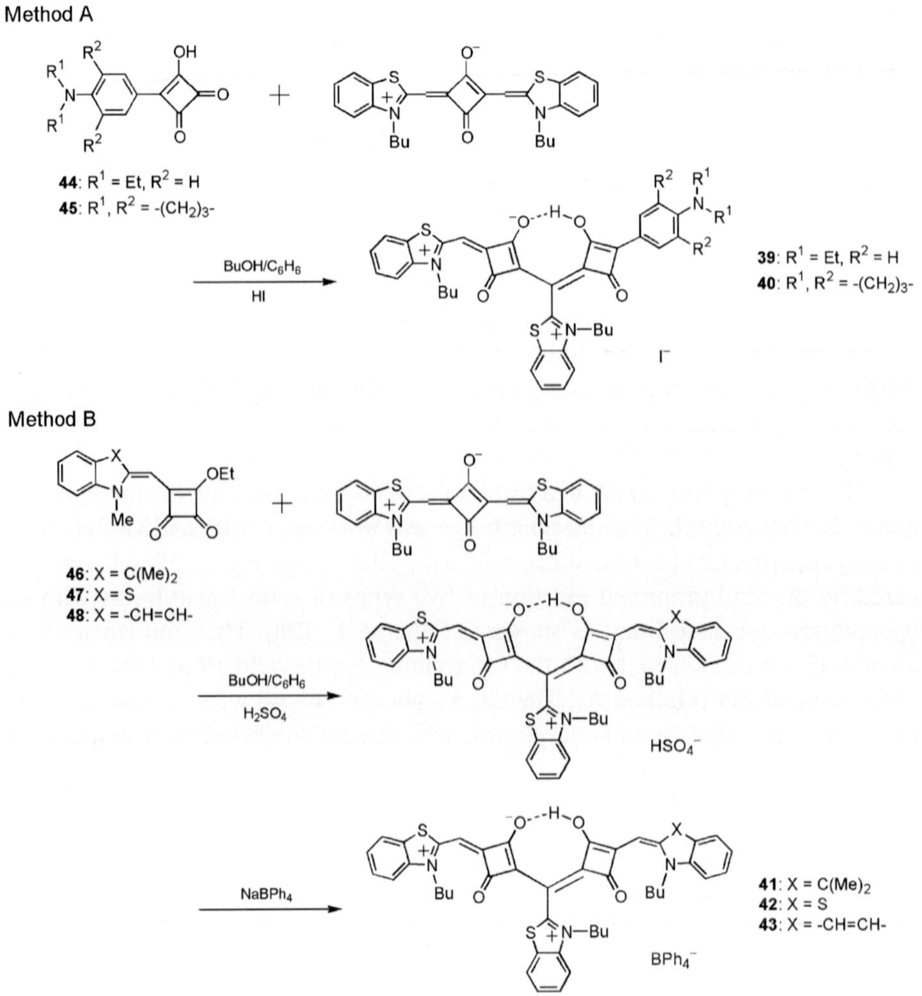

Scheme 12.

their carbonyl oxygen to attack on the methyne carbon of the squarylium dye to yield **41**–**43** in 10–17%, respectively (method B). Two types of acids were employed; HI for **44** and **45**, yielding **39** and **40** as the iodides, and sulfonic acid for **46**–**48**, yielding **41**–**43** as the tetraphenylborates after anion exchange. All these dyes absorbed near-IR light (771–815 nm) and exhibited large light-absorption coefficiencies as summarized in Table 4. Thus, the extensively π-conjugated structure from one semi-squarylium unit to the other should be maintained by the intramolecular hydrogen bonding between two cyclobutene moieties as confirmed by the X-ray structural analysis for **38**. Indeed, in the ^{1}H NMR study, the signal of

Table 5
Structures, yields, and electronic absorption data for unsymmetrical cationic bis(squarylium) dyes **39–43**

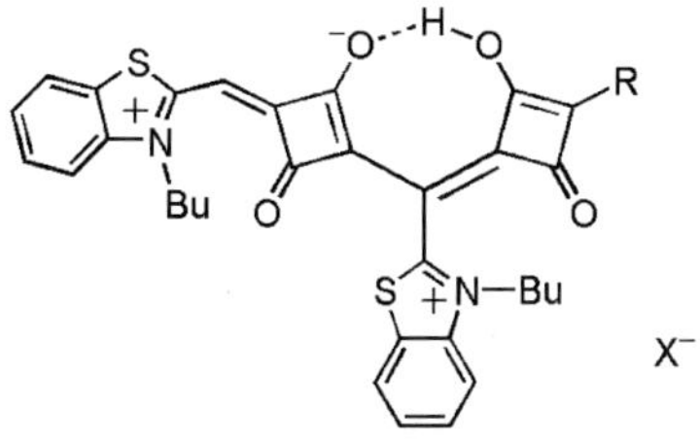

Compd.	R	X	Yield/%)	λ_{max}/nm (log ε) in $CHCl_3$
39	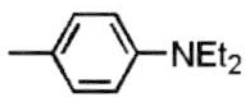	I	72	771 (5.02)
40	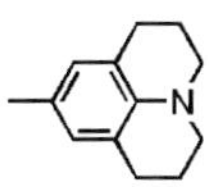	I	35	815 (5.13)
41	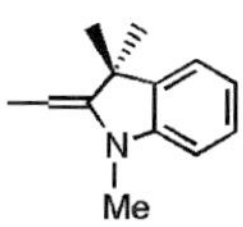	BPh_4	12	792 (5.40)
42	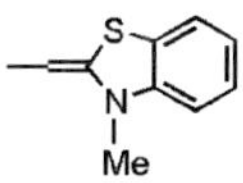	BPh_4	10	797 (5.48)
43	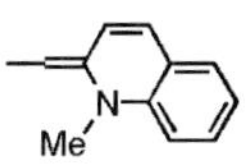	BPh_4	17	812 (5.43)

the OH proton of the dyes **39–43** was observed in the far low-magnetic field region (*ca.* 18–19 ppm), indicating the formation of strong intramolecular hydrogen bonding.

The methyne-bridged bis(squarylium) dye **49** possessing the analogous π-conjugation system to the cationic "Y-shaped" bis(squarylium)s **38–43** was also reported (Scheme 13) [29]. This "V-shaped" dye was obtained in quite low yield (3–10%) by the reaction of an anilino type of semi-squarylium in refluxed 2-propanol in the presence of an excess amount of tributyl orthoformate (10 eq.), although the detailed reaction mechanism has never been clarified. In comparison with the *cationic* "Y-shaped" dyes, the "V-shaped" dye possesses the electronically *neutral* structure, and the π-electronic conjugation extended through the whole molecule as clarified by the X-ray crystallographic analysis (discussed

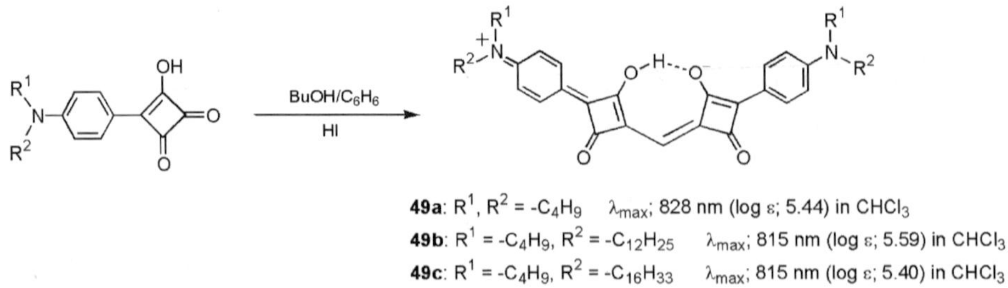

Scheme 13.

later), exhibiting intense light absorption band in the near-IR region (815–828 nm). The acidic OH proton in one of the cyclobutene units was also characterized to form the intramolecular hydrogen bonding to the oxygen in the other cyclobutene unit keeping high planarity of the π-conjugation structure. This was supported by the ^{1}H NMR spectrum where the largely shifted proton to the low magnetic field region was observed (18.9 ppm).

Recently, Yagi and Nakazumi *et al.* reported arene-bridged bis(squarylium) dyes **51–54** [30]. The palladium-catalyzed cross-coupling of the Liebeskind's tributylstannyl cyclobutenedione with diiodo-substituted arenes led to preparation of bis(squaric acid) precursors **50a–c** as shown in Scheme 14 [26]. Under the typical reaction conditions for squarylium synthesis, the bis(squarylium) dyes **51–54** were obtained in 8–90% yields as shown in Scheme 15. The light-absorption properties depended on both of the arene spacer and the heterocyclic components in the dyes (Table 6). Especially, the spacer significantly affected the π-conjugation structure: the *p*-phenylene-bridged bis(squarylium)s **51–53a** exhibited redshifted absorption maxima in comparison with *m*-phenylene-bridged dyes **51–53b**. This can be explained by the extensively conjugated π-electron system of the *p*-phenylene-bridged dye as shown in Scheme 16a. On the other hand, the π-electron resonance in the *m*-phenylene-bridged dyes might be terminated at the central phenylene moiety as shown in Scheme 16b, affording the blue-shifted absorption maxima. The 4,4′-biphenylene-bridged dyes **51c** and **52c** also exhibited redshifted absorption maxima in comparison with the *m*-phenylene-bridged dyes, but blue-shifted compared to the *p*-phenylene-bridged dyes. This was because the C1–C1′ bond rotation in the biphenylene moiety brought about partial breaking of the overlap of the p_z-orbitals between C1 and C1′. The fluorescence emission properties were also investigated for these dyes. These dyes exhibited fluorescence emission in the region from 571 to 736 nm although less emissive than the typical anilinosquarylium dye (Table 6). Some of them showed relatively large Stokes shifts. Thus, various combinations of the spacer and heterocyclic components yielded a wide range of variation of the optical properties.

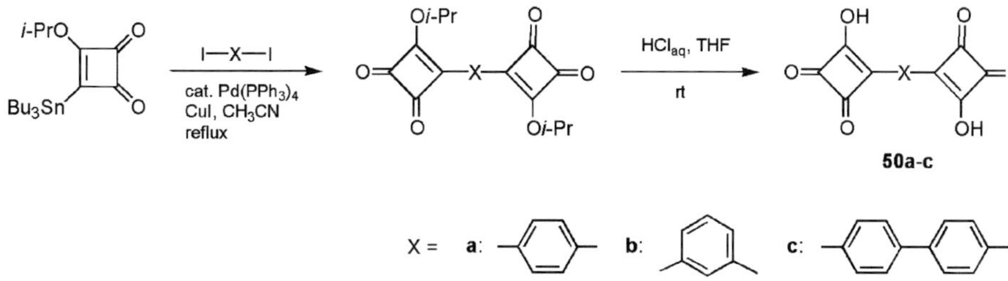

Scheme 14.

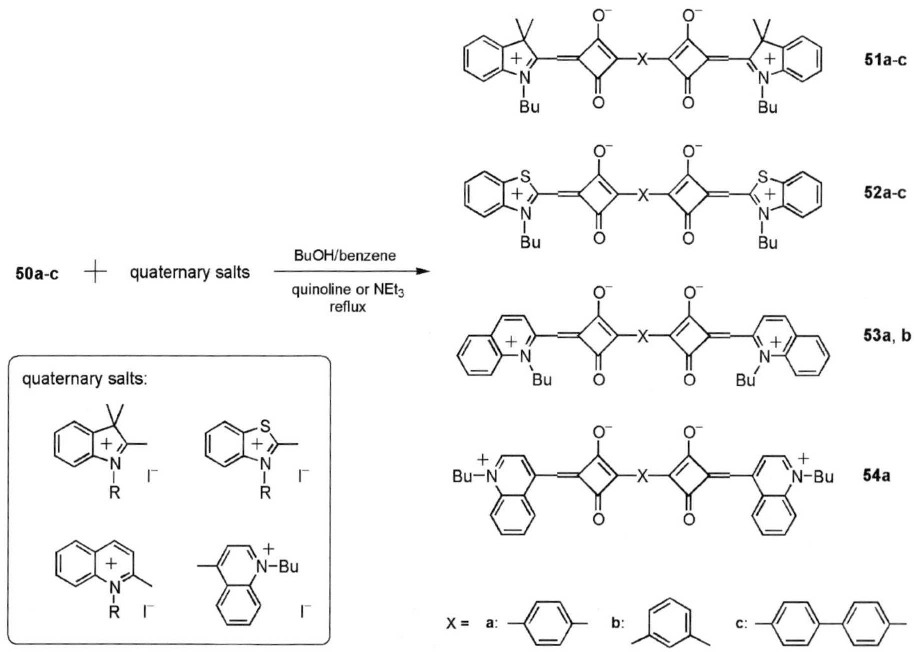

Scheme 15.

Another strategy to obtain π-extended electronic structures of squarylium dyes is to utilize aromatic or heterocyclic components possessing large π-conjugation systems. In this respect, stilbenoid π-conjugation systems have been appended to the diaryl-type squaryliums, as shown in Scheme 17. This is relatively an easy way to the purpose because the stilbene skeletons are available according to the Wittig-type diarylethene synthesis. Meier and Dullweber developed a series of bis(stilbenyl)squarylium dyes **55–60** (Scheme 17) [31]. Although

Table 6
Electronic absorption and fluorescence data of bis(squarylium) dyes **51–54** in $CHCl_3$

Compd.	Yield (%)	λ_{max}/nm (log ε)	λ_{max}/nm (relative intensity)	Stokes shift $\Delta\lambda$/nm
51a	75	699 (5.25) 635 (4.93)	714 (126)	22
51b	22	572 (5.27) 535 (4.87)	587 (56)	13
51c	45	618 (5.28)	656 (226)	37
52a	90	685 (5.28) 623 (4.99)	706 (206)	21
52b	30	543 (5.15) 479 (4.71)	586 (34)	15
52c	60	609 (5.01)	651 (157)	32
53a	8	768 (5.13) 692 (4.80)	809 (15)	45
53b	12	570 (5.30)	571 (3.4)	1
54a	8	704 (5.29) 641 (4.98)	736 (77)	24
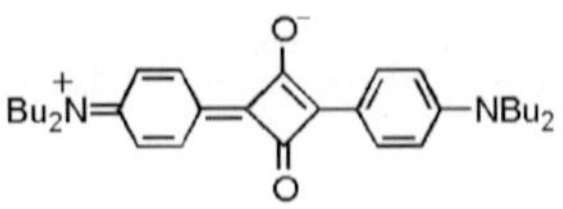		640 (5.50)	667 (2000)	12

normal homologous diaryl-type squarylium dyes exhibit light-absorption maxima at *ca.* 570 nm, the extension of the π-conjugation system at both ends brought about significant red-shift of light absorption; λ_{max} = 680–735 nm. This red-shift was owing to the electronic transition at the intramolecular charge-transfer state caused by the D–A–D charge-localized structure. The stilbene-based squarylium skeleton was applicable to the synthesis of a series of dendric oligomers with a phenylene core **61–65**, which led to the establishment of chromophoric assemblies with large light absorptivity [32]. Ajayaghosh *et al.* also prepared a stilbenoid squarylium dye exhibiting electrochromic activity [33]. The dye **66** showed its absorption maximum at 678 nm, which were oxidized to the radical cation and dication and reduced to the radical anion. The redox cycle occurred reversively, and the one-electron oxidation led to the generation of a new near-IR absorption band at 1000–1600 nm, whereas the absorption of **66** at λ_{max} significantly decreased according to one-electron reduction.

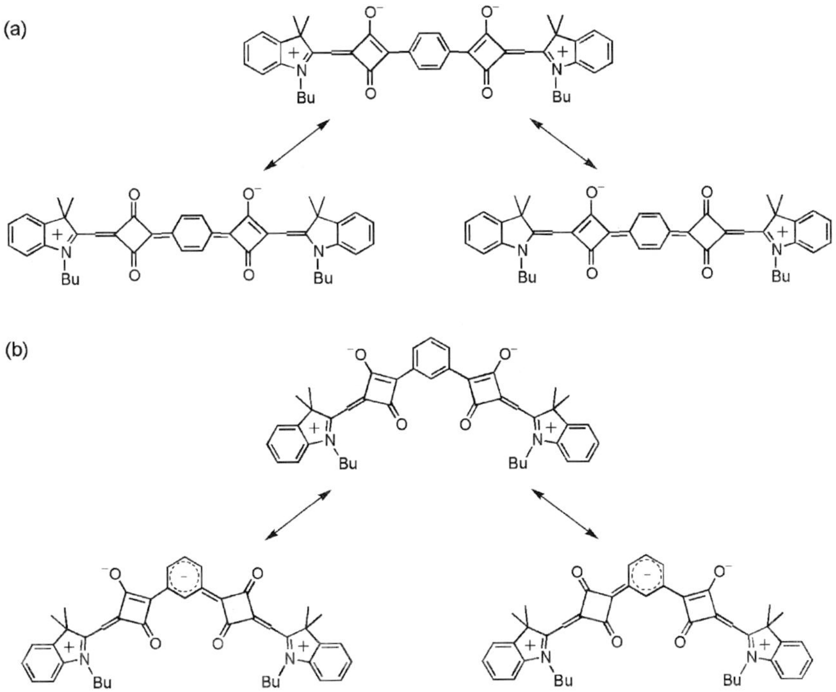

Scheme 16.

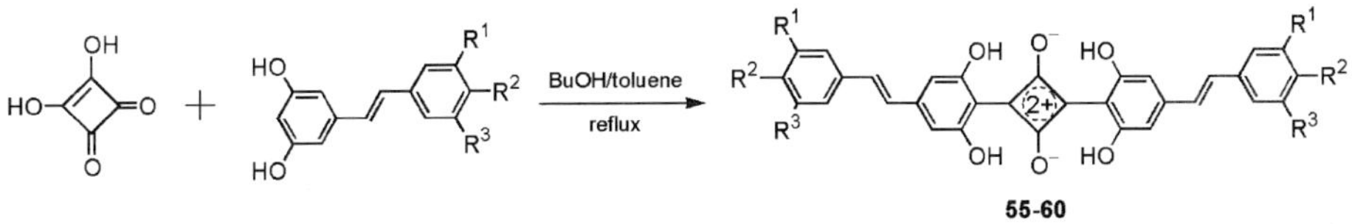

Compd.	R^1	R^2	R^3	Yield	λ_{max}/nm ($CHCl_3$)
55	H	OC_6H_{13}	H	33	714
56	H	OH	H	49	696
57	OC_6H_{13}	OC_6H_{13}	H	28	730
58	OC_6H_{13}	H_3	OC_6H_{13}	16	680
59	OC_6H_{13}	OC_6H_{13}	OC_6H_{13}	20	727
60	OC_6H_{13}	OH_3	OC_6H_{13}	21	735

Scheme 17.

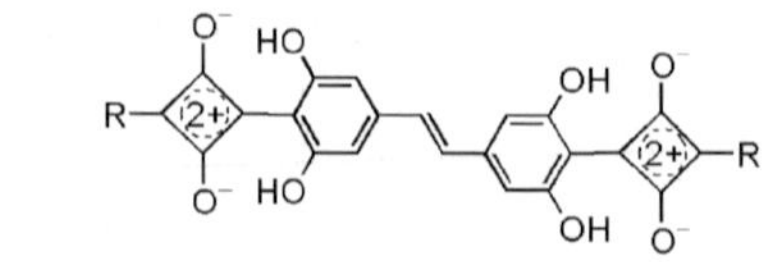

61

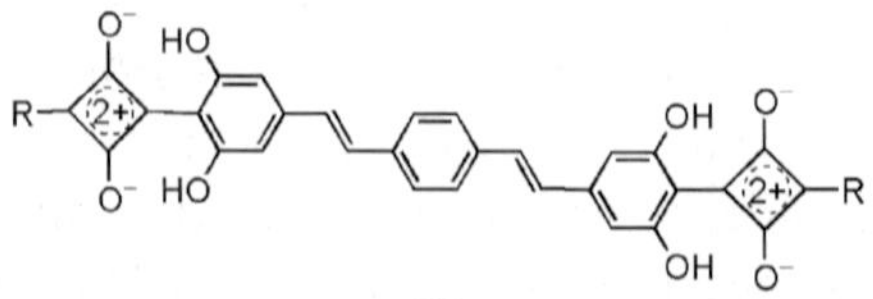

62

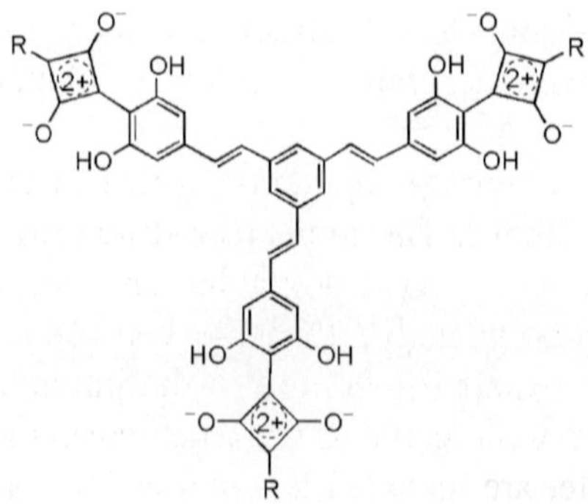

63

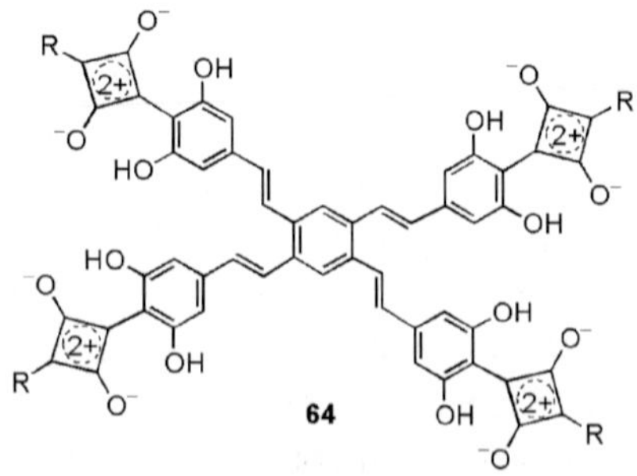

64

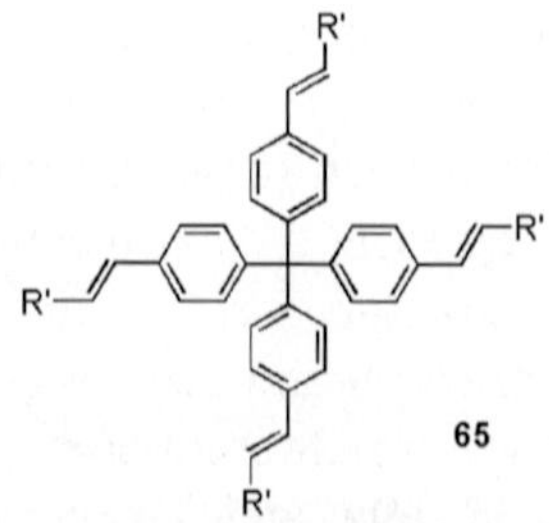

65

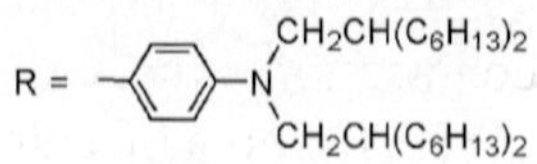

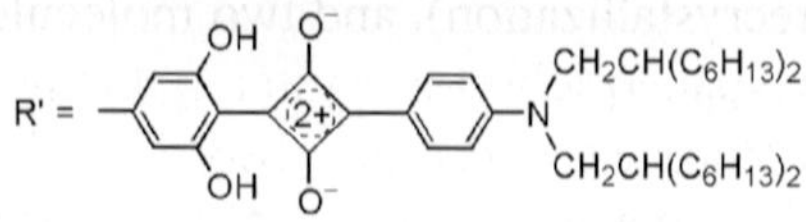

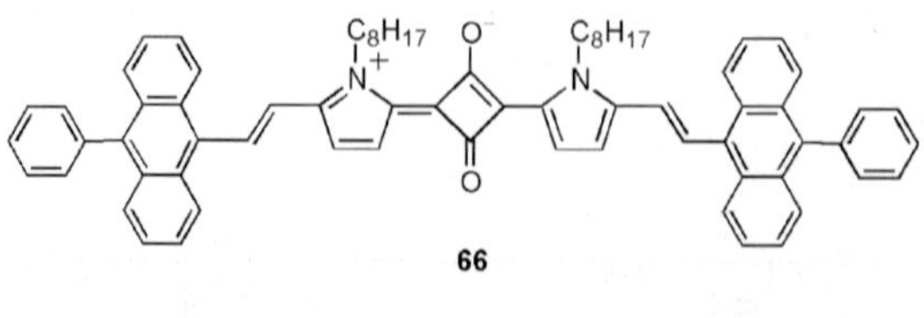

66

3. STRUCTURAL ANALYSIS OF SQUARYLIUM DYES

X-ray crystallographic analysis of dye molecules is one of the powerful methods to obtain information about structure–property relationships. In general, dye molecules possess planar structures due to their largely extended π-conjugation systems, and thus, often yield thin planar single crystals upon crystalline formation, which are not suitable to X-ray crystallographic analysis. A series of squarylium dyes are of course within this category, and for this reason, small numbers of examples of X-ray structural analyses have been reported for squarylium dyes. On the other hand, studies on the electronic structures of squarylium dyes have still been lying at the centre of discussion on squarylium chemistry because their special physicochemical properties such as intense light absorption from visible to near-IR regions, semiconductor properties at the solid states, and so on are caused by their unique π-conjugation structures and molecular packing.

X-ray structural analyses have been hitherto carried out for the squarylium dyes **67** [34], **4** (R^1, R^2 = Bu) [35], **9** (R = Me) [36], **12** (R = Bu) [37], and so on. Each of these dyes showed a highly planar structure. In most cases, the lengths of carbon–carbon bonds involved in squarylium π-cojugation systems showed the character between single and double bonds (two examples are shown in Fig. 3), and it could be concluded that significant intramolecular charge transfer from the aromatic or heterocyclic components to the central cyclobutene component resulted in delocalization of π-electrons over the whole molecules. Thus, it has been suggested that the zwitter-ionic D–A–D structure is more suitable to represent the squarylium electronic structure, rather than the cyclobutene diylium structure (see Scheme 1).

Nakazumi and Yagi prepared a series of new π-conjugation systems based on squarylium chromophores, as described above. The molecular structures of **18**, **25**, **38**, **49a**, and **51a** were clarified by X-ray crystallographic analysis, as shown in Figs. 4–8. The unsymmetrical squarylium dye **18** possesses a largely extended π-conjugation structure, from the benz[*c*,*d*]indolium to the julolidine moiety (Fig. 4) [13]. Two molecules of **18**, one molecule of benzene (solvent for recrystallization), and two molecules of water are included in a unit cell to form a stable inclusion crystal. Although any discussions about the bond lengths are not done because of relatively high *R* value (*R* = 0.123), this is the first report of the molecular structure of the near-IR absorbing unsymmetrical squarylium dye.

The molecular structure of the bisquarylium dye **25** possesses a quite planer structure with C_2 rotational symmetry [24]. As shown in Fig. 5, the bond lengths

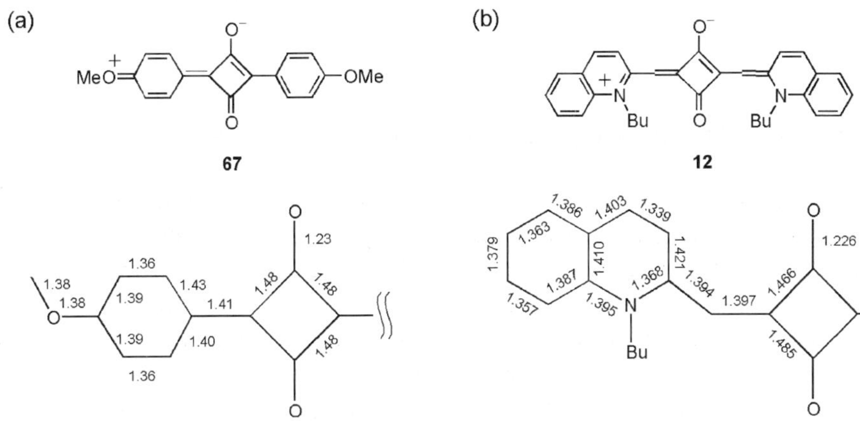

Fig. 3. Selected bond lengths (Å) of (a) squarylium dye **67** and (b) dye **12**.

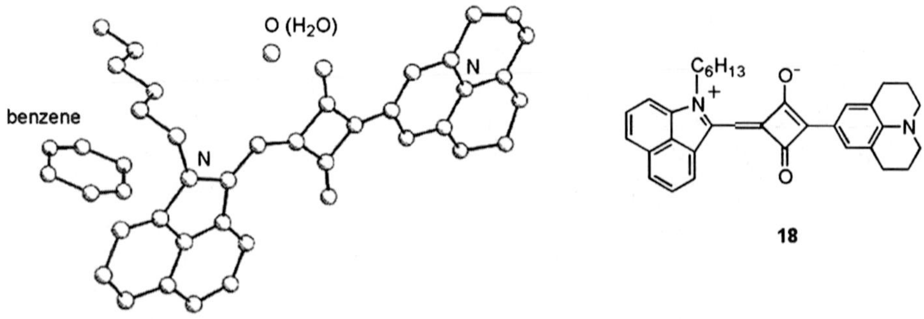

Fig. 4. The molecular structure of unsymmetrical squarylium dye **18**.

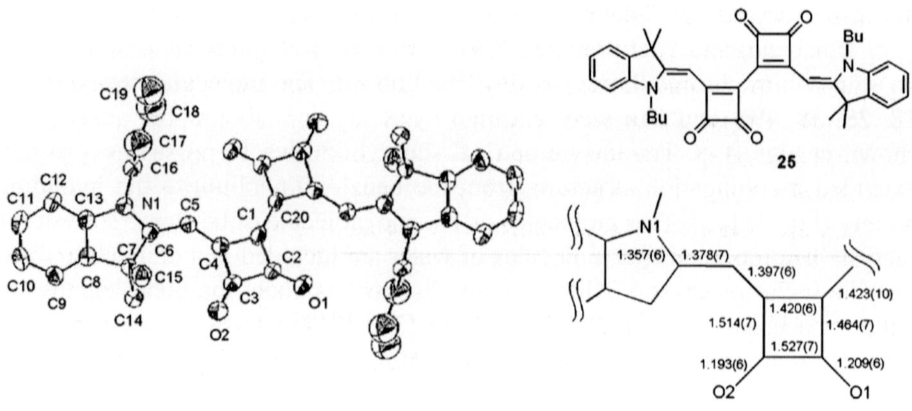

Fig. 5. The molecular structure of unsymmetrical bisquarylium dye **25**.

of N1–C6, C1–C4, C4–C5, C5–C6, and C1–C20 are within the range from 1.35 to 1.43 Å, showing double bond characters. On the other hand, the bonds C2–C3 and C3–C4 are long enough to show single-bond characters. In addition, the C3–O2 bond is relatively short in comparison with the length of the conjugated carbonyl C=O bond (e.g., C2–O1; 1.21 Å). Thus, the π-electron system of **25** is partially localized in the bisquaryl core.

The cationic bis(squarylium) dye **38** has a unique structure (Fig. 6) [27]. Two semi-squarylium moieties and bridging methyne carbon form a largely extended π-conjugation system, whereas one of three benzothiazolium moieties is connected to the methyne carbon in a pseudo-perpendicular manner. Thus, the positive charge delocalizes over the main π-conjugation system. Two shorter C–C bonds in each cyclobutene (1.43–1.46 Å) and the other C–C as well as C–N1 and C–N2 bond lengths also indicate a significantly resonanced polymethyne

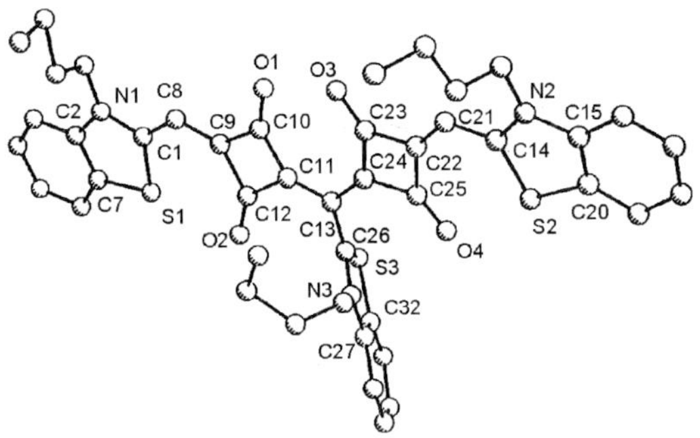

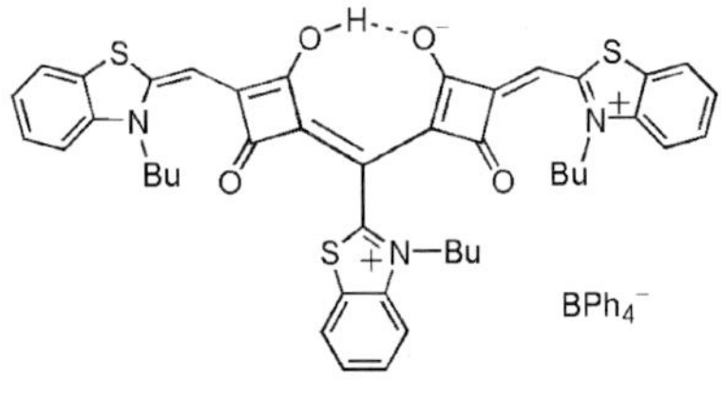

38

bond length /angstrom							
N1-C1	1.36(1)	C11-C12	1.49(1)	S2-C20	1.75(1)	C25-O4	1.20(1)
N1-C2	1.40(1)	C9-C12	1.50(1)	C14-C21	1.42(1)	C13-C24	1.41(1)
C1-C8	1.38(1)	C12-O2	1.20(1)	C21-C22	1.34(1)	C13-C26	1.43(1)
S1-C1	1.72(1)	C11-C13	1.41(1)	C22-C23	1.47(1)	S3-C26	1.68(1)
S1-C7	1.75(1)	C13-C24	1.41(1)	C22-C25	1.48(2)	S3-C32	1.74(1)
C9-C10	1.43(1)	N2-C15	1.41(1)	C23-O3	1.26(1)	N3-C26	1.32(1)
C10-O1	1.27(1)	N2-C14	1.35(1)	C23-C24	1.43(1)	N3-C27	1.38(1)
C10-C11	1.43(1)	S2-C14	1.72(1)	C24-C25	1.50(1)		

Fig. 6. The molecular structure of cationic bis(squarylium) dye **38**.

framework in **38**. The C10–O1 and C23–O3 bonds are longer than the other C–O bonds, and an intramolecular hydrogen bond is formed (O1–O3; 2.37 Å), yielding an effective π-electron delocalization over the polymethyne framework.

The neutral bis(squarylium) dye **49a** has similar structural characteristics to the polymethyne framework in **38**, as shown in Fig. 7 [29]. The bond distances of all C–C, C–O, and C–N bonds in the π-electron system except for C2–C5, C4–C5, C12–C15, and C14–C15 shows conjugated double bond characters, indicating a highly delocalized π-conjugation system over the whole molecule. The C3–O1 and C13–O3 bond lengths are longer than C5–O2 and C15–O4. The OH proton (H40) lies in the middle of the two oxygen atoms, O1 and O3, within van der Waals radii from both oxygen atoms (O1–H40, 1.29(6) Å; O3–H40, 1.15(5) Å; O1–O3, 2.433(4) Å; O1–H40-O3, 170(6) Å), forming a hydrogen bond to bridge the cyclobutene rings. The dye **49a** possesses a pseudo-C_{2v} symmetrical structure in the crystal packing. The least-squares planes of two anilinocyclobutene moieties omitting the alkyl chains (planes 1 and 2 consisting of O1, O2, N1, and C1–C11 and O3, O4, N2, and C12–C21, respectively) deviate from each other by 6.55°, whereas the mean deviations of the constituents atoms from the planes 1

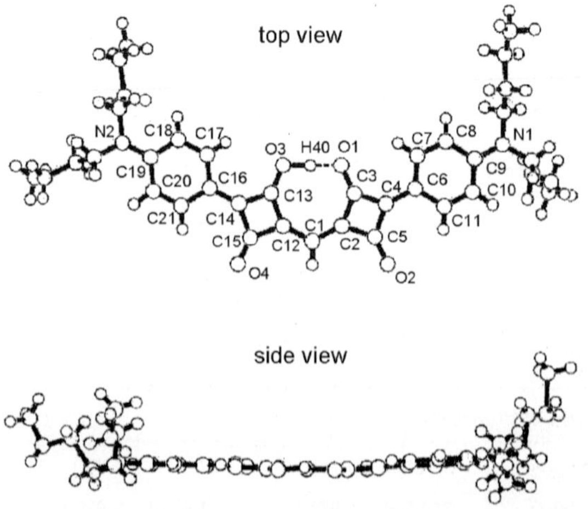

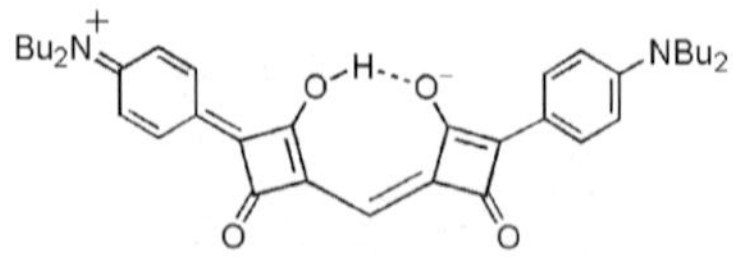

49a

bond length /angstrom							
O1-C3	1.280(4)	C1-C2	1.374(5)	C6-C11	1.408(5)	C14-C15	1.489(6)
O1-H40	1.29(6)	C1-C12	1.384(6)	C7-C8	1.351(6)	C14-C16	1.408(6)
O2-C5	1.215(5)	C2-C3	1.439(5)	C8-C9	1.407(6)	C16-C17	1.388(6)
O3-C13	1.269(5)	C2-C5	1.487(6)	C9-C10	1.392(6)	C16-C21	1.405(6)
O3-H40	1.15(5)	C3-C4	1.413(5)	C10-C11	1.359(6)	C17-C18	1.367(6)
O4-C15	1.209(5)	C4-C5	1.468(6)	C12-C13	1.445(6)	C18-C19	1.403(6)
N1-C9	1.368(5)	C4-C6	1.418(6)	C12-C15	1.494(6)	C19-C20	1.410(6)
N2-C19	1.336(5)	C6-C7	1.381(6)	C13-C14	1.430(6)	C20-C21	1.356(6)

Fig. 7. The molecular structure of neutral bis(squarylium) dye **49a**.

and 2 are 0.0645 and 0.0220 Å, respectively, and those from the bis(squarylium) skeleton consisting the planes 1 and 2 is 0.0950 Å. Thus, **49a** adopts high planarity of the π-conjugation system, tightly fixed by intramolecular hydrogen bonding.

The bis(squarylium) dye **51a** possesses another type of π-conjugation system in comparison with **38** and **49a**. As shown in Fig. 8, the structure clearly shows that two indolinylidenemethyl-substituted cyclobutene moieties are introduced to the central *p*-phenylene group, and the molecule of **51a** adopts a highly planar structure [30]: the mean deviation of the constituent atoms in the π-conjugation (C1–C16, N1, O1, and O2) from the least-squares plane is 0.0494 Å. Therefore, **51a** possesses an extensively conjugated π-electron system. The bond lengths of O1–C11, O2–C13, C10–C13, C11–C12, and C12–C13 exhibit conjugated C=O and C=C bond characters, respectively, whereas the C10–C11 bond

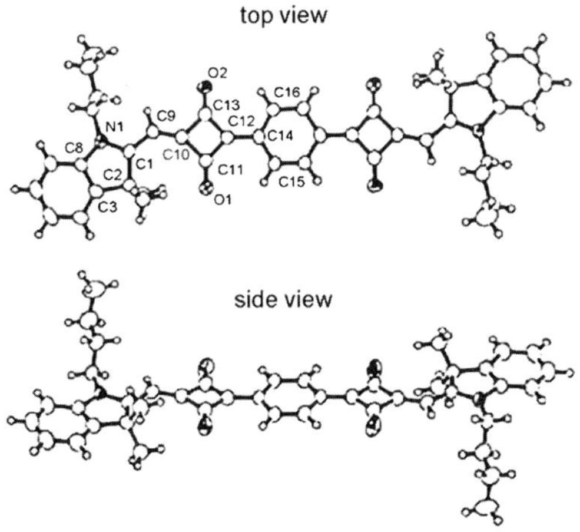

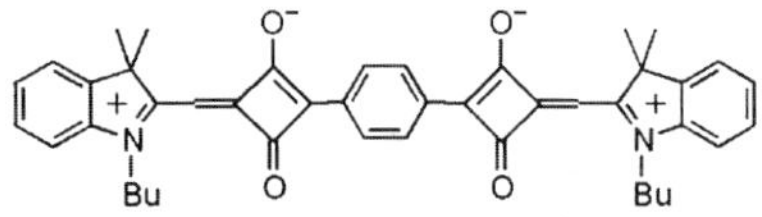

51a

bond length /angstrom			
O1-C11	1.230(5)	C10-C13	1.485(6)
O2-C13	1.236(5)	C11-C12	1.444(6)
C10-C11	1.520(6)	C12-C13	1.445(6)

Fig. 8. The molecular structure of bis(squarylium) dye **51a**.

is approximately a single bond. Thus, such a difference in the bond order indicates that the π-electrons in the cyclobutene rings localize at some extent.

4. PHYSICOCHEMICAL PROPERTIES AND APPLICATIONS OF SQUARYLIUM DYES

4.1. Photoconductive and electrochemical properties of squarylium dyes and their application

Since the first finding of photoconductivity of squarylium dyes [38], they have been receiving much attention with respect to applications in optoelectronics fields such as xerography and photoenergy conversion. The intense light absorption and highly polarized D–A–D structure of squarylium dyes make them applicable to photo-driven electronics materials based on generation of the electron–hole pair. The devices are often fabricated using the dyes in the solid state. The solid-state structure significantly affects the device performance, and thus, the solid-state chemistry of squarylium dyes have been eagerly investigated along with such applications [7,39,40]. In addition to the photoconductive property, the D–A–D intramolecular charge transfer of squarylium dyes have been giving opportunities to develop low-band-gap conductive materials. Recent advancement of squarylium dyes in electronics and optoelectronics fields is reviewed here.

For xerographic devices, anilino-type squarylium dyes have been mainly focused on, especially, applied as the charge generation layer (Fig. 9; the typical principle of xerography). Considering the light source used in the device, they are suitable because of large light absorption in the visible-to-near-IR region. Law

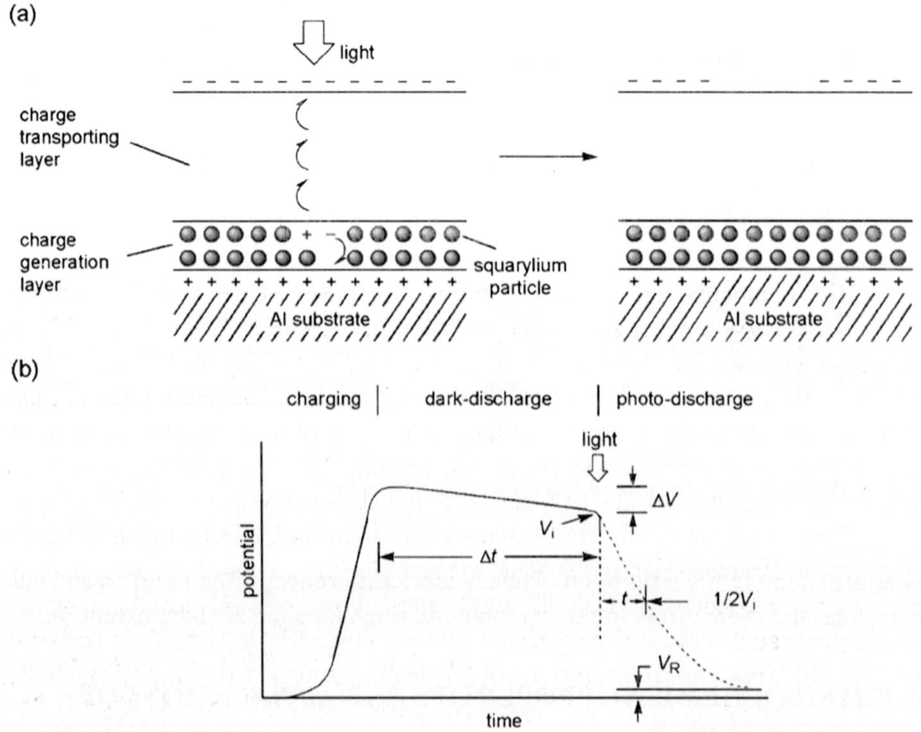

Fig. 9. (a) The configuration of a bilayer photoreceptor in a xerographic device. (b) The change of the electric potential of the xerographic photoreceptor during the photodischarge process: V_i, initial potential; V_R, residual potential, ΔV; dark-decay.

and coworkers synthesized a series of anilinosquarylium dyes and investigated their solid-state properties and xerographic performance [7,41–44]. Some of xerography-applicable dyes are **4a** and **68a–e**, and they exhibited broadened light absorption from the visible to the near-IR region in the solid state. From the X-ray powder diffraction and single-crystal structural analyses, any of these dyes showed tight packing of the dye molecules with *ca.* 3.5 Å spacing [45,46], and hence, intermolecular D–A interaction was suggested. A number of unsymmetrical squarylium dyes were also investigated toward xerographic application [17,18], and **69** exhibited a low dark-decay property ($\Delta V/\Delta t$; the decrease of the electric potential per unit time upon the dark-discharge process) and relatively high sensitivity, i.e., low $E_{0.5}$ ($E_{0.5}$; the energy required to discharge the initial electric potential to half upon photoirradiation): $\Delta V/\Delta t$ was –15 V/s, and the values of $E_{0.5}$ were 3.1 and 1.9 erg/cm^2 upon 600 and 790 nm light irradiation, respectively [47].

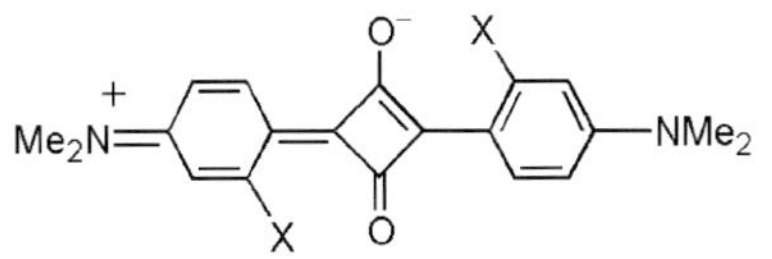

68a: X = -OH **68d**: X = $-CH_2CH_3$
68b: X = -F **68e**: X = -OMe
68c: X = $-CH_3$

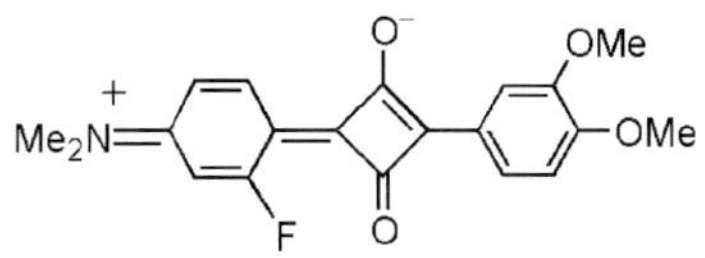

69

High photoconductivity of squarylium dyes as well as their large light absorptivity in the low-energy region of the solar-illumination wavelength is suitable for developing solar cell devices. At the early stage of the development of the light-to-power conversion, the devices were constructed mainly as Schottky-type thin film cells, taking advantage of photo-initiated generation of hole–electron pairs in the squarylium layer [48,49]. Nowadays, squarylium dyes are receiving increasing interest as sensitizers for the TiO_2 electrode in Grätzel-type solar cells [50–52].

If the successive arrangement of the squarylium D–A–D structure is obtained, high conductive materials are expected to be produced. Such materials are available by polymeric condensation of the highly electron-rich 2,5-unsubstituted pyrrole with squaric acid, which have been extensively studied by Ajayaghosh and coworkers [53–56]. They obtained two types of low band-gap polymers, each of which was selectively synthesized by choosing reaction conditions (Scheme18) [53,54]. The one is a zwitterionic 1,3-substituted squarylium polymer (**70** and **72**), and the other is a polymer containing diketonic 1,2-substituted squarylium repeating units in addition to the 1,3-zwitterionic polymer units (**71** and **73**). The latter polymer exhibited relatively high conductivity (8×10^{-6} S/cm for **71** and 2.3×10^{-5} S/cm for **73**) without any external chemical doping (conductivities of the other type of polymers; 2.4×10^{-6} S/cm for **70** and 2.5×10^{-6} S/cm for **72**). Upon doping iodine, the conductivity of the polymers were significantly improved; 2.5×10^{-4} S/cm for **70**, 6.0×10^{-4} S/cm for **71**, 1×10^{-3} S/cm for **72**, and 1.8×10^{-3} S/cm for **73**. Using a stylbenoid dipyrryl monomer, another type of polymer, the poly(phenylene-vinylene-squarylium) copolymer, was available (Scheme 19) [56]. This low-band-gap polymer, showing broadened light absorption from 600 to 1200 nm in the solid state, exhibited the conductivity that was affected by the length of the anchored alkyl groups (Table 7). The packing geometries of the polymer backbones were estimated by XRD analysis (Fig. 10), and the intrinsic conductivity decreased with the increase in the inter-column distance d_1. Exposure of the polymer samples to iodine vapor significantly improved their conductivity up to the 10^{-2} S/cm order.

4.2. Biological and environmental applications of squarylium dyes

For a few decades, organic functional dyes have been receiving increasing attention with respect to applications as labeling and chemosensory materials for environmentally and biologically important analytes such as typical and transition

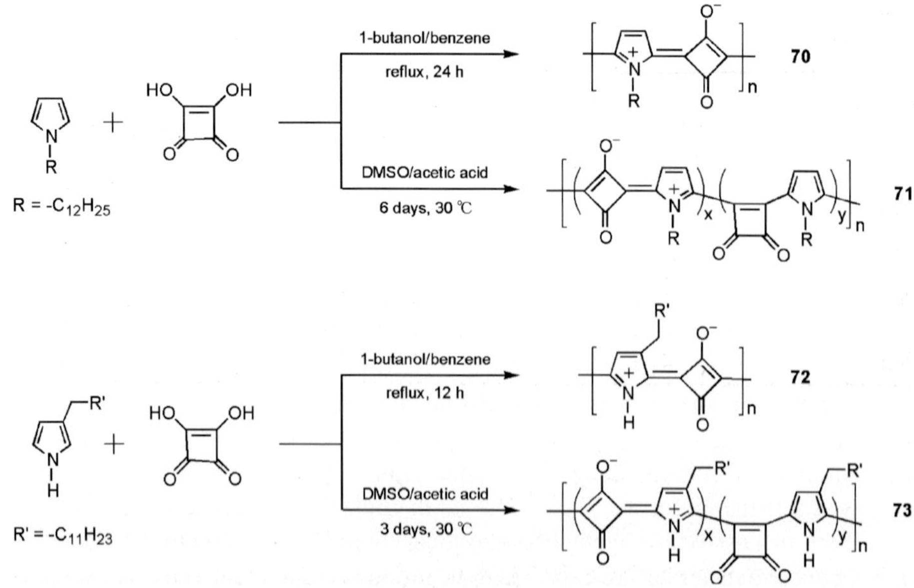

Scheme 18.

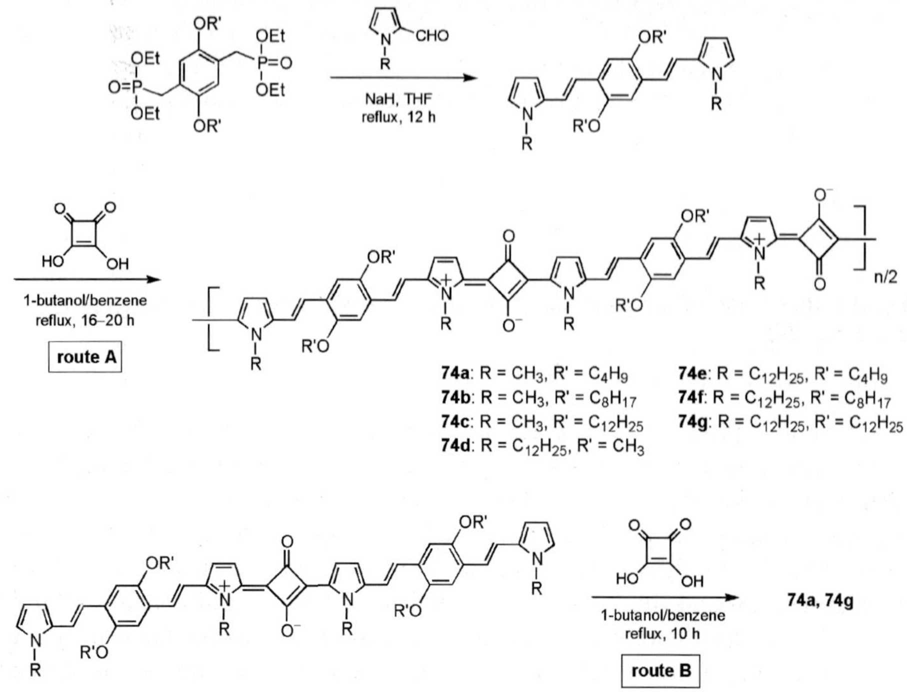

Scheme 19.

Table 7
XRD data and conductivity of poly(phenylene-vinylene-squarylium) copolymers **74a–g**

Compd.	inter-column distance d_l/angstrom	intrinsic conductitvity /S cm^{-1}	iodine-doped conductivity/S cm^{-1}
74a	11.0	5.30×10^{-4}	5.20×10^{-2}
74b	14.7	8.75×10^{-4}	2.98×10^{-2}
74c	19.2	4.21×10^{-6}	6.17×10^{-2}
74d	17.0	2.20×10^{-5}	1.10×10^{-2}
74e	18.4	4.40×10^{-6}	2.80×10^{-2}
74f	20.5	6.20×10^{-6}	5.10×10^{-2}
74g	21.0	6.30×10^{-7}	1.10×10^{-2}

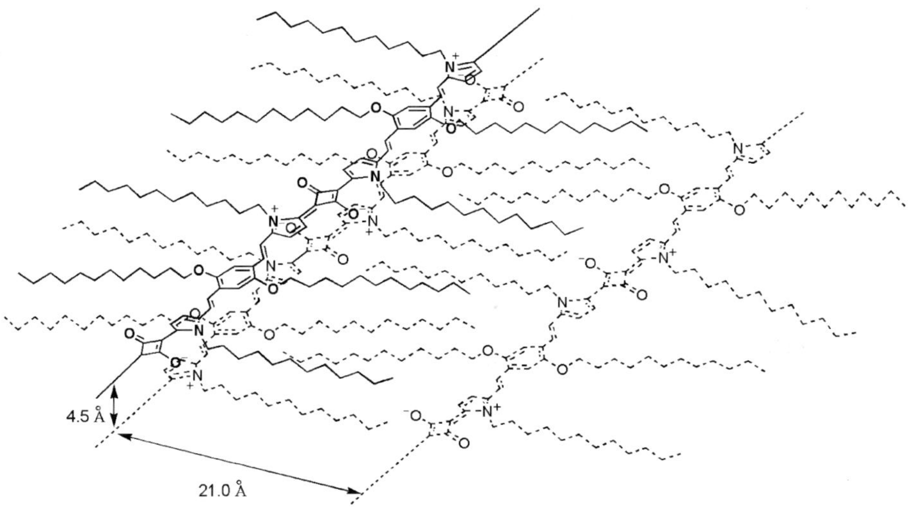

Fig. 10. Illustration of the molecular packing of the poly(phenylene-vinylene-squarylium) copolymer **74g**.

metal cations, proteins, nucleic acids, carbohydrates, and so on. Obviously, the techniques of the labeling and sensing of chemicals have been developing in parallel with the development of host–guest and supramolecular chemistry [57–59]. The basic concepts to produce such materials are shown in Fig. 11. The analyte binds to the dye (or dye appended receptor) to bring about the change of electronic environment around the chromophore, and thus, the spectral changes of light absorption and/or fluorescence emission occur to make possible the detection of the analyte. Hence, non-covalent interactions between the dye and the analyte lead to non-destructive detection. Squarylium dyes are suitable for such applications because they exhibit intense light absorption from the visible to the near-IR region

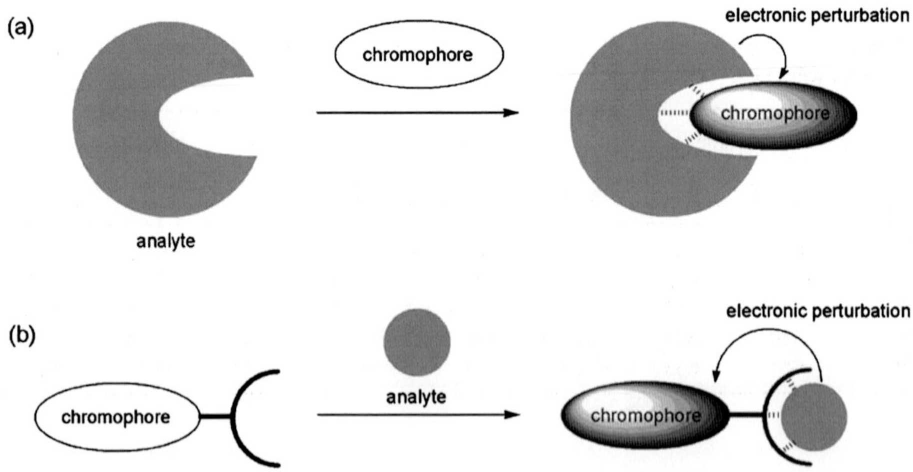

Fig. 11. Schematic illustration of (a) chromogenic labeling and (b) chromogenic chemosensing of an analyte.

and sometimes show fluorescence emission properties. In addition, when the π-electron of the squarylium chromophore joins in the complexation event with the analyte, significant electronic perturbation on their cyanine-like π-conjugation systems is expected to give rise to drastic spectral changes. Thus, the molecular design of labeling and chemosensory materials is based on how the interacting sites to the analytes are conjugated with the dye backbones. Recent development of squarylium-based labeling and chemosensory materials toward the detection of proteins and metal cations are shown below.

For non-destructive detection of proteins, squarylium-based labeling materials have been investigated [15,60,61]. Since squarylium dyes exhibit red-shifted absorption and emission spectra, the background absorption and emission caused by chromophoric residues in proteins are negligible, and the accurate analyses of proteins are possible. Terpetschnig *et al.* reported that symmetrical and unsymmetrical squarylium dyes **75**–**79** with high photostability bound to bovine serum albumin (BSA) to show slightly red-shifted absorption as well as drastic fluorescence emission enhancement, compared to those in alcoholic solvents (Table 8) [15]. These spectral changes are due to complexation of the dyes to the non-polar binding site in BSA. The indolenine-based squarylium dyes **75** and **76** exhibited largest emission enhancement (*ca.* seven times in quantum yields) and photostability (<10% bleaching) upon binding to BSA, and therefore, are suitable to the analytical use. The dye **77** showed less photostability (55% bleaching in BSA-H_2O), although it exhibited the highest quantum yield upon BSA binding. These squarylium dyes also showed longer fluorescence lifetimes upon complexation to

Table 8
Electronic absorption and fluorescence emission data of **75–79** in alcohols and BSA aqueous solutions

Compd.	absorption /nm		emission /nm (Stokes shift)		quantum yield		mean lifetime /ns	
	MeOH	BSA-H_2O	MeOH	BSA-H_2O	PrOH	BSA-H_2O	MeOH	BSA-H_2O
75	628	636	644 (16)	669 (33)	0.09	0.70	0.20	3.25
76	632	642	653 (21)	652 (10)	0.10	0.68	0.21	2.80
77	635	647	651 (16)	662 (15)	0.10	0.78	0.24	3.53
78	629	643	648 (19)	655 (12)	0.07	0.34	0.20	2.29
79	644	663	662 (18)	676 (13)	0.10	0.20	0.43	2.27

BSA than those in alcoholic solvents (Table 8). Such particular photophysical behaviors allow BSA detection by fluorescence lifetime measurement [62,63].

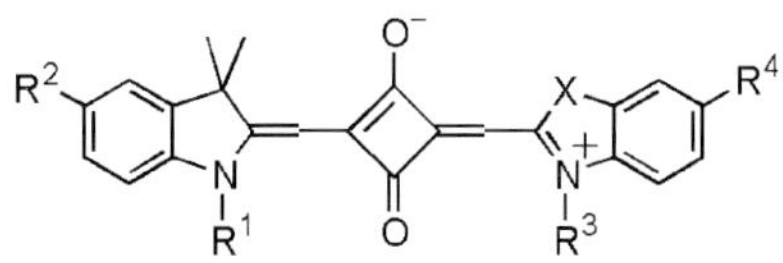

75: X = C(Me)$_2$, R^1 = R^3 = Et, R^2 = R^4 = H
76: X = C(Me)$_2$, R^1 = R^3 = Me, R^2 = R^4 = Cl
77: X = S, R^1 = Me, R^3 = Et, R^2 = R^4 = H
78: X = C(Me)$_2$, R^1 = R^3 = Me, R^2 = Cl, R^4 = H
79: X = Se, R^1 = Me, R^3 = Et, R^2 = Cl, R^4 = H

In practical uses of squarylium dyes toward detection of proteins, there is a serious problem: most of the organic dyes, including squarylium dyes, are quite insoluble in aqueous solutions due to their hydrophobic structural backbones. Nakazumi improved this problem by appending a carboxyl residue to squarylium skeletons [60]. The reported squarylium dyes **80a–d** and **82** are first dissolved in methanol and then can be diluted in an aqueous trizma buffer solution without precipitation. The unsymmetrical dyes were prepared in a similar manner to the stepwise synthesis of **16–21** [13]. The bathochromic shifts in electronic absorption spectra were observed for these dyes in the presence of human serum albumin (HSA), showing formation of dye–HSA complexes (Table 9). The complexed dyes exhibited their absorption maxima at 621–658 nm except for **80b**, applicable to the detection using a commercially available diode laser. Along with the absorption spectral changes, significant fluorescence enhancement was also observed upon the complexation with HSA. Job's analysis confirmed the complexation stoichiometry as dye:HSA = 1:1, the binding constants *K*s ranging form 1.7×10^5 to 5.8×10^6

Table 9
Absorption and fluorescence emission data of **79–82** in the absence and presence of HAS

Dye	Without HSA[a]		With HSA[b]			
	λ_{abs}/nm (log ε)[a]	Φ_f[c]	λ_{abs}/nm	λ_f/nm	Φ_f[c]	K/M^{-1}
80a	564 (4.81)	0.05	621	656	0.70	8.7×10^5
80b	550 (4.89)	0.05	550	618	0.39	1.1×10^6
80c	604 (4.96)	0.03	643	654	0.92	8.0×10^5
80d	558 (4.66)	0.04	656	677	0.06	1.7×10^5
81	658 (5.08)	0.01	658	680	0.34	5.8×10^6
82	622 (5.11)	0.02	640	645	0.97	3.8×10^6

[a] 5.0×10^{-7} M solution of the dye dissolved in trizma buffer (pH 7.4) at 298 K.
[b] 5.0×10^{-7} M solution of the dye dissolved in trizma buffer (pH 7.4) with 5.0×10^{-6} M HSA at 298 K.
[c] The Φ_f was determined relative to that of Rhodamine B (0.97); excitated at 630 , 550 , 630 , 630 , 658 , and 630 nm for **80a**, **80b**, **80c**, **80d**, **81**, and **82**, respectively.

M^{-1}. In the case of **80** and **81**, the fluorescence emission quantum yield reached up to 0.92, and the limit detection (signal-to-noise ratio; *ca.* 15) of 2.0 nm HSA was observed with **80c** by means of a conventional fluorescence spectrometer. It is well known that HSA has three substrate-binding sites; digitoxin, warfarin, and diazepam sites. Although warfarin possesses a coumarin fluorophore, the quantum yield of the warfarin–HSA complex is relatively low (0.11). Adding **80c** to the warfarin–HSA solution led to a decrease in warfarin–HSA emission (380 nm) as well as the generation of emission at 654 nm, indicating the squarylium dye was bound to the warfarin site. Although the unsymmetrical squarylium dye **82** did not have any carboxyl groups, it showed large binding constant and significant fluorescence emission enhancement; the dye **82** is also applicable to the HSA labeling. The emission enhancement of **80a** was also observed upon complexation with other proteins such as BSA, *β*-lactogloblin A, and tripsinogen [61]. Although the significant enhancement was observed for BSA as shown for HSA, the complexes of **80a**–*β*-lactogloblin and **80a**–tripsinogen were less emissive. Thus, the selective detection is possible between HSA (BSA) and -*β*-lactogloblin and between HSA (BSA) and tripsinogen. The practical detection was also carried out by using capillary electrophoresis with laser-induced fluorescence detection [61].

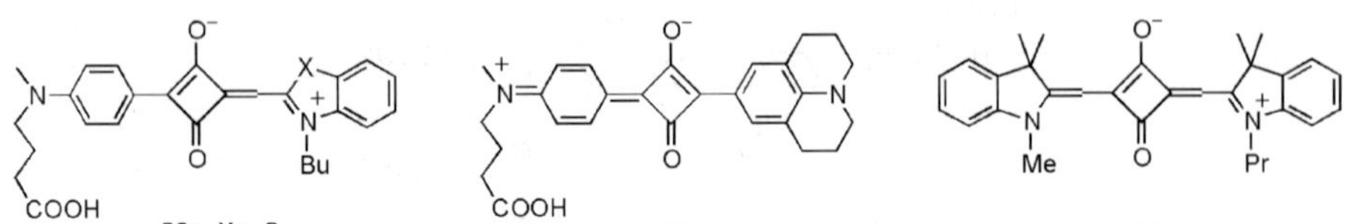

80a: X = S
80b: X = O
80c: X = $C(CMe)_2$
80d: X = -CH=CH-

81

82

Chemosensory systems based on squarylium chromophores have been eagerly investigated for typical and transition metal cations, employing crown ethers as ion-recognizing components. Akkaya and Oguz developed a squarylium-based chromoionophore **83**, in which azacrown moieties were integrated to the squarylium π-conjugation system [64]. The chromoionophore **83** was prepared by one-pot synthesis from phloroglucinol, aza-18-crown ether, and squaric acid, as shown in Scheme 20. Upon addition of a series of alkaline earth metal cations to an acetonitrile solution of **83** (λ_{max}, 640 nm; ε, 240,000 $mol^{-1}dm^3cm^{-1}$), blueshifts of the electronic absorption accompanied by hypochromic effects were observed. In parallel with the absorption spectral changes, the fluorescence emission also decreased. The decrease in the molar absorption coefficient was due to the decrease in the contribution of the nitrogen atoms to the π-conjugation system. Especially, Ba^{2+} significantly facilitated the blueshift as well as the hypochromism, and it was concluded that such spectral changes were caused by the formation of the H-aggregated dimer upon the complexation. Na^+ and K^+ cations were less effective on the spectral changes of the absorption and fluorescence emission. Hence, **83** is a potential chemosensor selective to alkaline earth metal cations, especially, Ba^{2+}.

For Ca^{2+} detection, a non-crowned chemosensor **84** was developed [65]. Instead of crown ethers, chelating units based on 1,2-bis(2′-aminophenoxy) ethanetetraacetic acid were employed as ionophores. Upon binding with Ca^{2+} in an aqueous buffer, hypochromism on electronic absorption as well as a decrease in fluorescence emission was observed. The addition of Mg^{2+} had no effect on the spectral changes. As Ca^{2+} and Mg^{2+} are the alkaline earth metal cations existing in the biological systems, **84** may be potentially applicable to bioanalysis and clinical use.

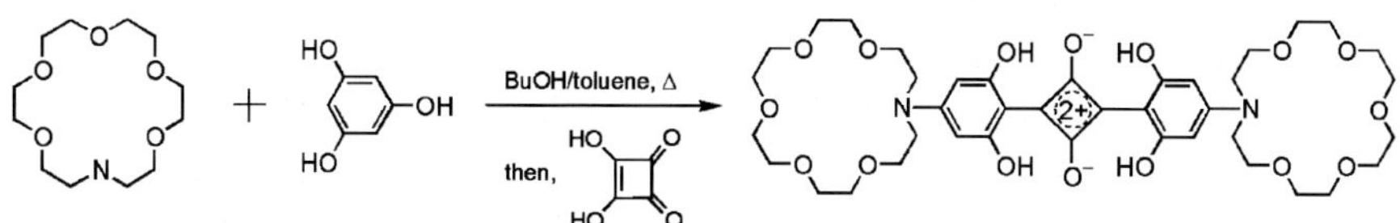

Scheme 20.

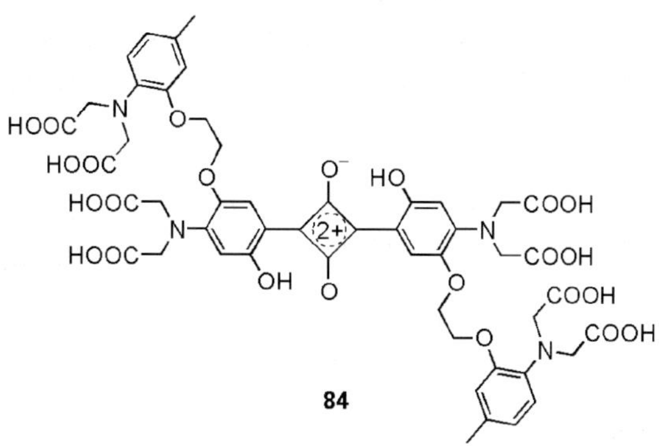

Akkaya and Oguz also developed another chemosensor **85** selective to Na^+ cation [66], which was prepared by the similar reaction to that for **83**. The diaza-crown moieties, which were selective to Na^+ and K^+ cations, were employed as ion-recognizing moieties. To increase the selectivity toward Na^+, i.e., to construct a more defined cavity for Na^+, lariat arms were appended to the crown rings. The detection of Na^+ by **85** was achieved by monitoring the bleaching of the original absorption band (λ_{max}; 642 nm) as well as the decrease in the fluorescence emission (λ_{em}; 651 nm). K^+ had no effect on the absorption nor on the fluorescence emission spectra of **85**, and thus, Na^+ selectivity was achieved by the lariat-crowned squarylium chromoionophore.

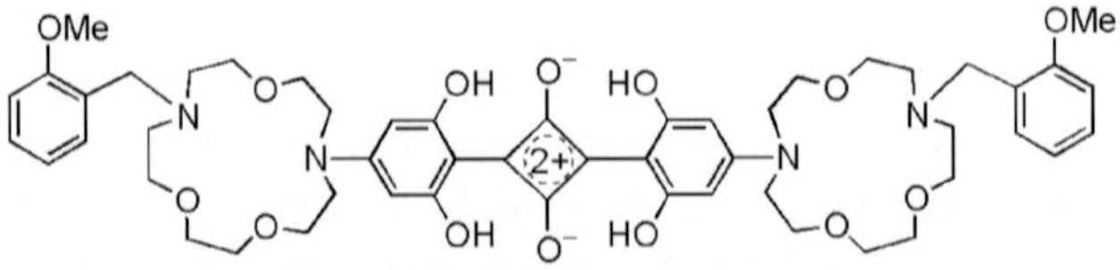

85

Ajayaghosh proposed another concept to develop a chemosensor for Ca^{2+}: the polyether-linked squarylium dimer **86** exhibited Ca^{2+}-selective spectral response caused by the intramolecular H-aggregate formation [67]. The stretched metal-free dimer was expected to adopt the face-to-face "dimeric" conformation upon complexation with a metal cation to exhibit different spectral properties from the "monomeric " form. Indeed, the addition of Ca^{2+} to an acetonitrile solution of **86** (λ_{max}; 630 nm) gave rise to the generation of the new electronic absorption band at the shorter wavelength (552 nm) accompanied by considerable quenching of the fluorescence emission (quantum yields; 0.03 and 0.008 before and after the Ca^{2+} addition, respectively), which was assigned to the complexation-induced intramolecular H-aggregate formation as shown in Fig. 12. The addition of Mg^{2+} induced modest spectral changes, and Na^+ and K^+ had neither any effect on the absorption nor on the fluorescence emission. Thus, the intramolecular H-aggregate formation was specific to Ca^{2+}, and the exitonic interaction between two squarylium chromophores achieved Ca^{2+}–signaling.

For sensing of a transition metal cation such as Hg^{2+}, Martínes-Máñes *et al.* developed a squarylium-based chemosensor **87** possessing two sulfur-containing azacrown moieties as ionophores [68]. The chromoionophore **87** exhibited the monomer (λ_{max}; 647 nm) and H-aggregate (λ_{max}; 560 nm) bands in its electronic absorption spectrum in an acetonitrile-H_2O solution that depended both on the concentration of **87** and on the volume ratio of the solvent mixture. Upon addition of Hg^{2+} to an H_2O-enriched solution (e.g., 20:80, MeCN/H_2O, v/v) of **87**, complete bleaching of both the monomer and H-aggregate bands was observed. This spectral change was fully reversible as confirmed by the addition of EDTA.

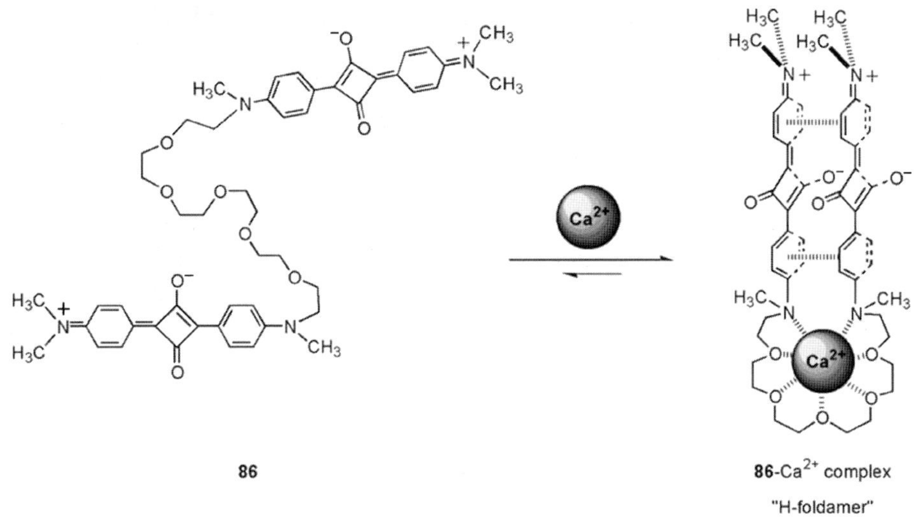

Fig. 12. Ca^{2+}-induced formation of the intramolecular H-aggregate of **86**, i.e., "H-foldamer."

On the other hand, the addition of Ag^+ induced the bleaching of the H-aggregate band and the enhancement of the monomer band. Other cations such as Li^+, Na^+, K^+, Mg^{2+}, Ca^{2+}, Ba^{2+}, Fe^{3+}, Ni^{2+}, Cu^{2+}, Zn^{2+}, and Cd^{2+} did not give rise to any noticeable spectral changes of either the monomer or H-aggregate bands. Therefore, employing the sulfur-containing ionophore as well as utilizing the monomer-H-aggregate equilibrium yielded an Hg^{2+}-sensing protocol in aqueous media (Fig. 13). It is worthy to note that Hg^{2+} could be detected at nanomolar levels in monomer containing solution (33:67, MeCN/H_2O, v/v). This analytical limit meets well the practical level of the analysis of drinking water.

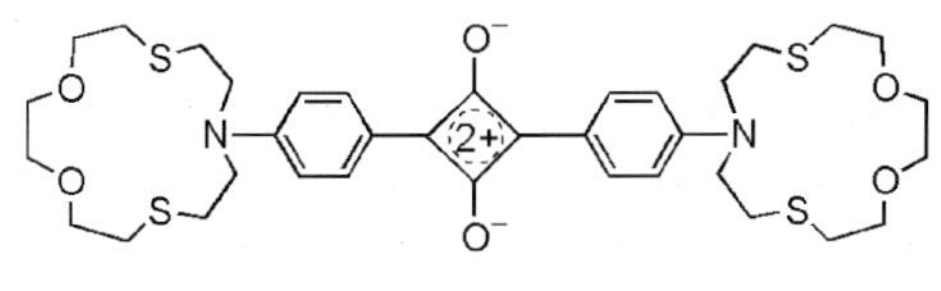

87

5. CONCLUDING REMARKS

In the present chapter, the authors reviewed recent development of squarylium dyes in the fields of synthesis and application. The synthetic procedure had been limited until 1980s: condensation of electron-rich aromatics or heterocyclic quaternary salts with squaric acid yielded symmetrical squarylium dyes. However, new protocols have been developed for the last two decades. The preparation of

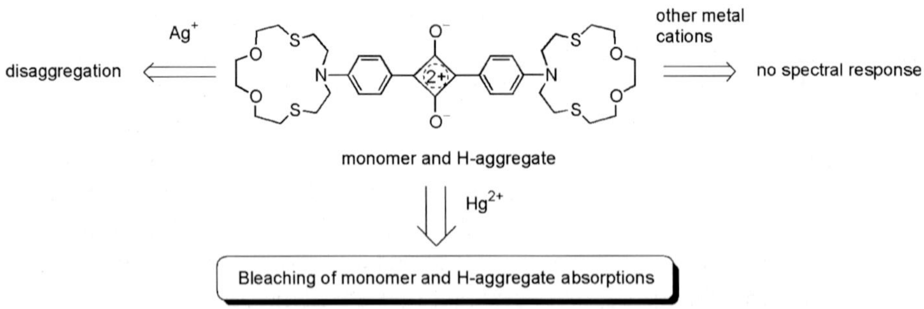

Fig. 13. Hg^{2+}-sensing protocol using the squarylium chromoionophore **87**.

unsymmetrical dyes has been achieved by stepwise synthesis employing semi-squarylium precursors. Thus, optical properties of squarylium chromophores have become tunable by choosing the aromatic and heterocyclic components at both ends of the dye skeleton. The development of reactions of squaric acid derivatives by Liebeskind have provided the opportunities to investigate new squarylium homologues, and the bisquarylium and bis(squarylium) dyes were prepared using bisquaryl and bi(squaric acid) with various spacers as synthetic precursors, respectively. The electrophilic addition of semi-squaryliums to the benzothiazolinosquarylium dye yielded a new class of squarylium homologues, i.e., the cationic "Y-shaped" dye, showing intense light absorption in the near-IR region. The analogous neutral dye, i.e., the "Y-shaped" dye, was also prepared, which possessed a similar π-conjugation system to the cationic dye. The X-ray structural analyses of squarylium dyes and their homologues were also mentioned, and it was clearly demonstrated how powerful the X-ray analysis is to investigate the electronic structures of squarylium dyes.

Applications of squarylium dyes were also reviewed. Their unique optical and electrochemical properties have provided great deal of opportunities to develop a variety of advanced materials ranging from electronics and optoelectronics to environmental and biological analyses. Outstanding achievements were found in such fields as xerographic devices, solar cells, and organic semiconductors. Intramolecular charge transfer as well as special stacking properties of some squarylium dyes in the solid state gave rise to the generation of photoconductivity applicable to xerographic and solar cell devices. Novel synthetic methods to produce polymeric squarylium backbone led to the development of low-band-gap conductive materials. Biolabeling and chemosensory materials have also been eagerly investigated. Focusing on the non-destructive analysis of proteins, squarylium-based labeling materials have been developed. Electric absorption and fluorescence emission of squarylium chromophores in the near-IR region were suitable for spectroscopic detection free from background signals caused by protein analytes. Some squarylium dyes were found to exhibit considerable fluorescence enhancement, potentially applicable to biological and clinical uses. The squarylium dyes bearing the ionophores integrated with their π-conjugation systems were applied to the sensing of metal cations. Selectivity and sensitivity were tunable by the combination of ionophores with squarylium chromophores. Electronic perturbation between the cations and the chromophores resulted in detectable spectral changes. The intra- and intermolecular aggregate formation was also employed to induce absorption and fluorescence emission spectral changes upon complexation with metal cations.

Although squarylium dyes themselves are old, they have great possibility to produce a variety of advanced materials. Indeed, the breakthroughs in the synthesis of squarylium dyes and their related compounds have been extending the range of functionality of the squarylium-based π-conjugation systems. The sophisticated material designs should be brought about not only by exploring novel synthetic protocols but also by understanding the physicochemical properties of squarylium dyes. Toward further development, more detailed studies on the relationship between the electronic structure and the functionality are expected.

REFERENCES

1. Mishra, A., Behera, R.K., Behera, P.K., Mishra, B.K. & Behera, G.B., *Chem. Rev.*, **100** (2000) 1973.
2. Treibs, A. & Jacob, K., *Angew. Chem., Int. Ed. Engl.* **4** (1965) 694.
3. Sprenger, H.-E. & Ziegenbein, W., *Angew. Chem., Int. Ed. Engl.*, **7** (1968) 530.
4. Ziegenbein, W. & Sprenger, H.-E., *Angew. Chem., Int. Ed. Engl.*, **5** (1966) 893.
5. Sprenger, H.-E. & Ziegenbein, W., *Angew. Chem., Int. Ed. Engl.*, **5** (1966) 894.
6. Maahs G. & Hegenberg, P., *Angew. Chem., Int. Ed. Engl.*, **5** (1966) 888.
7. Law, K.-Y., *Chem. Rev.*, **93** (1993) 449.
8. Fabian, J., Nakazumi, H. & Matsuoka, M., *chem. Rev.* **92** (1992) 1197.
9. Matsuoka, M., *Absorption Spectra of Dyes for Diode Lasers*, Bunshin, Tokyo (1990), pp. 51–59.
10. Sprenger, H.-E. & Ziegenbein, W., *Angew. Chem., Int. Ed. Engl.*, **6** (1967) 553.
11. Green, B.R. & Neuse, E.W., *Synthesis*, (1974) 46.
12. de Selms, R.C., Fox, C.J. & Riordan, R.C., *Tetrahedron Lett.*, **10** (1970) 781.
13. Yagi, S., Hyodo, Y., Matsumoto, S., Takahashi, N., Kono, H. & Nakazumi, H., *J. Chem. Soc., Perkin Trans. 1* (2000) 599.
14. Terpetschnig, E. & Lakowicz, J.R., *Dyes and Pigments*, **21** (1993) 227.
15. Terpetschnig, E., Szmacinski, H. & Lakowicz, J.R., *Anal. Chim. Acta*, **282** (1993) 633.
16. Jyothish, K., Arun, K.T. & Ramaiah, D., *Org. Lett.*, **6** (2004) 3965.
17. Law, K.-Y. & Bailey, F.C., *J. Chem. Soc., Chem. Commun.*, (1990) 863.
18. Law, K.-Y. & Bailey, F.C., *J. Org. Chem.*, **57** (1992) 3278.
19. Belluš, D., *J. Am. Chem. Soc.*, **100** (1978) 8026.
20. Kim, S.H., Hwang, S.H., Kim, J.J., Yoon, C.M. & Keum, S.R., *Dyes and Pigments*, **37** (1998) 145.
21. Reis, L.V., Serrano, J.P.C., Almeida, P. & Santos, P.F., *Synlett*, (2002) 1617.
22. Kim, S.H., Han, S.K., Park, S.H. & Park, L.S., *Dyes and Pigments*, **38** (1998) 49.

23. Kim, S.H., Han, S.K., Kim, J.J., Hwang, S.H., Yoon, C.M. & Keum, S.R., *Dyes and Pigments*, **39** (1998) 77.
24. Hyodo, Y., Nakazumi, H., Yagi, S. & Nakai, K., *J. Chem. Soc., Perkin Trans. 1*, (2001) 2823.
25. Liebeskind, L.S., Yu, M.S., Yu, R.H., Wang, J. & Hagen, K.S., *J. Am. Chem. Soc.*, **115** (1993) 9048.
26. Liebeskind, L.S. & Fengl, R.W., *J. Org. Chem.*, **55** (1990) 5359.
27. Nakazumi, H., Natsukawa, K., Nakai, K. & Isagawa, K., *Angew. Chem., Int. Ed. Engl.*, **33** (1994) 1001.
28. Yagi, S., Nakai, K., Hyodo, Y. & Nakazumi, H., *Synthesis*, (2002) 413.
29. Yagi, S., Fujie, Y., Hyodo, Y. & Nakazumi, H., *Dyes and Pigments*, **52** (2002) 245.
30. Yagi, S., Murayama, S., Hyodo, Y., Fujie, Y., Hirose, M. & Nakazumi, H., *J. Chem. Soc., Perkin Trans. 1*, (2002) 1417.
31. Meier, H. & Dullweber, U., *J. Org. Chem.*, **62** (1997) 4821.
32. Gerold, J., Holzenkamp, U. & Meier, H., *Eur. J. Org. Chem.*, (2001) 2757.
33. Buschel, M., Ajayaghosh, A., Arunkumar, E. & Daub, J., *Org. Lett.*, **5** (2003) 2975.
34. Farnum, D.G., Neuman, M.A. & Suggs, W.T., Jr., *J. Cryst. Mol. Struct.*, **4** (1974) 199.
35. Aswell, G.J., Bahra, G.S., Brown, C.R., Hamilton, D.G., Kennard, C.H.L. & Lynch, D. E., *J. Mater. Chem.*, **6** (1996) 23-26.
36. Kobayashi, Y., Goto, M. & Kurahashi, M., *Bull. Chem. Soc. Jpn.*, **59** (1986) 311.
37. Berstein, J., Tristani-Kendra, M. & Eckhardt, C.J., *J. Phys. Chem.*, **90** (1986) 1096.
38. Champ R.B. & Shattuck, M.D., U. S. Patent, 3,824,099 (1974).
39. Bernstein, J. & Goldstein, E., *Mol. Cryst. Liq. Cryst.*, **164** (1988) 213.
40. Law, K.-Y. & Bailey, F. C., *Dyes and Pigments*, **21** (1993) 1.
41. Law, K.-Y. & Bailey, F.C., *Can. J. Chem.*, **64** (1986) 1607.
42. Law, K.-Y. & Bailey, F. C., *Can. J. Chem.*, **64** (1986) 2267.
43. Law, K.-Y., *J. Phys. Chem.*, **91** (1987) 5184.
44. Law, K.-Y., Facci, J.S., Bailey, F.C. & Yanus, J.F., *J. Imaging Sci.*, **34** (1990) 31.
45. Wingard, R.E., *IEEE Ind. Appl.*, **37** (1982) 1251.
46. Tristani-Kendra, M. & Eckhardt, C. J., *J. Chem. Phys.*, **81** (1984) 1160.
47. Law, K.-Y., *Chem. Mater.*, **4** (1992) 605.
48. Merritt, V.Y. & Hovel, H.J., *Chem. Phys. Lett.*, **29** (1976) 414.
49. Morel, D.L., Ghosh, A.K., Feng, T., Stogryn, E.L., Purwin, P.E., Shaw, R.F. & Fishman, C., *Chem. Phys. Lett.*, **32** (1978) 495.
50. Chen, X.Y., Guo, J.H., Peng, X.J., Guo, M., Xu, Y.Q., Shi, L., Liang, C.L., Wang, L., Gao, Y.L., Sun, S.G. & Cai, S.M., *J. Photochem. Photobiol. A*, **171** (2005) 231.
51. Sayama, K., Tsukagoshi, S., Mori, T., Hara, K., Ohga, Y., Shinpou, A., Abe, Y., Suga, S. & Arakawa, H., *Sol. Energy Mater. Sol. Cells*, **80** (2003) 47.
52. Zao, W., Hou, Y.J., Wang, X.S., Zhang, B.W., Cao, Y., Yang, R., Wang, W.B. & Xiao, X.R., *Sol. Energy Mater. Sol. Cells*, **58** (1999) 173.
53. Ajayaghosh, A., Chenthamarakshan, C.R., Das, S. & George, M.V., *Chem. Mater.*, **9** (1997) 644.
54. Chenthamarakshan, C.R., Eldo, J. & Ajayaghosh, A., *Macromolecules*, **32** (1999) 251.
55. Chenthamarakshan, C.R. & Ajayaghosh, A., *Chem. Mater.*, **10** (1997) 1657.
56. Eldo, J. & Ajayaghosh, A., *Chem. Mater.*, **14** (2002) 410.
57. Vögtle, F., *Supramolecular Chemistry: An Introduction*, Wiley, Chichester (1989).
58. Balzani, V. & Scandola, F., *Supramolecular Photochemistry*, Ellis Horwood, Chichester (1991).
59. Lehn, J.M., *Supramolecular Chemistry: Concepts and Perspectives*, VCH, Weinheim (1995).
60. Nakazumi, H., Colyer, C.L., Kaihara, K., Yagi, S. & Hyodo, Y., *Chem. Lett.*, **32** (2003) 804.

61. Welder, F., Paul, B., Nakazumi, H., Yagi, S. & Colyer, C.L., *J. Chromatogr. B*, **793** (2003) 93.
62. Thompson, R.B. & Lakowicz, J.R., *Anal. Chem.*, **65** (1993) 853.
63. Szmacinski, H. & Lakowicz, J.R., *Anal. Chem.*, **65** (1993) 1668.
64. Oguz, U. & Akkaya, E.U., *Tetrahedron Lett.*, **38** (1997) 4509.
65. Akkaya, E.U. & Turkyilmaz, S., *Tetrahedron Lett.*, **38** (1997) 4513.
66. Oguz, U. & Akkaya, E.U., *Tetrahedron Lett.*, **38** (1998) 5857.
67. Ajayaghosh, A., Arunkumar, E. & Daub, J., *Angew. Chem., Int. Ed. Engl.*, **41** (2002) 1766.
68. Ros-Lis, J.V., Martínes-Máñes, R., Rurack, K., Sancenón, F., Soto, J. & Spieles, M., *Inorg. Chem.*, **43** (2004) 5183.

Functional Dyes
Sung-Hoon Kim (Editor)

Chapter 7

Fluorine-containing dyes

Masaki Matsui

Department of Materials Science and Technology, Faculty of Engineering, Gifu University, Gifu, Japan

1. INTRODUCTION

For fluorine-containing dyes in dye chemistry, a fluorine atom as a leaving group on pyrimidine and triazine moieties in reactive dyes was important. However, in the field of functional dyes, not only the chemical properties but also the physical properties of dyes are important. When a fluorine atom(s) is introduced into a dye molecule, the properties are changed. In this chapter, the physical properties and reactivity of fluorine-containing dyes are reviewed.

2. PHYSICAL PROPERTIES OF FLUORINE ATOM

A fluorine atom is introduced as a substituent in aromatic moiety and as a fluoroalkyl group. Their important physical properties are as follows:

(1) High electronegativity: A fluorine atom shows the largest electronegativity (Pauling, 4.0) among all elements; those of hydrogen and carbon atoms are 2.2 and 2.6, respectively. The Hammett's σ_p constants of fluoro, trifluoromethyl, triflouromethoxy, trifluoromethylthio, and trifluoromethylsulfonyl groups are reported to be 0.06, 0.54, 0.35, 0.50, and 0.93, respectively. Thus, fluorine atom and polyfluoroalkyl and perfluoroalkyl (R_f) groups show strong electron-withdrawing ability.

(2) Van der Waals radius slightly bigger than a hydrogen atom: The van der Waals radius of hydrogen atom is reported to be 1.20 Å and that of a fluorine atom is 1.35 Å. Thus, similar van der Waals radius of hydrogen atom to fluorine atom is known as mimic effect, being important in the field of medicinal chemistry. Since the C–F bond energy (116 kcal mol^{-1}) is larger than the C–H bond energy (99.5 kcal mol^{-1}), fluorine-containing compounds are not metabolized as the hydrogen derivatives. Meanwhile, in the case of R_f group, free rotation between the C–C bond is restricted due to slightly larger van der Waals radius of fluorine atom. Therefore, R_f groups are more rigid than the alkyl (R) groups.

Furthermore, the steric parameters (*Es*) of methyl and trifluoromethyl groups are reported to be -1.24 and -2.4, respectively. The size of trifluoromethyl group is similar to that of isobutyl group ($Es = -2.37$). Thus, though the size of a fluorine atom is similar to that of a hydrogen atom, that of fluoroalkyl group, especially R_f group, is much bigger than the R group.

(3) Lipophilicity: A fluorine and trifluoromethyl group show lipophilic property. Therefore, fluorine-containing compounds can easily penetrate into the cell wall. This property is also very important in the field of medicinal chemistry.

(4) Water and oil repellency: Teflon is a good example that shows water repellency. Polyfluoro polymers are used as coating materials. When the fluorine content in a molecule is higher than 70%, they are not soluble in water and organic solvents. These compounds are soluble in fluorous solvents such as perfluorohexane. Thus, fluorine shows both water and oil repellency.

Since a fluorine atom shows electron-withdrawing nature, fluorine chemistry is called mirror chemistry, which involves opposite concepts compared with the usual organic chemistry. For example, alkyl groups can stabilize a cationic center, whereas perfluoroalkyl group can stabilize an anion center. The details of fluorine chemistry are shown in the review [1].

3. FLUORINE-CONTAINING DYES

3.1. Azo dyes

Fluorine-containing azo dyes are shown in Scheme 1, 2, and 3. It is claimed that the photostability of trifluoromethyl-substituted azo dye **1** has improved. Electron-withdrawing and bulky *o*-trifluoromethyl group may inhibit photo-oxidation processes. Disazo dyes **2** substituted with R_f and R_fS group showed better solubility and dichroism than the R and RS derivatives [2]. R_f-Substituted ester disazo dyes **3** and benzothiazolyl disazo dyes **4** also showed improved solubility and dichroism than the corresponding R-substituted derivatives [3, 4]. These results can be attributed to the bulkiness of the R_f group. Since the R_f group is more rigid than the R group, dichroism in liquid crystals can be improved.

Azo dyes can act as second-order nonlinear optical chromophores (NLOphores). To obtain larger hyperpolarizability (β), strong intramolecular push-pull chromophoric system, bathochromic compounds, and a large molar absorption coefficient are required.

Azo dye **5** (Disperse Red 1, DR1) is usually used as an NLOphore. Thiazolylazo dyes are more bathochromic than the substituted phenylazo dyes. Therefore, dye **6** showed large nonlinearity due to electron-withdrawing R_f-substituted benzothiazolyl moiety. This compound shows improved relaxation behavior after poling [5]. The Hammett's σ_p constant of nitro group is reported to be 0.78. That of trifluoromethylsulfonyl group is reported to be 0.93. Therefore, R_fSO_2-substituted azo dye **7** can show larger than **5**. Furthermore, compound **7**

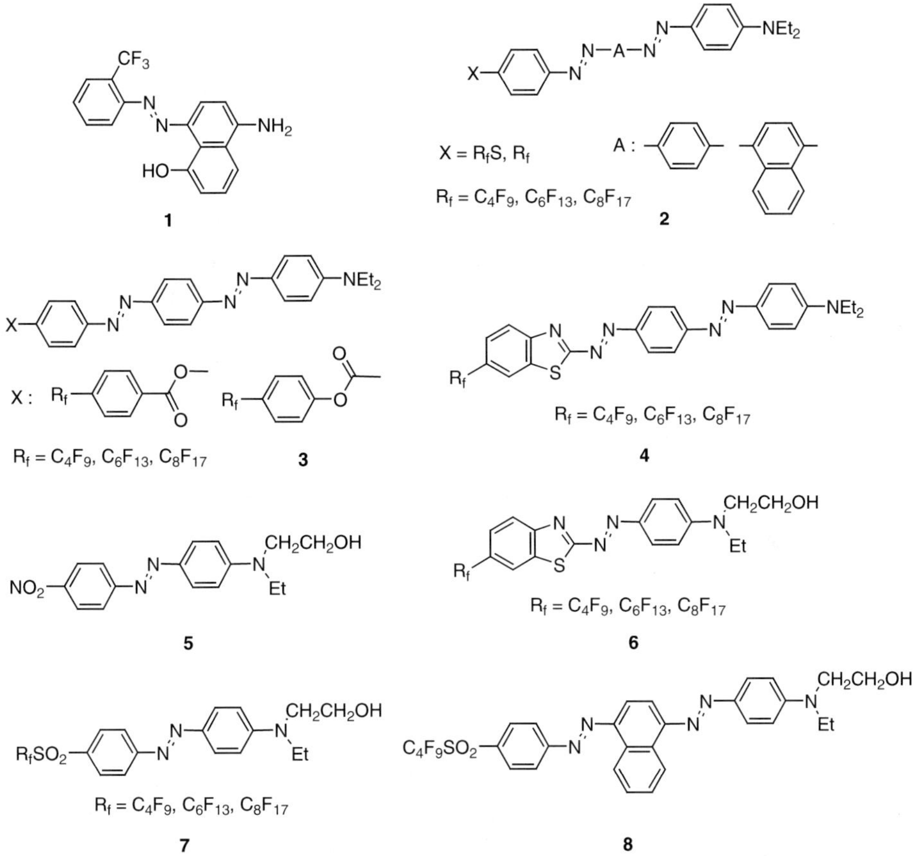

Scheme 1.

showed larger β value and solubility than **5** [6]. Bathochromic disazo dye **8** having a naphthylene moiety in the central part of the molecule was designed to improve the solubility and bathochromicity. This compound showed larger nonlinearity than the corresponding nitro derivative [7]. The methacrylate polymer pendanted with **8** showed larger nonlinearity and improved relaxation behavior compared with the methacrylate polymer pendanted with the DR1.

Azo dyes **9** having very bulky polyfluoroalkyl substituent, which was produced from perfluoropropene dimer and trimer were synthesized [8]. Azopyridone dyes **10** having perfluoroalkyl substituent(s) were synthesized as yellow dye diffusion thermal transfer dyes [9]. These compounds showed good photostability. Azo dye **11** substituted with polyfluoroalkyl group was also claimed to show good photostability on polyester [10].Water-soluble disazo dyes **12** having a polyfluoroalkyl group at the alkylamino moiety also showed good photostability [11].

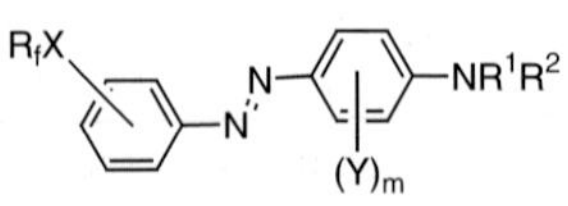

R^1, R^2 = H, C1-5-alkyl

R_f = C5-12-perfluoroalkenyl

X = -, O, S, $(CH_2)_p$, $(CH_2)_qO$

Y = COOH, SO_3H, C1-5-alkyl, halo; m = 0-2

9

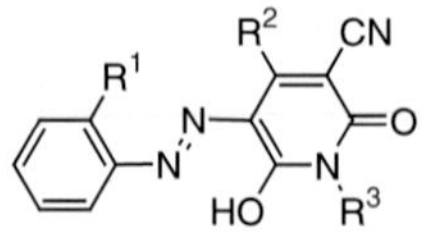

R^1 = CF_3, C_4F_9, C_6F_{13}, C_8F_{17}

R^2 = Me, CF_3

R^3 = H, Me, Pr, Bu, Hex, Oct

10

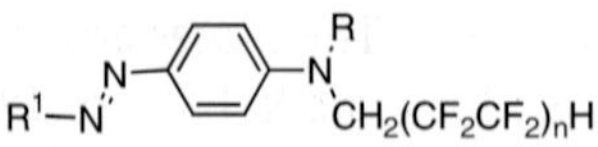

R = H, Me, Et, CH_2CH_2CN, CH_2CH_2OH

n = 1, 2 R^1 = substituted phenyl

11

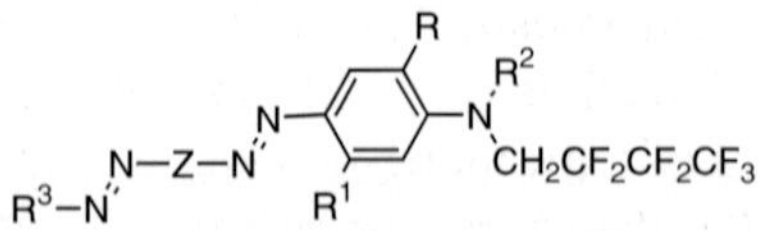

R = H, alkyl, alkoxy R^1 = H, Cl, alkyl, alkoxy, CH_3CONH

R^2 = H, alkyl, CH_2CH_2CN, hydroxyalkyl R^3 = R^4 or R^5

R^4 = 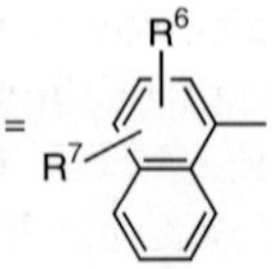 R^5 = 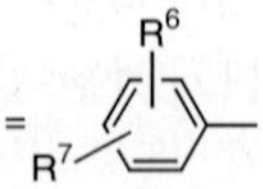

R^6 = SO_3H, PO_3H_2 R^7 = H, Cl, alkyl, alkoxy, NO_2

Z = *p*-C_6H_4, 1,4-$C_{10}H_6$, 1 or 2-alkyl or alkoxy

12

Scheme 2.

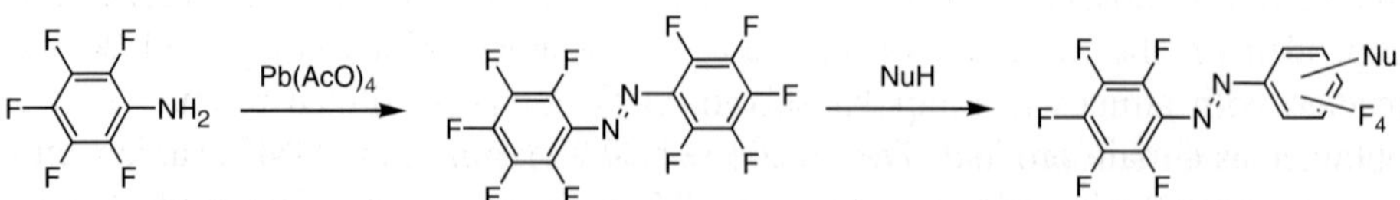

Nu = NR_2, OR

13   **14**   **15**

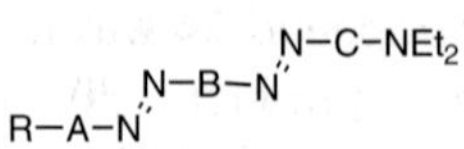

R : C_4H_9O, C_4H_9NH, C_4H_9S

A, B, C: 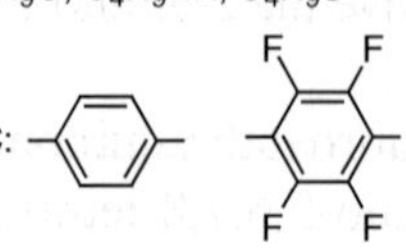

16

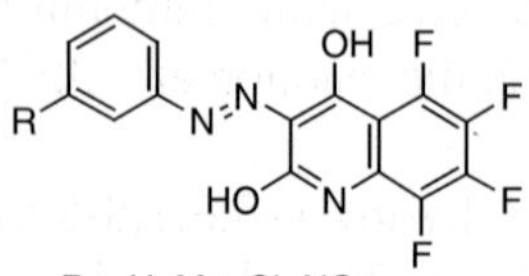

R = H, Me, Cl, NO_2

17

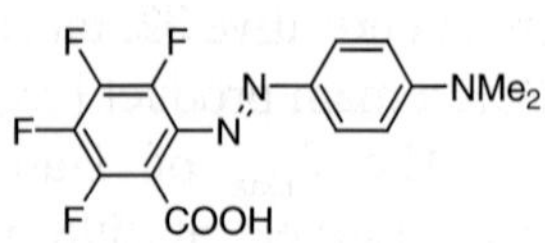

18

Scheme 3.

Oxidation of pentafluoroaniline (**13**) with lead tetraacetate gave decafluoroazobenzene (**14**) as a main product [12]. Compound **14** was planar by single X-ray crystallography [13]. Decafluoroazobenzene (**14**) easily reacted with nucleophiles such as amines and methoxide ion to give the substituted products **15**. By using this reaction, dichroic disazo dyes **16** having a perfluoro-*p*-phenylene moiety(ies) were synthesized [14]. The dichroism of **16** was slightly smaller than the fluorine-free derivative. However, the solubility of 4-(diethylamino)tetraphenylene derivatives was considerably improved. Yellow azo dye **17** was synthesized [15]. The azo and hydrazone tautomers of **17** showed the absorption maxima (λ_{max}) at 395 and 425 nm, respectively. The molar absorption coefficient of tetrafluorinated methyl red **18** under acidic and alkaline conditions was smaller than the fluorine-free derivative [16].

3.2. Quinone dyes

Fluorine-containing quinone dyes are shown in Scheme 4. Perfluoroalkanoyl peroxides **19** easily react with electron-rich substrates to give the perfluoroalkylated products [17]. R_f-substituted naphthoquinones **20** and anthraquinones **21** were obtained by using this reagent [18]. Compounds **21** showed hypsochromic shift compared with non-R_f derivatives. Blue anthraquinone dye **22** substituted with polyfluoroalkoxy group showed high dichroism and good durability [19]. In anthraquinone dyes **23**, no remarkable difference in λ_{max} between the $NHCH_2R_f$ and NH_2 derivatives was observed. The λ_{max} of OCH_2R_f derivative was *ca* 50 nm hypsochromic compared with that of the OH derivative [20].

The reaction of 1,2,3,4-tetrafluoroanthraquinone (**24**) with methoxide ion depended on the polarity of the solvent. In methanol, 1- and 2-methoxy derivatives were obtained in the ratio of 2:98, whereas in a methanol–benzene mixed solution, the ratio was observed to be 66:34 [21]. The reaction of **24** with amines also depended on the kinds of solvents and temperature. When compound **24** was reacted with amines in nonpolar solvent at 50°C, 1-substituted product **25** was obtained as a main product. The reaction of **24** with amines in DMSO at low temperature gave the 2-substituted derivative **26** as a main product. The product distribution also depended on the kinds of amines [22]. 1,2,3,4-Tetrafluoroquinizarine (**27**) reacted with aniline to give **28** and **29** [23]. When compound **24** was reacted with benzenethiol to give the 2,3-disubstituted product **30** [24]. Octafluoroanthraquinone **31** reacted with ammonia in toluene at 110°C to give1-amino derivative **32**, meanwhile the reaction at −35°C gave the 2-amino derivative **33** as a main product [25].

The λ_{max} of 1-amino-4-hydroxy-5,6,7,8-tetrafluoroanthraquinone (**34a**) was observed at 598 nm [26]. The λ_{max} of 1-amino-5,6,7,8-tetrafluoroanthraquinone (**34b**) was observed at 492 nm, being slightly more bathochromic than 1-aminoanthraquinone due to electron-withdrawing tetrafluoroanthraquinone moiety [27].

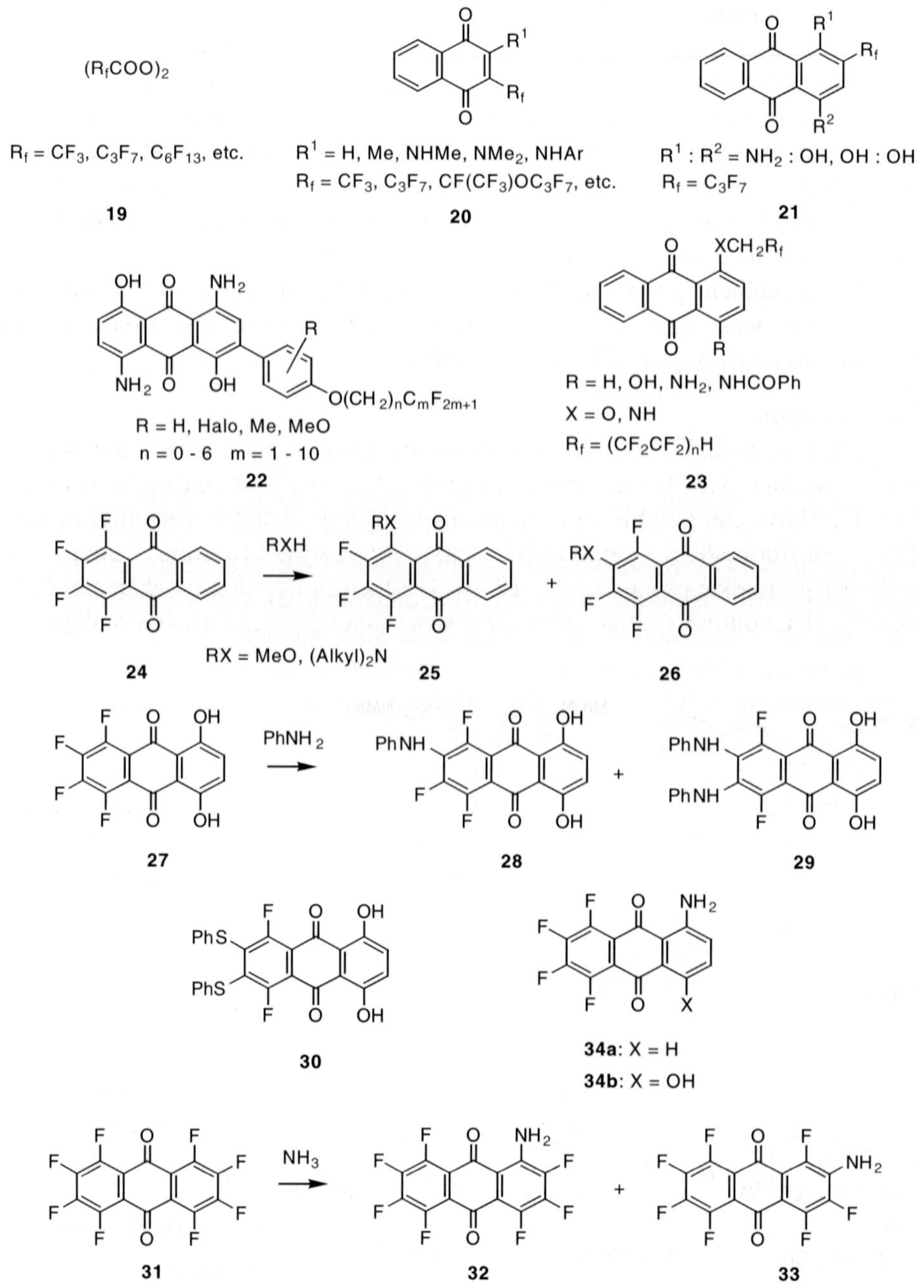
(RfCOO)2
Rf = CF3, C3F7, C6F13, etc.
19
R1 = H, Me, NHMe, NMe2, NHAr
Rf = CF3, C3F7, CF(CF3)OC3F7, etc.
20
R1 : R2 = NH2 : OH, OH : OH
Rf = C3F7
21
O(CH2)nCmF2m+1
R = H, Halo, Me, MeO
n = 0 - 6 m = 1 - 10
22
XCH2Rf
R = H, OH, NH2, NHCOPh
X = O, NH
Rf = (CF2CF2)nH
23
RXH
RX = MeO, (Alkyl)2N
24
25
26
PhNH2
27
28
29
30
34a: X = H
34b: X = OH
NH3
31
32
33

Scheme 4.

3.3. Arylmethane dyes

Fluorine-containing arylmethane dyes are shown in Scheme 5. The first and second absorption bands of pentafluoro malachite green **35** in 98% acetic acid were observed at 658 and 423 nm, respectively [28]. Those of fluorine-free derivative were observed at 621 and 428 nm, respectively. Thus, electron-withdrawing pentafluorophenyl group at the unstarred position in the alternant chromophoric system causes bathochromic shift. Trifluoromethylated derivative **36** showed better photostability than the unsubstituted one [29].

Diarylmethene precursor **37** whose aromatic moiety is substituted with fluorine atoms was synthesized [30]. The central cyanide ion dissociates in methanol to show λ_{max} at 500 nm ($\varepsilon > 70{,}000$).

3.4. Coumarins

Fluorine-containing coumarins are shown in Scheme 6. The reaction of ethyl trifluoroacetate with *m*-substituted phenols gave 4-(trifluoromethyl)coumarins **38** [31]. These compounds were more bathochromic than the 4-methyl derivative. 3-(Perfluoroalkyl)coumarins **39** were obtained by the reaction of coumarins with perfluoroalkanoyl peroxide **19** [32]. Electron-transfer mechanism from electron-rich coumarins to the peroxides was proposed. The 3-perfluoroalkylated

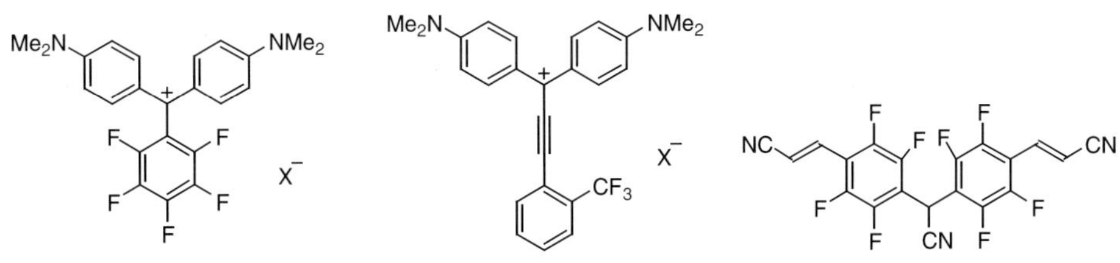

35 **36** **37**

Scheme 5.

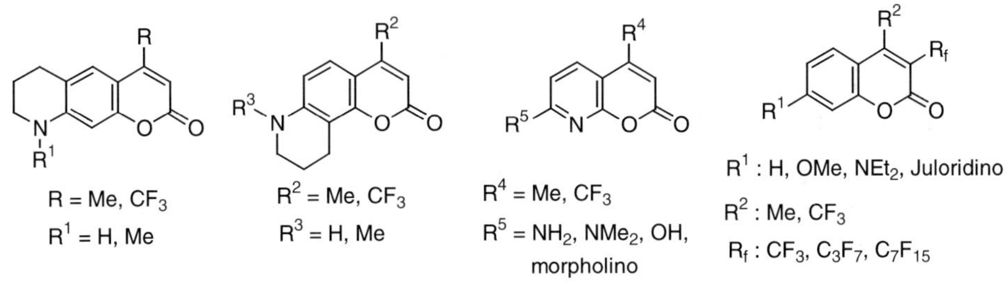

38 **39**

Scheme 6.

products **39** are more bathochromic than the unsubstituted ones. Coumarins show intramolecular charge-transfer chromophoric system from *o*-phenylene to lactone moieties. Therefore, the introduction of electron-withdrawing perfluoroalkyl group at the 3-position could cause bathochromic shift. 3-Perfluoroalkylated derivatives **39** were more stable than the unsubstituted one under UV-irradiation.

3.5. Cyanine dyes

Yagupol'skii et al. have synthesized a variety of cyanine dyes **40–44** as shown in Scheme 7 [33].

3.6. Other fluorine-containing dyes

The other fluorine-containing dyes are shown in Scheme 8. Nile red and quinolinone dyes reacted with perfluoroalkanoyl peroxide **19** to afford the corresponding

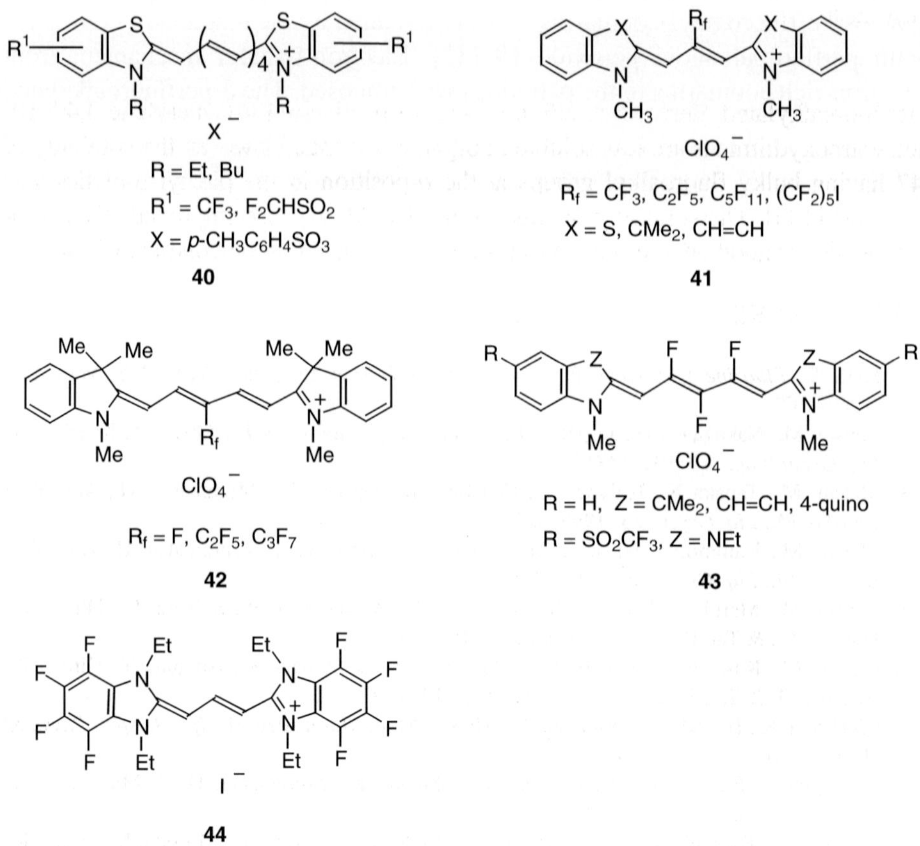

Scheme 7.

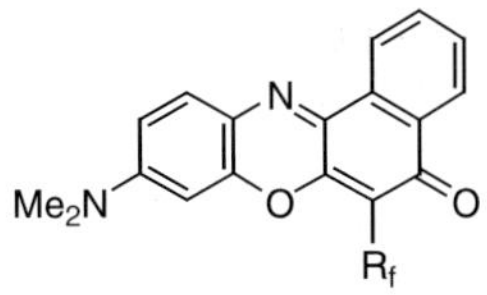

$R_f = CF_3, C_3F_7, C_6F_{13}$

45

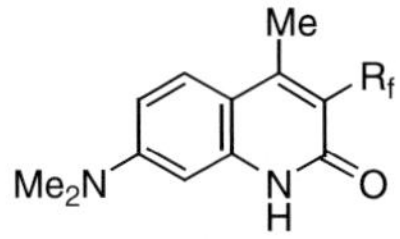

$R_f = C_3F_7, C_6F_{13}$

46

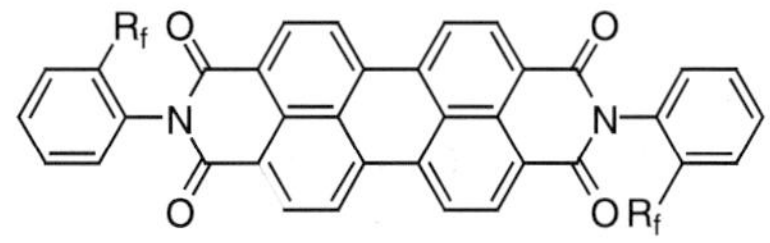

$R_f = C_3F_7C(CF_3)_2CH_2$

47

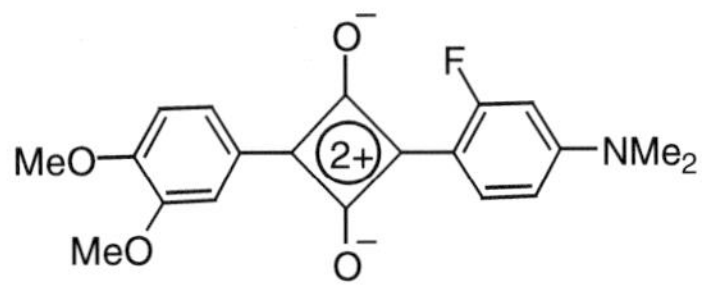

48

Scheme 8.

perfluoroalkylated derivatives **45** and **46**, respectively [34]. Perylene-3,4:9,10-tetracarboxydiimides are less soluble in organic solvents. However, the solubility of **47** having bulky fluoroalkyl groups at the *o*-position in the phenyl moieties was improved [35]. The solubility of squarylium dye **48** was also improved. This compound shows good charge-generation property in organic photoconductor [36].

REFERENCES

1. Paul, T., *Fluorine Chemistry Reviews*, Marcel Dekker Inc., New York, NY, Vols. 1–8, (1967–1977).
2. Matsui, M., Nakagawa, H., Joglekar, B., Shibata, K., Muramatsu, H., Abe, Y. & Kaneko, M., *Liq. Crystallogr.*, **21** (1996) 669.
3. Matsui, M., Tanaka N., Nakaya, K., Funabiki, K., Shibata, K., Muramatsu, H., Abe, Y. & Kaneko, M., *Liq. Cryst.*, **23** (1997) 217.
4. Matsui, M., Kamino, Y., Hayashi, M., Funabiki, K., Shibata, K., Muramatsu, H., Abe, Y. & Kaneko, M., *Liq. Cryst.*, **25** (1998) 235
5. Matsui, M., Marui, Y., Kushida, M., Funabiki, K., Muramatsu, H., Shibata, K., Hirota, K., Hosoda, M. & Tai, K., *Dyes Pigments*, **38** (1998) 57.
6. Matsui, M., Kawase, R., Funabiki, K., Muramatsu, H., Shibata, K., Ishigure, Y., Hirota, K., Hosoda, M. & Tai, K., *Bull. Chem. Soc. Jpn.*, **70** (1997) 3153.
7. (a) Hirota, K., Hosoda, M., Joglekar, B., Matsui, M. & Muramatsu, H., *Jpn. J. Appl. Phys.*, **32** (1993) L1811;
   (b) Joglekar, B., Miyake, T., Kawase, R., Shibata, K., Muramatsu, H. & Matsui, M., *J. Fluorine Chem.*, **74** (1995) 123;
   (c) Joglekar, B., Shibata, K., Muramatsu, H., Matsui, M., Hirota, K., Hosoda, M. & Tai, K., *Polym. J.*, **29** (1997) 184.
8. Tomota, H., Ando, M., Deyama, K., & Yano, K., JP 91-331774.

9. Matsui, M., Joglekar, B., Ishigure, Y., Shibata, K., Muramatsu, H. & Murata, Y., *Bull. Chem. Soc. Jpn.*, **66** (1993) 1790.
10. Gandel'sman, L.Z., Mostoslavskaya, E.I., Khomenko, L.A. & Yagupol'skii, L.M., *Dyes Pigments*, **2** (1981) 279.
11. Pechmeze, J.P.E. & Sureau, R.F.M., US 4287121.
12. Birchall, J.M., Haszeldine, R. N. & Kemp, J.E.G., *J. Chem. Soc. C.*, (1970) 449.
13. Chinnakali, K. & Fun, H.-K., *Acta Crystallogr.*, **C49** (1993) 615.
14. Matsui, M., Tanaka, N., Andoh, N., Funabiki, K., Shibata, K., Muramatsu, H., Ishigure, Y., Kohyama, E., Abe, Y. & Kaneko, M., *Chem. Mater.*, **10** (1998) 1921.
15. Ishikawa, N., *Nippon Kagaku Kaishi*, (1972) 1255.
16. Ishikawa, N., Tanabe, T. & Ohashi, T., *Nippon Kagaku Kaishi*, (1972) 202.
17. Sawada, H., *Chem. Rev.*, **96** (1996) 1779.
18. Matsui, M., Kondoh, S., Shibata, K. & Muramatsu, H., *Bull. Chem. Soc. Jpn.*, **68** (1995) 1042.
19. Takuma, H. & Kuroda, S., JP 94-45540
20. Gandel'sman, L.Z., Khomenko, L.A. & Yagupol'skii, L.M., *Dyes Pigments*, **4** (1983) 41.
21. (a) Fokin, E.P., Loskutov, V.A. & Vol'skii, L.N., *Zh. Org. Khim.*, **6** (1970) 1277;
(b) Coe, P.L., Croll, B.T. & Patrick, C.R., *Tetrahedron*, **23** (1967) 505.
22. (a) Loskutov, V.A., Nekrasova, L.N. & Fokin, E.P., *Izv. Sib. Otd. AN. Khim.*, (1970) 119;
(b) Fokin, E.P. & Loskutov, V.A., *Zh. Obsh. Khim.*, **38** (1968) 1884.
23. Masuda, K., Kaieda, O. & Tamaura, Y., EP 661350.
24. Gornostaev, L.M., Lavrikova, T.I. & Arnol'd, E.V., *Zh. Org. Khim.*, **28** (1992) 2291.
25. Inoue, H., Togano, T., Ikeda, K., Mihara, H. & Hida, M., *Nippon Kagaku Kaishi*, (1985) 2023.
26. Kim, H., Matsuoka, M., Yodoshi, T. & Kitao, T., *Chem. Express.*, **1** (1986) 129.
27. Tanabe, T., Ishikawa, N. & Chang, Y. S., *Nippon Kagaku Kaishi*, (1976) 797.
28. Hallas, G., Grocock, D.E., Hepworth, J.D. & Jones, A. M., *Tetrahedron*, **28** (1972) 893.
29. Muramatsu, H., Okumura, A., Shibata, K. & Matsui, M., *Chem. Ber.*, **127** (1994) 1627.
30. Sakai, T., Yoshio, O., Takayama, T, Ema, T. & Utaka, M., Abstract in 20th Japanese Symposium on Fluorine Chemistry, (1997) p. 144.
31. Atkins, R.L. & Bliss, D.E., *J. Org. Chem.*, **43** (1978) 1975.
32. (a) Matsui, M., Shibata, K., Muramatsu, H., Sawada, H. & Nakayama, M., *Synlett,* (1991) 113;
(b) Matsui, M., Shibata, K., Muramatsu H., Sawada, H. & Nakayama, M., *Chem. Ber.*, **125** (1992) 467.
33. (a) Pazenok, S.V., Kovtyukh, I.P. & Yagupol'skii, L.M., *Tetrahedron Lett.*, **32** (1991) 4595;
(b) Ushomirskii, M.N. & Yagupol'skii, L.M., *Dyes Pigments*, **16** (1991) 93;
(c) Ignat'ev, N.V., Datsenko, S.D., Pazenok, S.V. & Yagupol'skii, L.M., *Zh. Org. Khim.*, **26** (1990) 1740;
(d) Pazenok, S.V., Chaika, E.A. & Yagupol'skii, L.M., *Zh. Org. Khim.*, **23** (1987) 2021;
(e) Yagupol'skii, L.M., Pazenok, S.V. & Kondratenko, N.V., *Zh. Org. Khim.*, **22** (1986) 163;
(f) Yagupol'skii, L.M., Pazenok, S.V. & Kondratenko, N.V., *Zh. Org. Khim.*, **19** (1983) 2223;
(g) Lyubich, M.S., Tyurin, V.S., Yagupol'skii, L.M., Al'perovich, M.A., Troitskaya, V.I. & Larina, S.M., *Zh. Org. Khim.*, **19** (1983) 1976.
34. Matsui, M., Ishikawa, H., Hiramatsu, K., Shibata, K. & Muramatsu, H., *Dyes Pigments*, **27** (1995) 143.
35. Deyama, K., Tomoda, H., Muramatsu, H. & Matsui, M., *Dyes Pigments*, **30** (1996) 73.
36. Lee, K.Y., *Chem. Mater.*, **4** (1992) 605.

Subject Index